Burning Questions in Neuropsychology

Peter Brugger
CASA VERDE
Bergstrasse, Dielsdorf, Switzerland

CRC Press
Taylor & Francis Group
Boca Raton London New York

CRC Press is an imprint of the
Taylor & Francis Group, an **informa** business
A SCIENCE PUBLISHERS BOOK

First edition published 2027
by CRC Press
2385 NW Executive Center Drive, Suite 320, Boca Raton FL 33431

and by CRC Press
4 Park Square, Milton Park, Abingdon, Oxon, OX14 4RN

CRC Press is an imprint of Taylor & Francis Group, LLC

Library of Congress Cataloging-in-Publication Data (applied for)

ISBN: 978-1-032-47186-0 (hbk)
ISBN: 978-1-032-47188-4 (pbk)
ISBN: 978-1-003-38499-1 (ebk)

DOI: 10.1201/9781003384991

Typeset in Times New Roman
by Prime Publishing Services

To the women in my life—the one I met on the boat,
the one who followed me river-up and river-down,
the mother of my children and the one who was with my mother
when she died—to Susan.

Acknowledgements

My teachers, especially Thedi Landis, Marianne Regard, and Rüdiger Wehner.

My students, especially Alexia Anagnostopoulos, Fatma Aras, Shahar Arzi, Daniel Bächtold, Anna Burrack, Christine Busch, Silvia Cavelti, Noemi Dannecker, Marion Funk, Lorena Gianotti, Leonie Hilti, Philipp Katumba, Ira Kurthen, Andreas Kyriakou, Rebekka Meier, Christine Mohr, Tobias Pflugshaupt, Neda Rashidi-Ranjbar, Yannick Rothacher, Ladina Schlosser, Corinne Tamagni, Kirsten Taylor, Evelyn Unterburger, Olivia Zindel-Geisseler and all those who contributed questions to this book.

My collaborators, especially Reto Agosti, Atsushi Aoyama, Olaf Blanke, Stefan Bracha, Gabriella Bottini, Markus Christen, Chris Dijkerman, Sara Fabrikant, Arko Ghosh, Melitta Giummarra, Roger Graves, Thomas Grunwald, Marie-Claude Hepp-Reymond, Spyros Kollias, Peter Krummenacher, Hans-Peter Landolt, Bigna Lenggenhager, Ryan McKay, Ervin Macas, Tamar Makin, Angelo Maravita, Thomas Metzinger, Andreas Monsch, Lorimer Mosley, René Müri, Veit Mylius, Masao Okura, Eraldo Paulesu, Peter Sandor, Sophie Schwartz, Maggie Shiffrar, Werner Surbeck, Devin Terhune, Isabelle Viaud-Delmon, Patrik Vuilleumier, late Volker Henn, late Hansruedi Isler and late Tom Troscianko.

My friends, especially Pad Alexander, Fredy Baumann, Susan Brugger, Fabio Conti, Georg Egli, Martin Fischer, Roger Gassert, Dimitris Georgiadis, Manuela Gaggiotti, Brigitte Hofer, Andreas Horst, Lukas Imbach, Claudine Kallenberger, Heidi Kämpfer, Daria Knoch, Peter Kramer, Tobias Loetscher, Sohee Park, Kostas Priftis, Hariklia Proios, Gianluca Saetta, Joachim Scheuerer, Armin Schnider, Margitta Seeck, Markus Weber, late Norman Cook, late Thomas Hauschild, late Peter Schell, late Urs Schwarz, and late Aiha Zemp.

The many unnamed patients from Valens, Victoria (CA), and Zurich.

(Many of the named teachers, students, and collaborators could also have been listed under “friends”)

Contents

Where Can We Find Which Question?

A selection of the burning questions answered in this book is listed in the table below. Most questions are not literally duplicated here. They are arranged topically. Not according to the chapters in which they appear and are addressed, but according to some grand themes of neuropsychology.

FRONTAL ISSUES

Preliminary note: almost all burning questions in neuropsychology touch on issues of the frontal lobes. We may add one more question here: "Is there any human behavior which would NOT be affected by frontal lobe functions?" We answer it by "No" and suggest that you contemplate on this question/answer after having read the entire book.

HEMIPATIAL NEGLECT

FROM BODY TO SELF

CHAPTER 1

What Is Neuropsychology, and What Is Special About This Book?

Neuropsychology: The Multidisciplinary Study of Behavior

Neuropsychology is the study of behavior in close relation to the central nervous system (CNS), especially the brain.

"Behavior" encompasses an enormously broad variety of facets of living: from perception and sensation to cognition and emotion, and from intention to action—including interaction with conspecifics and other species. After all, brains live in people, and people live in societies with different cultural traditions, diverging mentalities, and vastly different behavioral norms.

? **Is neuropsychology's claim to describe behavior in terms of underlying neural machinery not futile in the face of these vast differences due to a person's cultural background?**

The study of brain and behavior does not exclude comparative analyses that gauge the relative influences of biology and culture. In fact, describing the coevolution of these two factors is a core aim of cultural neuroscience. In the domain of perception, the dynamic relationship between neuropsychological processing and cultural context has long been recognized (Nisbett and Miyamoto 2005).

Before reading on, please draw a rough sketch visualizing the sentence "The girl pushes the boy." Done? If you come from a culture with a script oriented from left to right, your drawing is likely to place the girl to the left of the boy, such that the action flows rightward. In a series of elegant experiments, Maass and Russo (2003) demonstrated that people from cultures that read and write from right to left do not exhibit a comparably strong reversal. A combination of scanning habits and lateral action initiation explains the asymmetry observed across study populations. While scanning habits are culturally determined, motor action initiation is a shared biological factor, reflecting a left-hemisphere bias and the depiction of agency from left to right.

DOI: 10.1201/9781003384991-1

Ready et al. (2025) analyzed the number of articles published in 13 neuropsychology journals between 2010 and 2020 that included cultural or multicultural topics in their titles. Regardless of journal impact factor, the figures ranged from 1% to 13%, and they increased over the decade. So yes—your critical stance is justified. But there is hope that future neuropsychologists will be more attuned to the interplay between biology and culture.

Experimental Neuropsychology

Who shows the behavior studied by experimental neuropsychologists? Is it only people with brain damage? The renowned cognitive neuroscientist Tim Shallice, professor of neuropsychology in London, opened the abstract of one of his articles by stating: "Cognitive neuropsychology is characterized as the discipline in which one draws conclusions about the organization of the normal cognitive systems from the behaviour of brain-damaged individuals" (Shallice 2015). Indeed, neuropsychologists often work with people who have neurological problems. But imagine if the field were confined to such a highly specialized population!

Let me provide some examples of simple behavioral studies[1] across different groups of research participants who all have one thing in common: a CNS—or, to avoid an overly narrow definition, an organized organic machinery underlying their behavior. These groups include:

Healthy People

A simple reaction time task can reveal how the two cerebral hemispheres communicate, and how long it takes for information to cross the corpus callosum (see Chapter 11, p. 251). Looking at a face for a while and then shifting your gaze to another will demonstrate cortical adaptation—a key principle of short-term plasticity (see Figure 3 in Chapter 7).

People With Special Bodies or Ways of Seeing the World (No Overt CNS Lesions)

Behavioral testing in amputees with phantom limbs has revealed many laws of cortical organization—some of which have later been explored with neuroimaging (Chapter 10). Higher-order phantom experiences, such as out-of-body or near-death states, also fall within the scope of neuropsychological research. Another example is people with digit-color synesthesia. They perceive specific digits in specific colors. Simple behavioral tasks can distinguish genuine synaesthetes from those who merely pretend (Chapter 7, p. 160). No neuroimaging method can do this alone—but it can help pinpoint the level(s) at which sensory modalities converge.

[1] Evidently, structural and functional neuroimaging, electrophysiology and, in animals various invasive methods may complement the behavioral experiments.

People With CNS Lesions

Some stroke patients exhibit a selective deficit in recognizing objects only in motion, while still recognizing them when static, highlighting separate motion-processing pathways (Chapter 7, p. 152). Patients who neglect the left side of space may nevertheless "unconsciously" avoid a stimulus with threatening information in the neglected field (Chapter 9). In patients with *pure alexia*, reading is impaired while writing is preserved—a dissociation that puzzled early researchers and helped define the visual word form system (Chapter 4). Early autopsy findings in such patients revealed a lesion location involving the most posterior part of the corpus callosum (Chapter 4, p. 71).

Healthy Animals

Some ant species navigate by detecting the polarization of skylight. If you cover a specific part of the compound eye, this ability is lost. Honeybees can differentiate between human faces. A mealworm's memory span can be determined by its behavior in a T-maze (Chapter 6, Figure 1). Dogs show lateralized tail wagging: a right-side bias (left hemisphere) reflects approach; a left-side wag (right hemisphere) indicates withdrawal (Chapter 11, p. 266). Even single-celled organisms, such as amoebae, use coordinated pseudopod activity to swim toward or away from chemical stimuli—depending on factors like temperature, pH, or friction.

Comparative neuropsychology reminds us that the human brain is not the "crown of evolution." Every surviving species is highly adapted to its niche. If extinction threatens them, it is not due to their "inferior cognition," but due to a lack of humane regard by humans. This field should make us more modest and less anthropocentric.

Animals With CNS Lesions

Much of what we know about human action, perception, and memory derives from studies in animals with targeted CNS lesions. That is a difficult truth—but a foundational one. Animal research has contributed to the development of new medications and preventive strategies for humans. At the same time, ethical oversight is crucial. Ethical Review Boards (ERBs) have gained influence, but some have degenerated into bureaucratic gatekeeping and may even act in an anti-scientific spirit (see Chapter 11, footnote 2).

Ironically, ERBs do not consider all animals equal: studies on insects, for instance, often bypass ethical review entirely.[2] *Poor drosophila! Poor ants! Poor butterflies!*

[2] At least this holds for most European countries. In Switzerland, zebra fish are animals. But only from 5 days post fertilization on. I am afraid that bureaucracy determines what they are during days 1 to 4.

? Animal neuropsychology is purely experimental. Should clinical neuropsychologists care about it at all?

I agree that there is no such thing as a proper "veterinary neuropsychology." However, animals can—and sometimes do—suffer from naturally occurring CNS lesions. A scattered but fascinating literature documents the range of resulting behavioral abnormalities.

One striking example is that of an ant with a brain tumor, described by Brun (1925), an entomologist known locally for his expertise in ant behavior. He one day received a letter from a colleague. Enclosed was a meticulous account of a single ant in his artificial nest colony that exhibited an unusual movement disorder. Rather than walking straight along scent trails like the others, the "patient ant" moved in persistent spirals, always to the right. Its right antenna and the distal segment of its right foreleg appeared paretic. It failed to engage in normal antennal contact with its conspecifics and instead opened its mandibles, appearing to exhibit an act of aggression. In response, the other ants would grab it and carry it a few centimeters away before rereleasing it.

The colleague also noted a "white mass" protruding from the ant's mouth, preserved the animal in alcohol, and sent it to Brun for further investigation. Brun performed an autopsy and found an infiltrating tumor occupying a large portion of the left side of the ant's most rostral ganglion. This lesion could account for the ant's rightward spiraling, the partial hemiparesis, and, as Brun speculated, the breakdown of its social behavior.

For a modern overview of tumors in invertebrates, see Tascedda and Ottaviani (2014). For human circling behavior, see Schaeffer (1928) and the references in Leuenberger et al. (2015).

The amoeba example mentioned earlier shows that behavior is not necessarily dependent on a brain—or even on a CNS in the traditional sense. But amoebae do have sensory receptors. These allow them to detect whether their chemical environment (the world outside themselves)[3] is acidic or neutral, warm or cold, light or dark.

They also exhibit motility: through coordinated pseudopod activity, they can swim from sour to sweet water. Each time we place a drop of lemon juice into the microscopic pond beneath the binocular microscope and document where the amoeba's first pseudopod emerges on the cell membrane, we are conducting neuropsychological research.

Having a head is no prerequisite for being selected for neurocognitive research (Baluška and Levin 2016).

[3] "Outside"—what a provocative word in this context! Can unicellular organisms differentiate between inside and outside? Do they possess something like a *corporeal awareness* (see Chapter 10)? Do they have interoceptive capabilities?

? Is there something like a "plant neuropsychology"? Do plants have minds?

A provocative double-question!

To the first part: No, there is no proper "plant neuropsychology." Plants do not have neurons. Yet, they unquestionably behave. And behavior, in all cases, requires an underlying substrate. The prefix "neuro" is not appropriate when discussing plant behavior. That is why, in 2009, the "Society for Plant Neurobiology" (founded in 2005) renamed itself the "Society for Plant Signaling and Behavior."

This society operates an open-access journal dealing with issues such as botanic sensorimotor processing, memory, communication, decision-making, and many other domains that overlap with traditional neuropsychology. To give you a small taste: Mimosa plants close their leaves rapidly when touched (a sensorimotor act). However, if the stimulus is repeated and harmless, the plant shows habituation—a basic form of learning—and gradually stops closing its leaves. Systematic studies have examined how long this memory of harmless touches persists (Gagliano et al. 2016). Other forms of plant memory include priming, and some are involved in transgenerational epigenetic inheritance (Harish and Osherov 2022). Such work contributes to general conceptions of memory (see Chapter 6).

Now, to the second part: Do plants have minds? I am personally reluctant to offer a definite answer. The very definition of "mind" seems a bit wobbly to me, even in human psychology (see Maher, 2017, for specific objections regarding plant minds). Yet the idea is taken seriously by some. Kawano et al. (2025), for instance, propose that each plant may, in fact, possess two minds—comparable, in a way, to the dual mental systems hypothesized in humans and nonhuman animals (see Chapter 11; regarding humans, Schechter 2018).

In any case, the field of plant biopsychology—focusing on the relationship between behavior and biological substrate—is mind-boggling. Every neuropsychologist should take an interest. Let your own ideas about it take root!

Clinical Neuropsychology

The work of clinical neuropsychologists is as multifaceted as that of their experimentally working colleagues.

While experimental neuropsychology draws on cognitive neuroscience, biology, anatomy, anthropology, and even sociology to understand brain-behavior relationships, clinical neuropsychology requires a firm grounding in diverse areas of medicine and psychology. These include test psychology, physiology, rehabilitation medicine, developmental sciences, neurosurgery, geriatrics, and, depending on specialization, applied fields such as forensic neuropsychology.

But if experimental neuropsychologists value behavioral paradigms, clinical neuropsychologists should take pride in their observational skill. They are often the ones who *see* what no test can quite capture—discrepancies, anomalies, or moments that only appear when context, body, task, and awareness intersect. Consider the following examples, each showing how behavior, when carefully observed, becomes diagnostic insight:

* Some patients can read a word but not say what it means—until you ask them to draw it. (A form of optic aphasia: the meaning is there, but inaccessible to language.)
* Some patients with visual extinction will notice a touch on their left hand—unless you also touch their right. Then the left disappears. (Extinction is not a deficit of sensation but of attention when stimuli compete.)
* Some people with apraxia cannot mimic brushing their teeth when asked, but they will do it spontaneously if you hand them a toothbrush. (Volitional imitation and object-based action are not always governed by the same neural pathways.)
* A patient with left-sided neglect might feel a tap on the left hand if it is placed on the right side of space, but not when it is on the left. (Body and external space are not always aligned in the brain. "Left" is not always left!)
* Some split-brain patients can name objects explored blindfolded with the right hand but must draw them with the left to express what they palpated with that hand. (The hemispheres still communicate—just not via the usual verbal routes.)
* Some patients cannot point to their own nose, but can point to yours. (Disturbances of body schema may spare interpersonal representations.)

These are not rare curiosities. They are the core material of clinical neuropsychology. Observation is not a passive act—it is a skill, and one that distinguishes the best clinicians from those who merely score and tabulate. Observations can lead to novel findings, which must pass the test of time: Is the observed pattern coincidental, or can it be observed repeatedly by others as well?

> In many years of clinical practice, I've occasionally encountered a striking phenomenon I'd like to call *hetero-anosognosia*. Anosognosia, classically, is the failure to recognize one's own deficits—say, denying a hemiplegia that is otherwise obvious (Chapter 10, p. 226). But what about the failure to recognize someone else's?
>
> In close relationships—almost always, in my experience, between spouses—it is sometimes the *partner* who denies or downplays the patient's cognitive deficits. A few examples stand out: "*He was never good at drawing*," says the wife, confronted with her husband's attempt to copy a complex figure—so rudimentary it can barely be labeled as "recognizable." "*Her memory hasn't been any better since we married*," says the husband, after his wife recalls none of ten words just five minutes after presentation.
>
> In some of these cases, the patient is more aware of their own impairments than the spouse is. In rare instances—perhaps 1 in 500?—the pattern is so striking that the non-recognition on the partner's side overshadows the patient's own insight. Interestingly, the presence of hetero-anosognosia sometimes appears inversely correlated with the patient's own anosognosia.

> Whenever possible, I try to speak with the spouse alone. The reactions vary: from sudden breakdowns to a doubling down of denial. Describing these conversations would fill half a book ... Look out for such observations yourself. As far as I know, they are yet unpublished.

The point I wish to make here is: without the clinical observation of hetero-anosognosia (let us assume the phenomenon is "replicable" in the sense of potentially confirmed by similar observations by the readers of this book), no prospective research on the phenomenon is ever possible.

? When I think of the plethora of technical instrumentation available to (neuro)psychologists, I am inclined to ask: Do you not overestimate the status of observation in our modern times?

Appreciating and depreciating the value of observational data reflect two opposing worldviews within science. May I recommend a timeless paper that, as early as 1973, lamented young researchers' impression that "observation is merely of historical interest" (Hailman and Sustare 1973)?

The authors describe how they inspired enthusiasm for the observational method by having students pull the string of a stuffed toy—Horton, a talking elephant—10,000 times. With each pull, Horton emitted one of ten vocalizations in seemingly random order. The task: by observation alone, infer the mechanism behind Horton's speech.

From the fact that each sentence occurred with roughly equal frequency, that immediate repetitions were rare but possible, and that certain sequences tended to follow others preferentially (i.e., there were unequal transitional probabilities of utterances), the students deduced that a rotating circular device with a one-way pointer must underlie the behavior.

Crucially, they were asked not to manipulate the mechanism (e.g., pulling the string halfway), but to rely solely on observation and logical analysis. By the end of the course, their hypothesis about Horton's inner workings was confirmed through autopsy.

As the example of hetero-anosognosia above shows, clinical neuropsychologists sometimes also need to care for the patient's relatives. They may offer practical advice on how to handle new, often puzzling, behaviors brought about by brain damage. While they do not need to be trained psychotherapists, their guidance can be invaluable—because it is based on direct, first-hand observation.

Finally, one of the clinical neuropsychologist's key tasks is to write a report for the referring physician or healthcare provider. This is more than just a summary of test scores. While a tabular overview of results may be helpful, it cannot replace a narrative account—one that integrates observations across the entire assessment.

Too often, standardized report templates reduce rich behavioral findings to phrases like "Function X is well preserved/moderately/severely impaired." These mechanical statements may fulfill administrative needs, but they risk ignoring what was actually seen. Clinical neuropsychologists should be proud to be narrative.

The word narrative may carry negative connotations in science—"only narrative" as opposed to meta-analysis or statistics—but in this context, it means something else. It means formulating an informed judgment after integrating diverse impressions, including scores, spontaneous behavior, body language, patient self-perception, reactions to feedback, and how a test was approached rather than just how it was completed.

Being narrative does not mean being wordy, vague, or poetic. It means being clear. It means daring to formulate your own view—briefly, thoughtfully, and with the patient in mind. A good report is not just for the file. It can help the referring colleague, and ultimately, it can help the patient.

? There is a big problem with defining neuropsychology as "the field that studies behavior in relation to the brain." There are occasional case reports of individuals who appear largely normal, but whose brain scans incidentally reveal no brain. What can you say to such cases?

You are likely referring to a widely cited commentary in *Science* titled *"Is your brain really necessary?"* (Lewin 1980). Though tongue-in-cheek, the title caused quite a stir—especially in widespread press coverage. It summarized the work of John Lorber, a British professor of pediatrics, who reported the case of a socially well-integrated university student with an IQ of 126 and a first-class honors degree in mathematics. He had been referred to Lorber because of an abnormally large head.

"*When we did a brain scan on him, we saw that instead of the normal 4.5-centimeter thickness of brain tissue between the ventricles and the cortical surface, there was just a thin layer of mantle measuring a millimeter or so. His cranium is filled mainly with cerebrospinal fluid.*" (Lewin 1980)

The student had hydrocephalus (see Chapter 2, p. 39), a condition in which cerebrospinal fluid accumulates in the ventricles, expanding them and exerting pressure on the surrounding brain tissue and skull. When this occurs abruptly in adulthood, it can cause headaches, balance problems, and gait disturbance. Neuropsychologically, it is often associated with slowed psychomotor speed, executive dysfunction, and memory impairments.

However, when hydrocephalus develops gradually in early life—as in this case—it may remain unnoticed for years. Symptoms are minimal, and headaches may be explained away. To the untrained eye, an unusually large head does not necessarily signal a neurological condition—especially not in someone who is thriving academically.

Juvenile hydrocephalus with largely preserved cognitive and emotional development illustrates two key points. First, it demonstrates the adaptive plasticity of the CNS—how brain function can be preserved despite massive anatomical distortion. Second, it shows how the media can latch onto rare findings and turn them into neuromyths.

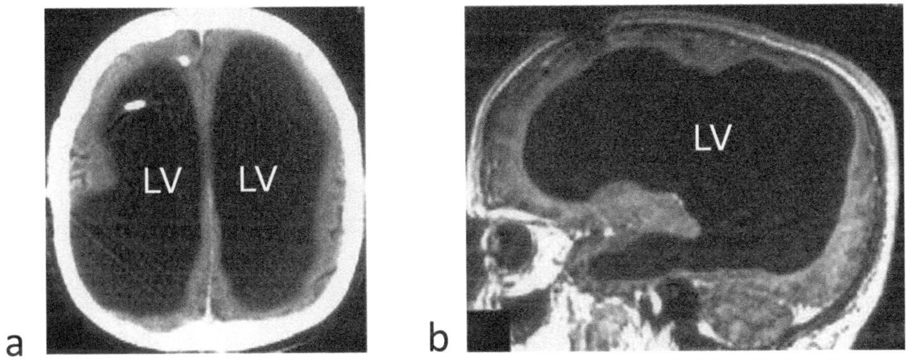

Figure 1: Massive ventricular enlargement in a patient with normal social functioning. Compare Figure 5, Chapter 2, for regular ventricular size.
a: CT-scan (transverse section) shows both left and right lateral ventricles to take up the space of almost the entire hemisphere. b: MRI Scan shows the same for one hemisphere, in a sagittal section. LV: lateral ventricle. Reprinted, with permission from The Lancet, from Feuillet et al. 2007.

So-called "no-brain" cases (Figure 1) do not show the absence of brain tissue. Instead, they show that slow and progressive compression of the brain can sometimes be remarkably well tolerated—at least up to a point. Instead of casting doubt on the importance of the brain, these cases highlight just how astonishingly adaptable it can be.

Such cases are not limited to hydrocephalus. In clinical work, I have seen both sides of the dissociation:

(1) patients with striking posterior cortical atrophy on MRI, yet intact visual functions;

(2) others with severe visual-perceptual deficits and age-appropriate imaging.

Both patterns are rare, but they caution us against assuming a strict one-to-one correspondence between structure and function. They remind us that damage visible on a scan does not always predict behavior, and conversely, that functional deficits may exist even when the anatomy appears preserved. Neuropsychology must therefore remain a science of observation, not just of imaging.

What Is Special About This Book?

This book is a textbook, but not quite like the others. It is neither a complete introduction nor a comprehensive reference. Some chapters go deeper than beginners might expect; others may omit topics that are covered elsewhere (you will not find a separate chapter on dementias, for instance). It does not aim to replace standard textbooks, but to complement them by raising questions they often leave unasked.

It is special in at least five ways.

First, it is driven by questions. Real questions. Often asked by students, sometimes strange, sometimes brilliant, always sincere. You will find them scattered throughout the chapters, not as polished FAQ boxes, but as interruptions, because that is what actual questions do: they interrupt our flow, making us reconsider. They invite thought and more questions. Learning, here, is not just about obtaining answers, but about understanding how deep even a simple question can go.

Second, it takes historical texts seriously. Neuropsychology has a short history, but a dense one. Reading older work is not an antiquarian habit—it is a way of tracing how today's frameworks emerged, and how alternatives were once imagined. Many modern neuropsychological ideas are less new than they seem.

Third, the book dares to cross boundaries. Between human and animal research. Between brain and body. Between science and what is often considered pseudoscience. The chapter on apraxia, for instance, briefly discusses paranormal phenomena, not because they are taken at face value, but because their study reveals how action and intention are structured in the brain. The book occasionally adopts the perspective of a "devil's advocate" to examine science from its edge. As Seymour Kety (1960) once remarked, "It requires just as much oxygen to think an irrational thought as to think a rational one."

Fourth, it treats observation with the same respect as measurement. While test scores have their place, they are not the whole story. Clinical neuropsychologists are trained observers, and what they witness cannot always be reduced to a number. Narratives matter, not only because they explain, but also because they facilitate the integration of information across various sources. This book tries to preserve that spirit.

Finally, *Burning Questions in Neuropsychology* recognizes that not everything must be simplified. Sometimes, the text may be more detailed or technical than what is typically found in introductory material. Readers are encouraged not to be discouraged. Skip what you find too dense; return to it later if you wish. Some references are there for scholars who enjoy following threads; others are essential for showing how conclusions were reached.

This book, then, is an invitation: to think, to wonder, to notice what others overlook, and to ask better questions. That, more than any list of syndromes or test scores, is what makes a neuropsychologist.

Where Can We Read More?

Bellone, J.A. and van Patten, R. Becoming a Neuropsychologist. Advice and Guidance for Students and Trainees. Springer: New York, 2021.

Block, C. (ed.). The Neuropsychologist's Roadmap. A Training and Career Guide. American Psychological Association: New York, 2021.

Boulton, A.A., Baker, G.B. and Hiscock, M. Neuromethods Vol. 17: Neuropsychology. Humana Press: Totowa, 2013 (1st edition 1990).

Heilman, K.M. and Valenstein, E. (eds.). Clinical Neuropsychology (5th edition). Oxford University Press: Oxford, 2011 (1st edition 1979).

Parsons, M.W. and Braun, M.M. (eds.) (4th edition). Clinical Neuropsychology. A Pocket Handbook for Assessment. American Psychological Association: New York, 2024.

CHAPTER 2

How Much Neuroanatomy?

You cannot know enough about the anatomy of the nervous system to make sense of neuropsychology. It is, however, all a question of balance. Knowing simply where each structure is situated and with what other structures it is connected does not offer an understanding of its functionality. The study of the substratum on which all cognition, emotion, and behavior rest must be conducted in conjunction with the study of cognition, emotion, and behavior themselves. Functional neuroanatomy is fascinating. On your way to becoming an expert neuropsychologist, whether clinical or experimental, you will automatically gravitate more toward brain structure or more toward the organization of the mind. Never forget that both mindlessness and brainlessness are dead ends in the study of behavior (Eisenberg 1986).

Functional neuroanatomy must not be the reason why you are holding this book in your hands. You will only find the need-to-know basics here. Table 1 will guide you to some introductory literature, and Figure 1 offers advice to those among you who are enrolled in university studies: take dissection courses! To handle brains means, almost literally, to get a grip on neuroanatomy. Whether you want to take that grip as a springboard to behavior is then up to you.

? Figure 1 appears a bit provocative to me. Does it not foster a "materialistic" or reductionist view of neuropsychology?

Without delving too deeply into philosophy, materialism would say that matter is primary and all non-matter—i.e., mind, "soul," ideas—secondary. Idealism would say the opposite. The picture of the dead brains in a bath does not intend to suggest either philosophical stance. It merely conveys that neuropsychology requires the study of both structure and emergent function. And indirectly, it promotes the view that neuropsychology concerns the functioning of living tissue; all assumptions about "out-of-body" consciousness (Chapter 10, p. 231) and "life-after-death" issues (Chapter 3, p. 47) belong to the esoteric realm and are ultimately a mere matter of belief (pun intended…).

As for reductionism, I do not think cognition or emotion can be reduced to the physiology of neural tissue. Mind springs from complex interactions between different aspects and levels of life—material, psychological, and social. This book may adopt a mechanistically oriented spirit, but it does not defend a reductionist one.

DOI: 10.1201/9781003384991-2

Figure 1: Getting a grasp on brain anatomy begins with dissection courses. Here, an assistant from the Institute of Anatomy at the University of Zurich grabs a human brain for dissection. Photography by Josef Stücker. Reprinted with permission of Josef Stücker from Hunziker and Mazzola (1990).

Table 1: Recommended primers of functional neuroanatomy for neuropsychologists.

References	**Major Focus**	**Nice**	**Not So Nice**
Fix (2002) *Neuroanatomy*	Comprehensive overview, rich in illustrations. Contains examination questions and answers.	Can be borrowed for free (see the Reference Section of this book).	At times, more for students of neurology than neuropsychology.
Vanderah and Gould (2025) *Nolte's The Human Brain*	Comprehensively describe functional neuroanatomy, including neurological disorders.	An in-depth account with clinical content for medical students. Almost 700 pages.	More advanced-level than announced on the back flap.
Society for Neuroscience (2005) *Brain Facts*	Provide glimpses into the study of the nervous system, from neuron to systems to disease and therapy.	Really simple and brief. Broad scope, from neurons to systems.	At times, too simple; illustrations are more artistic than technical.
Schoenberg et al. 2011 *Neuroanatomy Primer…*	Introduce the nervous system, including its peripheral and spinal components, with gross pathologies.	Didactically appealing "rule of thumb" boxes; well-illustrated.	At times, too much detail and overloaded figures; disturbing typos.
Carter (2019) *The human brain book*	Comprehensive guide to brain structure and function and neuropsychiatric disorders. Primarily a reference book.	Mixture of CGI illustrations and MRI scans.	At times, it is too structurally oriented at the expense of functional context.
Henley (2021) *Foundations of Neuroscience*	Introduces systems neuroscience and neuroanatomy to undergraduates of various subjects	Well-structured, open-access, with multiple-choice tests at the end of each section.	-
Ray (2024) *Fundamentals of Brain and Behavior*	Introduces human neuroscience, including its history. Evolutionary and social neuropsychology	Comprehensive; contains learning objectives and checks.	More on behavior than on neuroanatomy. Expensive.

Basic Facts About Brain Anatomy and Neural Functions

The human brain measures approximately 14 × 16 × 9 cm and weighs about 1.5 kilograms. According to various popular sources, it contains around 100 billion neurons. These are the basic units of communication in the central nervous system, but they are by no means the only cells occupying our skull. They are surrounded by at least as many glial cells, which support, hold in place, and protect and nourish the neurons.

Neurons communicate with each other either electrically or chemically via synapses. It is estimated that the human brain contains approximately 100 trillion synapses.

What do all these numbers mean? Not much—unless we begin to think in terms of relationships:

* How does the size of the human brain compare to the brains of other animals?
* How does brain size relate to body size—or to neuron count?
* How does the number of neurons relate to the number of synapses?

And perhaps most importantly:

* How are these synapses interconnected, and how plastic are the connections?

? **What is the "encephalization quotient," and are bigger brains better?**

The encephalization quotient (EQ) is a measure of the ratio between the observed brain mass of a species and the brain mass expected from its body size. It allows comparisons across species—whales, mice, and humans—with regard to the relationship between cognition and brain size. The EQ was introduced in the 1970s (Jerison 1973) and is based on the finding that, across species, the correlation between brain and body size can be described mathematically. This allows us to predict brain size from body size for any species. An EQ of 1 indicates that observed and expected values match. EQs greater than 1 indicate larger-than-expected brains. Humans have an EQ of about 7 compared to all mammals, and about 3 compared to other primates.

However, there are caveats regarding the usefulness of EQ as a predictor of cognitive capacity. One major finding is that brain size is not a reliable indicator of neuron count (Herculano-Houzel et al. 2007). As a result, several long-held neuroanatomical views on the uniqueness of the human brain and mind require re-examination (see Herculano-Houzel (2009) for an overview). A meta-analysis by Deaner et al. (2007) suggested that absolute brain size, rather than relative size or EQ, is the best predictor of cognitive abilities.

However, even the absolute number of neurons may not be decisive. The nematode *C. elegans*, with only 302 neurons, shows remarkable associative learning abilities (Zhang et al. 2005). Neural network studies reveal that complex navigation and numerical abilities in insects depend on surprisingly small numbers of neurons (Cruse and Hübner 2008; Dehaene and Changeux 1993). In their review paper "Are bigger brains better?", Chittka and Niven (2009) concluded that modularity and connectivity are more informative predictors of adaptive cognition than either neuron number or brain size, even when scaled to body size.

A Brief History of Structure-Function Relations in the Human Brain

The name of Franz Joseph Gall (German anatomist, 1758–1828) is today associated with phrenology, the belief that the shape of an individual's skull reflects their mental strengths and weaknesses. As a result, Gall's contributions to neuropsychological thinking are often judged as having hindered "real progress." Yet we should remember that the ideas of any historical period must be evaluated against the general level of knowledge at the time—"general" meaning the state of the art in natural science, philosophy, and medicine.

It is quite possible that we would all feel deeply embarrassed if we were allowed to time-travel just 100 years into the future and look back at the large number of today's "false beliefs" about brain-behavior relationships!

It is true that Gall was an ardent proponent of phrenology. But he was also among the first to argue that the brain—and particularly its surface, the cerebral cortex—was far more than a homogenous mass contributing diffusely to psychological function. He proposed that there is functional specialization in different cortical regions and that functional maps could be overlaid onto the human cortex. He was even quite specific in localizing what he called the language function to the anterior segments of the frontal cortex. In fact, Gall initiated fundamental work on the functional specialization of the two cerebral hemispheres (Harris 2019), which are referred to in this book as FUHEMAS (see Chapter 11). While his craniometric mappings have long since been abandoned by science, his early attempts to establish cerebral maps relating to mental function proved highly influential.

Korbinian Brodmann (1868–1918), another German anatomist and psychiatrist (see Mueller and Kanis-Seyfried (2019) for an overview of his life and work), became world-famous for his cytoarchitectonic parcellation of the human cerebral cortex (Brodmann 1909). Even today, his meticulous nomenclature—comprising more than 50 cortical regions—is still widely followed. ("Brodmann area" should perhaps be listed under *B* in the ABC of neuropsychology-related eponyms—see Chapter 4, Appendix.)

Keep in mind that, in Brodmann's time, the function of a particular cortical region could be inferred from two main sources: first, from the post-mortem analysis of brains with visible structural alterations in circumscribed regions, and second, from electrical stimulation of defined cortical areas, mostly in nonhuman brains.

? **Should neuropsychologists care about Brodmann's areas?**

Neuropsychologists who limit themselves to reading patient charts and neuropsychological reports will rarely encounter the Brodmann nomenclature. Those who—laudably—also consult detailed neurosurgical reports or read publications in the field of neuroimaging, however, will come across it regularly. Table 2 provides a list of all Brodmann Areas (BAs). You do not need to memorize them. Some are mentioned more frequently in connection with well-defined neuropsychological deficits, such as BA44 and BA45, commonly associated with expressive aphasia. Still, the specific functions of individual areas and the deficits caused by their lesions are only coarsely defined, often based on a relatively small number of neuroimaging studies.

Brodmann's meticulous parcellation of the cortical surface was based on the cytological characteristics of the underlying tissue. Whether all of the Brodmann Areas (BAs) are also distinct in a functional sense remains debatable. Some voices argue for a reduction in the number of areas, while others call for an expansion to allow for a more nuanced functional segregation.

Table 2: List of the 52 Brodmann areas (BAs), their location, and attributed function.

Brodmann Area (BA)	Location and Labeling, Keywords to Function(s)	If Lesioned, Cardinal Symptoms Involve
BA 01-03	The somatosensory cortex is on the postcentral gyrus and posteriorly adjacent. S1, S2, S3. Tactile perception and integration, proprioception.	Inability to perceive and localize touch/pressure in specific contralateral body parts.
BA 04	Primary motor cortex on the precentral gyrus. Motor execution, voluntary movement control.	Weakness or paralysis of specific contralateral body parts.
BA 05	Somatosensory association cortex around the superior parietal sulcus (SPS).	Astereognosis (nonrecognition of objects by touch).
BA 06	Premotor and supplementary motor cortex (SMA); motor sequencing and planning.	Deficits in planning and sequencing a motor action.
BA 07	The somatosensory association cortex is on the superior parietal gyrus (SPG). Spatial and corporeal awareness.	Deficits of spatial attention (e.g., hemineglect), corporeal illusions, and bodily disintegration.
BA 08	Pre-SMA, containing frontal eye fields. Eye movements and visual attention.	Difficulties in shifting eye gaze properly, attention deficits.
BA 09	Dorsolateral prefrontal cortex (DLPFC). Attention, working memory, and executive functions.	Impaired decision making, "dysexecutive syndrome."
BA 10	Frontal pole with abstract thinking, self-control, decision making, evaluation of risk and conflict, and other higher cognitive functions.	"Dysexecutive syndrome" with impairments in the functions enumerated in column 2.
BA11-12	Orbitofrontal cortex/superior frontal gyrus with regulation of emotion, reward processing, and social cognition. Taste perception. Mental time travel.	Disinhibited, socially inappropriate behavior, poor decision making, and personality changes.
BA 13,14*, BA15*,16	Insular cortex. Interoception, temperature, pain, itch, sensual touch, hunger, thirst (posterior insula), and decision-making, love, craving, and subjective time experience (anterior parts).	Interoceptive disorders, deficits in emotion regulation, socially maladaptive behaviors, and distortions of time.
BA 17-18	Striate (V1) and extrastriate cortex (V2) with brightness, color, elementary visual form, and visual depth analyses.	Cortical blindness (V1), impairment in spatial frequency discrimination, and pattern recognition.
BA 19	V3–V5 of the extrastriate visual association cortex. Differentiation point between the dorsal and ventral stream of visual processing (Chapter 7, Figure 4).	Cerebral achromatopsia (V4), Visual motion blindness (V5), and visual agnosias.
BA 20	Inferior temporal gyrus (ITG). Object recognition and visual object memory.	Object agnosia.

Table 2 contd. ...

...Table 2 contd.

Brodmann Area (BA)	Location and Labeling, Keywords to Function(s)	If Lesioned, Cardinal Symptoms Involve
BA 21	Middle temporal gyrus, MTG. Multisensory integration and semantic memory.	Learning and retention deficits, prosopagnosia (especially RH).
BA 22	Superior temporal gyrus (STG) with Wernicke area. Analysis of verbal (LH) and nonverbal sounds (RH).	Receptive aphasia; auditory agnosia for environmental sounds.
BA 23	(Ventral) posterior cingulate cortex (PCC). Visual processing, memory, and emotional experience.	Impaired emotion processing.
BA 24	(Ventral) anterior cingulate cortex (ACC). Regulation of attention and motivation, error monitoring.	Lack of motivation, emotional blunting, and general apathy.
BA 25	Cingulate cortex (subgenual region). Regulation of mood, emotion processing.	Affect incontinence, emotional instability, depression.
BA 26	Retrosplenial cortex. Episodic memory, spatial navigation.	Forgetfulness, difficulties in topographic orientation.
BA 27	The "Piriform cortex" on the parahippocampal gyrus contains cells that are sensitive to the facing direction of the head.	Difficulties in orientation.
BA 28	(Ventral) entorhinal cortex with major connections to the hippocampus. (Spatial) memory encoding.	Deficits in sensory integration, insufficient (spatial) memory.
BA 29, BA 30	Part of the retrosplenial cortex, on the isthmus of the cingulate gyrus. Autonomic control, spatial navigation.	Topographagnosia (espec. RH), autonomic disturbances.
BA 31-33	Cingulate cortex. Emotional connotations of sensory input, response to pain, regulation of aggressive impulses, impulse control, decision-making, and attention allocation.	Attentional disorders and deficits in the heterogeneous functions mentioned in column 2.
BA 34	Entorhinal cortex/superior temporal gyrus (STG). Olfactory perception and memory.	Anosmia of the ipsilateral nostril.
BA 35-36	Perirhinal cortex. Object recognition and memory, especially in BA 36, involves the consolidation and retrieval of contextual declarative memories.	Object agnosia (especially LH), deficits in contextual understanding.
BA 37	Occipital-temporal cortex with "fusiform face area" (FFA).	Prosopagnosia (bilateral or RH lesions).

Table 2 contd. ...

...Table 2 contd.

Brodmann Area (BA)	Location and Labeling, Keywords to Function(s)	If Lesioned, Cardinal Symptoms Involve
BA 38	Temporal pole (rostral part of BA 21, 22). Semantic memory, social cognition. Strongly linked to orbitofrontal areas by the uncinate fasciculus.	Deficits in socially appropriate behavior, (emotional) semantic memory disturbances.
BA 39	The angular gyrus, located at the end of the lateral fissure in the parietal lobe. Reading, writing, arithmetic, body schema, and spatial orientation.	Gerstmann syndrome (agraphia, acalculia, finger agnosia, and difficulties in discriminating left from right).
BA 40	Supramarginal gyrus/intra-parietal sulcus. Tactile processing, writing, language processing (LH), and "reading" emotional facial expressions (RH).	LH: deficits in complex linguistic functions, working memory.
BA 41-42	Primary (A1) and secondary (A2) auditory cortex.	Cortical deafness; deficits in recognition and processing of complex sounds.
BA 43	Postcentral gyrus (PCG) with primary gustatory cortex.	Ageusia.
BA 44-45	Inferior frontal gyrus (IFG) with Broca's area. Speech production, imitation of hand movements, and sensory-motor integration.	Expressive aphasia.
BA 46	Also, "middle frontal area 46", corresponding to the dorsolateral prefrontal cortex (DLPFC). Sustaining attention, working memory, and executive functions.	Dysexecutive syndrome.
BA 47	The orbital part of the inferior frontal gyrus (IFG) has a curved appearance that also comprises the orbitofrontal cortex (OFC). Syntax of spoken language and music; linguistic processing, emotion processing.	Deficits in language tasks and musical analysis, and insufficient discrimination of different emotions.
BA 48, BA 49	Retro- (BA 48) and parasubicular (BA 49) areas are positioned between the hippocampus and entorhinal cortex. Not well-explored functions of spatial processing and regulation of the hypothalamic-pituitary-adrenal axis activity.	Somatosensory-spatial deficits, inappropriate social conduct
BA 50-51	Posterior insula. Interoception (BA 50), emotion and social cognition (BA 51).	Deficits in socially directed behavior.
BA 52	Parainsular area along the border of the insula and temporal cortex. Attention, auditory processing.	(Mild) attention disorder.

Abbreviations/comments: LH = left hemisphere; RH = right hemisphere;
*area originally found exclusively in the monkey brain by Brodmann (1909). According to Damiani et al. (2020), only 43 of the 52 BAs were initially found in man.

Impressively, contemporary neuroimaging continues to rely heavily on the Brodmann classification. Today, BAs primarily serve as a common code through which neuroscientists communicate findings with one another.

? Is Brodmann's parcellation of the cortex into minute functional fields not reminiscent of phrenology, if not on the skull, then on the brain's surface?

Brodmann was far from viewing his areas as "functional fields." What he did was to divide the cortical surface (and some subcortical regions) into areas that differed in cellular architecture. He even explicitly criticized the tendency among some of his contemporaries to localize mental functions to the brain in a simplistic way:

"In fact, the more recent period was abundant with theories that, in the manner of phrenology, sought to localize complex mental activities such as memory, will, imagination, intelligence, or spatial qualities like sense of form and sense of space in different cortical districts." (Brodmann 1909)

He regarded such ideas as "naïve conceptions" and concluded his book with a clear warning:

"One thing must be emphasized with all urgency: a functional localization of the cerebral cortex, of man as well as of animals, will remain absolutely impossible without guidance from anatomy." (Brodmann 1909)

Brodmann would almost certainly find today's neuroscientists' habit of localizing a BOLD signal to one of his areas a rather bold idea. Not that he thought of himself as a connectionist—he was, through and through, an anatomist.

Neuroimaging methods gained prominence in the 1970s. Techniques such as cerebral angiography and electroencephalography (EEG) were then complemented by X-ray computed axial tomography (CAT scans) and, later, by Positron Emission Tomography (PET) and Magnetic Resonance Imaging (MRI). PET and functional MRI (fMRI) were revolutionary, as they allowed scientists to observe a person's brain in action during the execution of specific tasks, whether motor or cognitive. For a brief history of brain imaging techniques, see Oldendorf (1978).

There are precursors to modern neuroimaging. The mathematical foundations for three-dimensional X-ray scanning were already described before 1920, the Austrian mathematician Johann Radon (see Parks 1986 for an English language translation). The technique of tomography dates back to the Italian radiologist Alessandro Vallebona, who called it *stratigrafia* at the beginning of the 20th century (Vallebona 1955).

Ideologically, functional neuroimaging was anticipated by the Russian physiologist Ivan Pavlov, who, in 1913, speculated about the possibility of localizing human brain functions by visualizing those neuronal structures activated during task performance (Pavlov 1913/1928; cited in Brugger 1997a). Pavlov wrote:

"If we could look through the skull into the brain of a consciously thinking person, and if the place of optimal excitability were luminous, then we would see playing over the cerebral surface, a bright spot with fantastic, waving borders constantly fluctuating in size and form, surrounded by a darkness more or less deep, covering the rest of the hemispheres."

Brain imaging techniques that allow us to visualize white matter pathways emerged relatively late in the history of neuroscience. They only became possible once the focus of interest began to shift—from asking what the individual places on the map do, to exploring what the connections between them might reveal about cognitive functioning.

Admittedly, there are important precursors to connectionist thinking in the 19th century, particularly in the context of disconnection syndromes. These will be discussed under the heading Connectomics (see below, p. 41). For now, suffice it to say that, although we are currently living in an era shaped by connectionist thought, we must not neglect the points on the map. Certain landmarks are essential for basic orientation. Others are unquestionably central—key hubs of grey matter that do matter for neuropsychology. These will be addressed in the following sections.

? **Your words about places on the map versus connections between them make me ask: What is more important to neuropsychologists, grey matter or white matter?**

This is a moot question. Thinking depends on grey matter—the sites where neurons fire and where activity can be measured with traditional neuroimaging techniques such as functional MRI. But thinking equally depends on the white matter tracts that connect those grey matter regions. If you think of grey matter as the chips and white matter as the wires, it becomes clear that only when both work in concert can the brain—as a system—function properly. Chips and wires are structurally and functionally distinct, but neither functions without the other. What matters most when dealing with brain topographies is to resist the urge to "localize" a function solely to the region whose damage produces its dysfunction. As John Hughlings Jackson already pointed out in the 19th century:

"To locate the damage which destroys speech and to locate speech are two different things." (Jackson 1874)

It is equally important to remember that brains exist within persons, and persons are part of societies and cultures. Social matters do matter, and no neuropsychologist should ever close their eyes to this fact (Brugger et al. 2018).

Topography: Major Cortical Landmarks

When learning a written list of ten actions—hammering, knitting, sawing, drawing, and so on—our recall after an hour will be better if we pantomime the actions rather than simply read the words (the enactment effect; see Chapter 6, p. 123). In the same spirit, your ability to orient yourself on the surface of the brain (the cortex) and in the underlying structures (subcortical nuclei, brainstem, and cerebellum) will profit from drawing a rough sketch of the different regions yourself.

Let us begin by sketching the outline of the left cerebral hemisphere. Start near the middle of your page, draw downward to the left, and continue counterclockwise until you close the loop and return to your starting point. The sheet in front of you should now contain something like this:

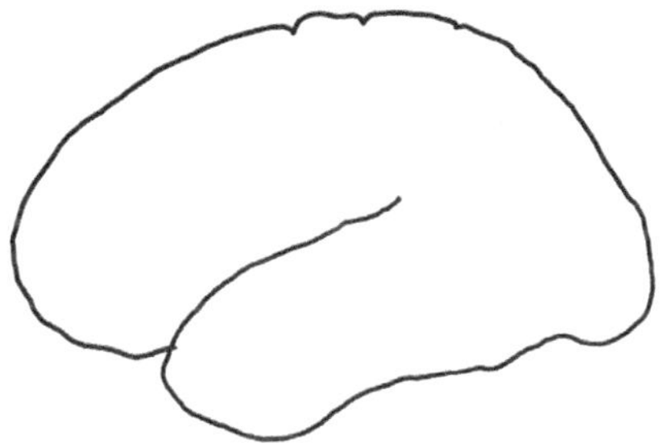

The line you began with is the lateral sulcus, also known as the Sylvian fissure. Sulci (plural of sulcus) or fissures are grooves on the cortical surface, with fissures simply being deeper, as in the case of the longitudinal or interhemispheric fissure, which separates the two hemispheres (though this one is not visible in your current drawing). A gyrus (plural: gyri) refers to the ridge-like elevation between two sulci.

Now add the following six lines to indicate some of the most important sulci and gyri:

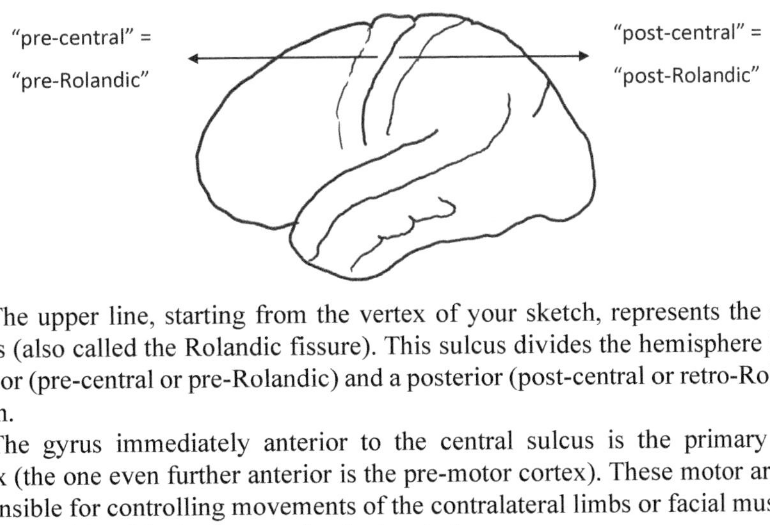

The upper line, starting from the vertex of your sketch, represents the central sulcus (also called the Rolandic fissure). This sulcus divides the hemisphere into an anterior (pre-central or pre-Rolandic) and a posterior (post-central or retro-Rolandic) region.

The gyrus immediately anterior to the central sulcus is the primary motor cortex (the one even further anterior is the pre-motor cortex). These motor areas are responsible for controlling movements of the contralateral limbs or facial muscles—that is, the side of the body opposite the hemisphere.

The gyrus immediately posterior to the central sulcus is the primary somatosensory cortex, which processes touch and other somatosensory sensations from the contralateral half of the body .

Add a few labels to your drawing, and you now have a very crude lateral view of the left hemisphere:

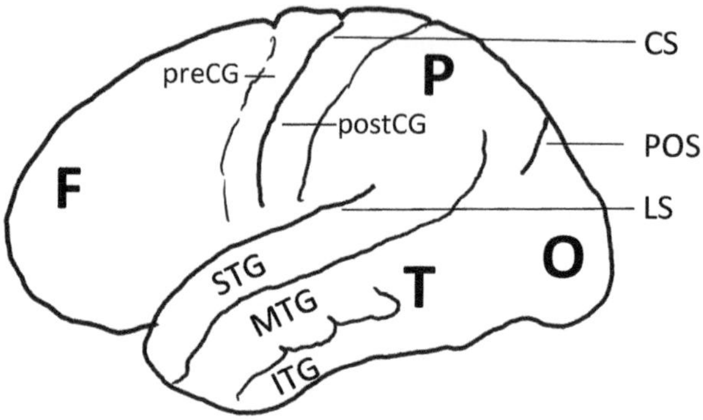

F = Frontal lobe, demarcated by CS (central sulcus) and LS (lateral sulcus);

P = Parietal lobe, demarcated by CS and POS (parieto-occipital sulcus) and indistinctly toward its caudal side from T;

O = Occipital lobe, demarcated by POS from P and indistinctly toward its anterior side from T;

T = Temporal lobe, demarcated indistinctly toward its posterior side from O and its cranial side from P;

preCG: precentral gyrus; postCG: postcentral gyrus; STG: superior temporal gyrus; MTG: medial temporal gyrus; ITG: inferior temporal gyrus.

A Very Brief Description of the Functions of the Four Lobes

There are entire books devoted to each lobe, so what follows is not only brief but also necessarily highly incomplete.

? Why only *four* lobes? I have read about the "limbic lobe" and seen the "insular cortex" discussed as yet another lobe.

It is indeed debatable how many lobes we want to distinguish within the cerebrum. In this book, I subsume the limbic system, including the hippocampus, amygdala, cingulum bundle, and some related structures, under the medial temporal lobe (see p. 25). Similarly, the insula, which lies buried beneath a lid-like structure at the lateral fissure (the operculum), is typically considered part of the temporal lobe (Kolb and Wishaw 2008). There are also close phylogenetic ties between limbic and insular structures.

What matters is that we acknowledge both limbic and insular functions as neuropsychologically significant—involving emotion and memory in the case of the former, and interoception, social cognition, and time perception in the case of the latter (see Craig (2009a,b), for an overview).

So, while I began this section by mentioning that there are entire books dedicated to the four classical lobes, I may now add that there are also academic monographs devoted to these "fifth" lobes: one to the limbic cortex (Isaacson 2013), and one to the insular cortex (Turgut et al. 2018).

The Frontal Lobes

The frontal lobe, located at the front of the brain, is the largest of the four classical lobes. It houses the motor cortex, the strip just in front of the central sulcus (see p. 21 above). The motor cortex is responsible for all voluntary movements, from walking and talking to fine motor skills. Just anterior to it lies the premotor cortex, involved in planning and controlling movements.

The prefrontal cortex, the most anterior part of the frontal lobe, is often described as the part of the primate brain that distinguishes humans from other primates. It plays a central role in complex thought processes such as reasoning, problem-solving, and decision-making. It is also crucial for regulating emotions and promoting socially appropriate behavior.

Damage to the frontal lobe can have profound consequences, resulting in inflexibility, perseveration, personality changes, and difficulties with concentration and task completion. Lesions in motor or premotor areas may result in paralysis, loss of voluntary movement in specific body parts, or difficulty coordinating sequential behaviors. For more on the diverse functions of the frontal lobes, see Chapter 8.

The Parietal Lobes

The parietal lobe, located near the top and rear of the brain, integrates somatosensory information from across the body and plays a key role in spatial awareness and coordination. The postcentral gyrus, located just behind the central sulcus, receives input from sensory receptors in the skin, muscles, and joints. This enables the brain to process sensations such as touch, pressure, pain, and temperature.

The somatosensory cortex also creates a map of the body (homunculus in males, femincula in females, see Chapter 10), allowing individuals to localize sensations on the skin and within the muscles. Additionally, the parietal lobe contributes to spatial reasoning, navigation, tool use, pantomiming (see Chapter 5), and cognitive tasks such as reading, writing, and understanding numerical concepts.

Damage to the parietal lobe can lead to sensory deficits, spatial neglect, and difficulty directing attention to the contralateral side of space (see Chapter 9). Its proximity to the occipital lobe also makes it a frequent site of visual agnosias and related visual-spatial disorders (see Chapter 7).

One remarkable work devoted entirely to the parietal lobes is MacDonald Critchley's 1953 monograph *The Parietal Lobes*. Though more than 70 years old, it still offers invaluable insight for clinicians and neuroscientists. It remains unmatched in its coverage of morphology, comparative anatomy, and vascular supply. For those interested in the rich clinical symptomatology of these lobes, encompassing tactile misperceptions, bodily distortions, visual-spatial and language disorders, as well as reflections on psychiatric symptoms and hysteria, the book remains a refreshingly and necessarily non-imaging-centered classic.

? Are you seriously recommending reading a book that is more than 70 years old?

I am absolutely serious. MacDonald Critchley (1900–1997) was one of the most brilliant writers in the field of neurology. He wrote *The Parietal Lobes* at around 50, by which time he had already developed formidable expertise—not only in neurology, but in its behavioral dimensions as well. The cases he presents in that monograph, particularly those involving impairments in the perception of body and space, remain instructive—especially for neuropsychologists.

For those who would prefer to read more broadly about brain-behavior relations in a more accessible style, I recommend *The Divine Banquet of the Brain* (Critchley 1979). This collection of essays touches on nearly everything covered in the present book, from aphasia to misoplegia.

I have now answered more than you asked—but let me add one final thought: yes, we should read the classics. Some neuropsychology journals feature "historical sections." These are sometimes dismissed as a "Jurassic Park of long-forgotten papers and fossilized ideas" (Goldenberg 2001). However, as Goldenberg himself notes, reviving old ideas and elaborating on them from new perspectives can spark scientific progress. Especially when it comes to clinical insight, MacDonald Critchley's writings remain a rich source for the rejuvenation of old ideas.

The Occipital Lobes

The occipital lobe, located at the posterior end of the brain, is the first cortical region to receive visual input via the lateral geniculate nucleus of the thalamus. The primary visual cortex (V1) is responsible for the initial stages of processing visual information, including light, color, and shape. Motion perception, on the other hand, is a largely independent function, handled by area V5.

Visual information is then transmitted to other brain regions for further processing. The ventral route, which supports object and face recognition, is directed toward the temporal lobe, while the dorsal route, which enables visually guided action, is directed toward the parietal-frontal cortex.

Damage to the occipital lobe may lead to a wide range of visual disorders, including selective blindness for color or motion, visual field deficits (from scotomas to quadrantanopia or hemianopia), and difficulty recognizing objects or faces (different visual agnosias; see Chapter 7). Visual hallucinations resulting from occipital lesions are usually unimodal, but may become multimodal when adjacent cortical or subcortical areas are also involved.

The Temporal Lobes

The temporal lobes are located on the sides of the brain. The name temporal refers to the region near the temples, although it is an intriguing coincidence that the insula, buried deep beneath the lateral sulcus, is centrally involved in time perception.

The temporal lobe is primarily concerned with auditory processing, language comprehension, and memory. The auditory cortex enables us to hear, recognize, and understand speech, sounds, and music. Wernicke's area, located posterior to the auditory cortex in the left hemisphere, is essential for language comprehension (see Chapter 4). Its homologue in the right hemisphere analyzes nonverbal environmental sounds but also contributes to language, specializing in rare or non-prototypical expressions, whereas Wernicke's area proper handles more standard linguistic forms (Passeri et al. 2015).

The medial temporal lobe (MTL), situated beneath the cortical surface, comprises the hippocampal formation, the parahippocampal region, and the amygdala. It plays a central role in memory formation (see Chapter 6), spatial navigation, and emotion processing. The amygdalae, located adjacent to the hippocampi, are involved in associative learning, especially when emotional salience is high. They also play a key role in fight-or-flight responses and are implicated in the genesis of phobias.

The Limbic System

As the term "system" implies, the limbic system has no fixed anatomical boundaries. At its core are structures of the medial temporal lobe, especially the hippocampus and amygdala. Some authors include the hypothalamus, which connects the nervous and endocrine systems. Others expand the system to encompass the cingulate gyrus, thalamus, and parts of the basal ganglia.

A common mnemonic describes the five Fs of limbic function:

Feeding (hunger, thirst, satiety; hypothalamus);

Forgetting (memory; hippocampus);

Fighting (or Fleeing, emotional reactivity; amygdala);

Family (reproduction and parental behavior);

Fornication (sexual drive).

The functional overlap with the insula is striking (see Table 3).

A more concise way to remember the breadth of limbic function is through Jaffe's (2000) memorable definition: "Psychiatry is but neurology of the limbic system."

The dazzling behavioral symptomatology that follows limbic disturbances often extends beyond the well-known memory deficits of hippocampal damage. Particularly notable is the high epileptogenicity of limbic structures. Seizures in temporal lobe epilepsy (TLE) are not the classical convulsive episodes with full loss of consciousness. Instead, they are associated with a behavioral syndrome involving intense spirituality, cosmic preoccupations, and a belief in paranormal forces (Waxman and Geschwind 1975; Roberts et al. 1990).

? I have read that the temporal lobes also house a "god module"?

I assume you came across this in a popular science text... As simplistic as it is to frame an entire lobe as being *responsible* for the experience of God, the idea captures an important truth about the unique role of the (medial) temporal lobes in a semiology that includes bliss, cosmic serenity, and encounters with divine presence.

The association between temporal lobe epilepsy (TLE) and religious experiences has been documented as far back as Babylonian times (Reynolds and Kinnier Wilson 2008). Based on a model of sensory-limbic hyperconnection, Bear and Fedio (1977) developed a standardized questionnaire to assess the experiential phenomena of TLE. Later, Persinger and Makarec (1987) created an inventory to measure similar experiences in healthy individuals.

Persinger's research demonstrated that functional hyperconnectivity in cortical–limbic circuits is accompanied by elevated signs of TLE, including hyperreligious and other irrational beliefs (Persinger 1987). Conversely, in epileptic patients, these signs may become exaggerated, resulting in what Crowe and Kuttner (1991) described as "schizophrenia-like episodes of TLE" during the interictal phase.

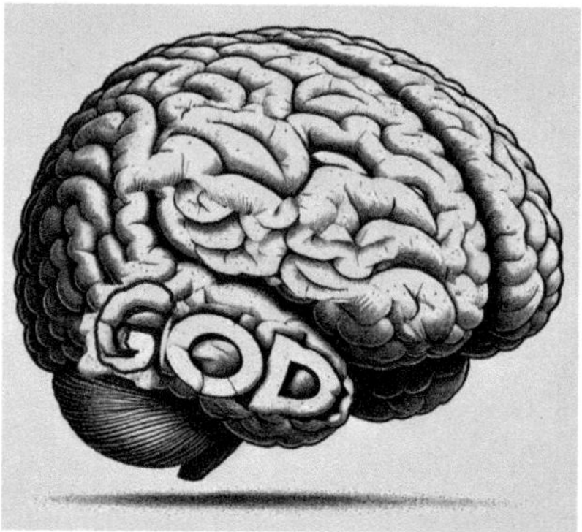

The popular concept of a "god module" in the right temporal lobe is based on observations that patients with temporal lobe epilepsy often report mystical experiences, a sense of a divine presence, and sudden conversions from atheism to strong religious convictions.[1]

The Insula

This part of the temporal lobes is bilaterally represented, lying deep within the lateral fissures, and covered by the cortical opercula (or "lids"). The diversity of insular functions is mirrored in the multifaceted personality of Johann Christian Reil (1759–1813), the German physician who first described it. Reil was not only a popular general practitioner and renowned anatomist, but also a psychiatrist—in fact, he coined the term "psychiatry." For a long time, the insula was referred to as the "island of Reil."

Functionally, the insular cortex appears to be an "island for everything." The mnemonic in Table 3 will help you recall the six major functional domains it supports. For an accessible overview, I recommend Gogolla's (2017) primer and the comprehensive review by Craig (2009a).

[1] I thank ChatGPT, an AI model developed by OpenAI, for assistance with this figure. However, I must state here that its form of intelligence does not include the distinction between left and right nor that between the temporal and parietal lobes.

Table 3: INSULA: A mnemonic of the core functions of the insular cortex.

I	**Interoception** is at the top of all associations with the insular cortex and at the bottom of many of the emotional and cognitive functions that build upon it. Interoception enables us to feel hunger and experience satiation, monitor bowel filling, and track our heartbeat. It mediates the sensual aspects of touch and sexual arousal and the experience of orgasm. The insula integrates autonomous functions, visceral sensations, and gut feelings into a general awareness of feeling good or bad. It thus links individuals' inner homeostasis to a balanced emotional and social equilibrium (Adolfi et al. 2017). The insular cortex also processes vestibular information, which serves the balance of much more than just the physical body in space (Lopez 2016).
N	**Nociception**. As an interoceptive feeling, pain is a prominent stimulus to elicit insular responses. Pain studies show that noxious stimulation applied to one hand activates the contralateral posterior insula. Empathic feelings for a loved one receiving the painful simulation, however, were associated with activation of the anterior parts of both insulae. A general rule is that the posterior insula is more responsive to the physical aspects of noxious stimulation, whereas the anterior insula is more responsive to more abstract forms of pain such as suffering from social isolation (Eisenberger (2015) for critical review). Nociceptive experiences related to pain include itch and disgust, the latter being a prime emotion of the insula, which houses the primary gustatory cortex. Again, disgust, as the nausea-related feeling about a physical stimulus, is coded in the posterior; its more abstract form, "moral disgust," is in the anterior insula.
S	**Self-awareness** includes the **sense of time.** The insula's interoceptive expertise is accompanied by a general interest in the self. What are its bodily boundaries, how do we feel today, and toward whom should we open ourselves up? People who believe that one of their fully functional limbs does not belong show structural alterations in the insula, and many less dramatic disorders of the bodily self have an interoceptive component (Brugger and Lenggenhager (2014) for review). Over time, one's self is experienced as a constant. Our sense of time enables us to connect early versions of our autobiographic self to the current one and to anticipate the future. Time perception is critically dependent on the insula (Craig 2009b; Naveh and Arzy 2019).
U	**Urges** imply strongly encouraging someone to do something. They are mostly experienced as negative impulses, and the U in INSULA is closely related to the A for addiction. While urges to consume a substance are better known as cravings, we often use the word in a non-substance context, such as the urge to gamble, read, or write. Damage to the insular cortex was found to reduce the urges associated with nicotine use. "Premonitory urges" are uncomfortable sensations of mounting inner tension that precede a tic in Tourette's syndrome and are temporarily alleviated by tic execution. Patients find these sensory aspects of the disorder as troublesome as the motor tics themselves (Cox et al. 2018). Interestingly, high levels of interoceptive awareness are associated with more frequent premonitory urges to tic.
L	**Love 'n' Lust**. Loving friendship and sexual desire are neighbors in the insula. In accordance with the posterior-to-anterior gradient, posterior insular regions code for the more corporeal, sensory aspects of attraction, while anterior regions code for the more long-lasting component of "true" (also platonic and maternal) love and social bonding (Cacioppo et al. 2012). There is quite a literature on the insular involvement in orgasm. The review and meta-analysis by Stoléru et al. (2012) showed that, apart from the insula, a dozen other areas are also activated. Also, sexual arousal is most frequently brought about by having subjects watch erotic movie clips, such that the research focus is relatively narrow. In a larger context, neuroimaging work with romantically involved partners has shown that purely anticipatory representations of tactile caressing activated posterior insular cortex regions in the same way as the anticipation of reward (Ebisch et al. 2014).
A	**Addiction**. The insula has been labelled "the hidden island of addiction" (Naqvi and Bechara 2009). This holds for substance abuse, where drug-induced interoceptive states translate into overwhelming urges, which can only be satisfied by further drug intake. However, excessive use of online technology is likewise associated with insular (dys) function. A systematic review (PRISMA protocol) identified more than 20 neuroimaging studies published between 2013 and 2023, which suggested that the insula is invovled in Internet and smartphone addictive behavior (León Méndez et al. 2024).

Before going to describe selected areas and structures covered by the cortical surface we have drawn so far, we add the brain stem and cerebellum to reach the following view:

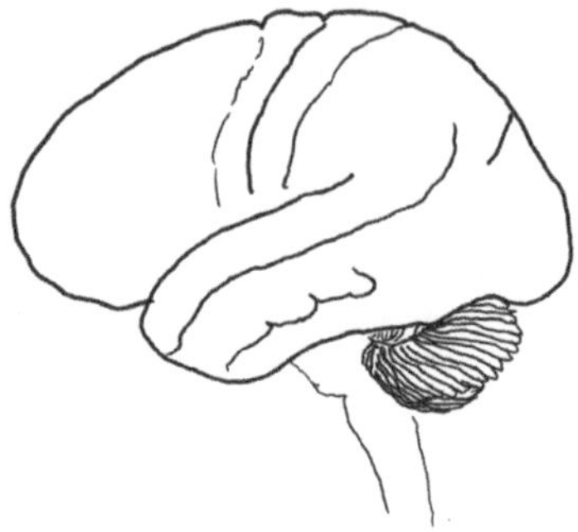

And before we "cut the brain open," we must familiarize ourselves with the methods we can use to do this.

Anatomical Planes

The brain is a 3D object. There are thus three planes that separate (1) left and right, (2) front and back, and (3) upper and lower portions, respectively.

(1) *The Sagittal Plane*

It is perpendicular to the ground and splits left and right hemispheres. If it is laid through the middle of the corpus callosum, it is labelled the "medial" or "midsagittal" plane (see, e.g., Figure 2). It is "parasagittal" when it is shifted laterally, i.e., goes through one hemisphere.

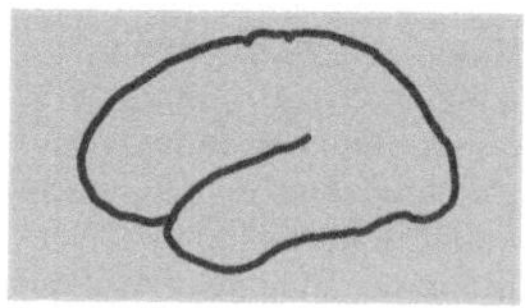

(2) *The Coronal Plane*

It is also perpendicular to the ground, but splits the front and back parts of each hemisphere. These parts are referred to as "anterior" and "posterior," respectively. The cut displayed here is a coronal section through the posterior part of the brain.

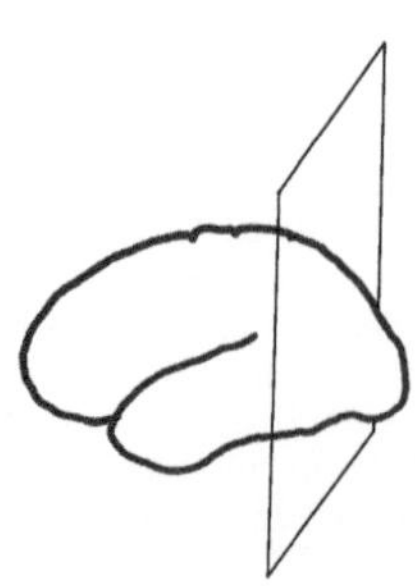

(3) *The Transverse (Also "Axial" or "Horizontal") Plane*

splits the top and bottom parts of the brain. These parts are referred to as "cranial" and "caudal," respectively. The cut displayed here splits the brain approximately at the level of the basal ganglia. See Figure 3 for a transverse section at the height of thalamus and basal ganglia.

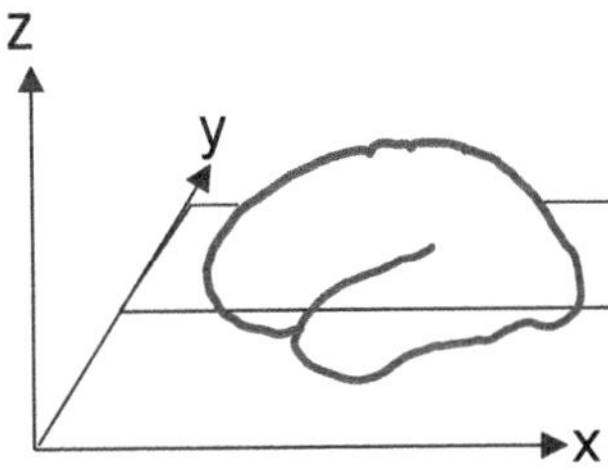

You will need these terms to interpret MRI slices. For instance, some parasagittal planes, shifted along the *y*-axis, show the hippocampus (of either the left or right hemisphere) in its anterior–posterior extension. Many coronal planes show the hippocampi of both hemispheres in a particular cross-section along the *x*-axis, and many transverse sections, each presenting them in cross-section along the *z*-axis.

Note that, according to radiological convention, brains in a transverse section are viewed from below—so left and right are reversed in the image.

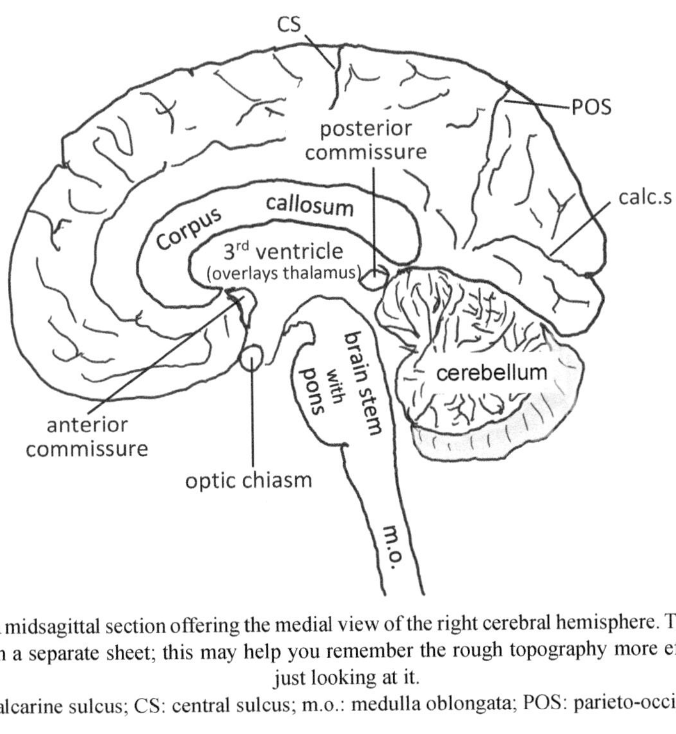

Figure 2: A midsagittal section offering the medial view of the right cerebral hemisphere. Trace the figure or copy it on a separate sheet; this may help you remember the rough topography more effectively than just looking at it.

Calc.s.: calcarine sulcus; CS: central sulcus; m.o.: medulla oblongata; POS: parieto-occipital sulcus.

Selected Subcortical Structures and the Cerebellum

Subcortical structures are as functionally specialized as cortical regions. Their importance for neuropsychology is often underestimated, in part because the cortex is traditionally seen as the seat of "higher functions," and research is correspondingly biased.

Moreover, standard techniques such as EEG and TMS are inherently surface-oriented—they can record from or stimulate cortical areas much more effectively than subcortical ones. (Stereotactic EEG with depth electrodes is not considered here; see Gholipour et al. 2020, for an overview.)

Table 4 provides a brief overview of the frequency with which selected subcortical structures are mentioned in the PubMed database in combination with the term "neuropsychology" during the decade from 2015 to 2024 (last accessed December 26, 2024).

Table 4: Number of publications listed in MedLine/PubMed that contain the keyword "neuropsychology" in title, abstract, or text in combination with a subcortical structure.

Structure	# Named Without the Term "Neuropsychology"	# Named in Combination With Term "Neuropsychology"	Percentage (Column 3/ Column 2) * 100%
Thalamus	18,078	237	1.31
Basal ganglia	13,352	176	1.32
Striatum	25,900	275	1.06
Caudate	7,859	146	1.90
Putamen	6,729	125	1.86
Accumbens [nuclei]	8,279	62	0.75
Subthalamic [nucl.]	4,060	61	1.50
Substantia nigra	8,988	31	0.35
Hippocampus	72,636	652	0.90
Amygdala	18,708	306	1.64
Cingulate [gyrus]	18,294	422	2.31
Brain stem	10,364	29	0.28
Cerebellum	25,609	269	1.05

Note: Search period 2015–2024. To control for publication frequency independent of neuropsychology, the percentage of publications with vs. without the keyword "neuropsychology" is also provided. Search terms did not include the words in square brackets.

Table 4 allows for the following tentative inferences: the percentage of publications containing the keyword "neuropsychology" relative to the total number of publications mentioning a given subcortical structure—without reference to neuropsychology—is surprisingly consistent across the different target structures. The proportion ranges from 0.3% for the brainstem to 2.3% for the cingulate gyrus.

Interestingly, if the same analysis is applied to the four major cortical lobes, the percentages are only slightly higher: 3.16% for the frontal, 3.61% for the parietal, 0.54% for the occipital, and 3.73% for the temporal lobe.

The following ultra-brief descriptions of the main functions of the subcortical structures listed in Table 4 aim to illustrate their contributions to neuropsychologically relevant processes.

Anatomy and Functions of the Thalamus

The functional organization of the central nervous system is hierarchical in nature. Primary sensory areas deliver information about the external world, while primary motor centers initiate responses according to current demands. Between these two poles lies an immense amount of integrative processing—a task largely orchestrated by the thalamus.

Often described as a gatekeeper or relay station, the thalamus filters incoming signals and routes them to the appropriate cortical regions. In this sense, the role of the thalamus is crucial for all neuropsychological functioning, as every cognitive or emotional act ultimately builds upon sensory input, whether exteroceptive or interoceptive.

A broad distinction may be drawn between three functional types of thalamic nuclei:

(1) Relay-type processors, which transmit afferent signals from primary sensory pathways and project to specific cortical targets;

(2) Association-type nuclei, which convey preprocessed cortical signals to associative cortical areas;

(3) Nonspecific thalamic neurons, which receive input from the cortex, other thalamic nuclei, or the brainstem‘s ascending reticular activating system (ARAS) (see Chapter 3), and project diffusely back to various cortical areas.

Lesions to the thalamus may cause diffuse neuropsychological impairments, such as disturbances in the sleep–wake cycle, hypersomnia, and fatigue. They can also result in more focal deficits, particularly frontal lobe-related cognitive impairments, which are especially persistent after bilateral or left-sided infarctions (Hermann et al. 2008).

When disrupted, each of the three thalamocortical loops (outlined in Figure 1, Chapter 8) may result in distinct neuropsychological deficit patterns—targeting motivation/arousal, intellectual processing, or affect regulation and inhibition.

A left-sided thalamic stroke can lead to thalamic aphasia, as well as dysarthric and hypophonic speech (see Chapter 4). In contrast, right-sided infarctions may result in severe hemispatial neglect (see Chapter 9).

Amnestic syndromes are another frequent outcome of thalamic pathology. These often co-occur with disorientation to time and place, as well as a pronounced lack of motivation or drive. The thalamic pain syndrome, while not cognitive in nature, is indirectly relevant to neuropsychology, as chronic pain can broadly impair both cognitive and motivational performance.

Figure 3, following the next paragraph, shows the location of the thalami in a transverse section, along with the surrounding basal ganglia.

Anatomy and Functions of the Basal Ganglia

The basal ganglia are bilateral structures that surround the two thalami. They form a highly integrative network consisting of the striatum (which includes the caudate nucleus, putamen, and nucleus accumbens), the pallidum, the subthalamic nucleus, and the ventral tegmental area/substantia nigra complex. Their anatomical and functional unity was already recognized in the 17th century (Parent 1986).

For neurologists, the basal ganglia are primarily associated with movement disorders—both hyperkinetic and hypokinetic—which often result from disrupted modulation of thalamo-cortical projections. Clinical syndromes include Tourette syndrome, Obsessive–Compulsive Disorder, and a broader spectrum of addictive behaviors. Also, Huntington's disease, like Tourette's, is a hyperkinetic disorder caused by the gradual degeneration of the basal ganglia (especially the caudate nucleus and putamen).

Parkinson's disease is the key motor disorder associated with basal ganglia dysfunction. It typically involves dopaminergic underproduction in the substantia nigra and frequently affects not only movement but also cognition and affect regulation. The dense functional connectivity between basal ganglia structures and the cortex, especially the frontal lobes, as well as with limbic structures (notably between the striatum and the hippocampus–amygdala complex), supports rich motor-cognitive interactions.

? With respect to neuropsychology, what are the major differences between Parkinson's and Huntington's disease?

Parkinson's disease (PD) is characterized by bradyphrenia (slowed thought processes) and a movement disorder affecting voluntary actions, including resting tremor, rigidity, and bradykinesia. Memory deficits primarily manifest as retrieval difficulties, while encoding remains relatively preserved; patients can often benefit from cueing. As the disease progresses, a dysexecutive syndrome emerges, affecting planning, flexibility, and working memory (see Chapter 8). Some but not all patients develop dementia, with a profile similar to dementia with Lewy bodies. Depression and apathy are common, and some cases exhibit medication-induced impulse control disorders.

Huntington's disease (HD), by contrast, is an autosomal-dominant disorder caused by a mutation in the huntingtin gene. It presents with a triad of motor, cognitive, and psychiatric symptoms. Unlike PD, executive dysfunction often precedes motor symptoms and is characterized by marked deficits in attention, cognitive flexibility, and problem-solving. Procedural learning is severely impaired, and retrieval deficits cannot be compensated for by cueing. HD features choreiform involuntary movements, which clearly differentiate it from PD even for a relatively unexperienced observer. Psychiatric symptoms—such as irritability, aggression, and apathy—often precede motor signs. HD patients also show greater anosognosia for their movement disorder, even when

> controlling for cognitive decline (Sitek et al. 2011). Sleep disturbances differ between the diseases: PD is associated with REM sleep behavior disorder (acting out dreams), whereas HD is characterized by a drastic reduction in REM sleep.

Of particular relevance to neuropsychology are the behavioral consequences of deep-brain stimulation (DBS) in Parkinson's disease. While DBS is often highly effective in alleviating severe tremor, stimulation of the subthalamic nucleus may lead to perplexing disinhibition at the behavioral level.

> Pathological gambling and other forms of non-substance addiction are frequently reported, particularly following deep brain stimulation (DBS), even when the intervention is successful in alleviating motor symptoms. Moral reasoning may also deteriorate considerably.
>
> In one striking case of otherwise successful treatment with subthalamic nucleus stimulation, the patient began to exhibit intolerable sexual disinhibition (unpublished observation). He would invite two or three prostitutes to his home—where he lived with his wife—and openly engage in novel sexual practices. Eventually, the neurologist and the patient's wife agreed to reduce the stimulation frequency. The patient—formerly known for his empathy and insight—gradually regained emotional sensitivity and expressed genuine remorse. He ultimately provided written consent to discontinue the treatment, despite the fact that his motor symptoms (tremor) had again become severely disabling.

The dorsal striatum, divided into the caudate nucleus and putamen, is involved in sensorimotor activities. Its ventral counterpart, which includes the nucleus accumbens, is connected to the limbic system via the dopaminergic mesolimbic pathway. There are extensive bidirectional connections between the striatum and pallidum.

Pallidal neurons operate according to a principle of disinhibition: they fire at high baseline rates in the absence of input, and striatal signals reduce their firing rate. Because pallidal neurons exert inhibitory effects on their targets, the net effect of striatal input is a reduction of tonic inhibition. These complex interactions—between limbic and subcortical areas, or across different neurotransmitter systems—may help explain seemingly inexplicable behavioral changes in cases of basal ganglia disease or intervention.

Among the often-neglected questions in neuropsychology are those linking paranormal belief to brain function. Schizophrenia—and its milder, nonclinical variant schizotypy—is associated with a marked tendency to believe in scientifically unsubstantiated forms of causation (Brugger et al. 1994b). Acute psychotic states typically coincide with hyperdopaminergic activity. Indeed, administration of levodopa can induce schizophrenia-like perceptual aberrations and belief patterns even in healthy individuals (Krummenacher et al. 2010b).

Strikingly, Parkinson's disease is associated with lower levels of religiosity in cross-sectional studies. Conversely, pronounced religiosity has been found to reduce the risk of developing Parkinson's disease (Otaiku 2023).

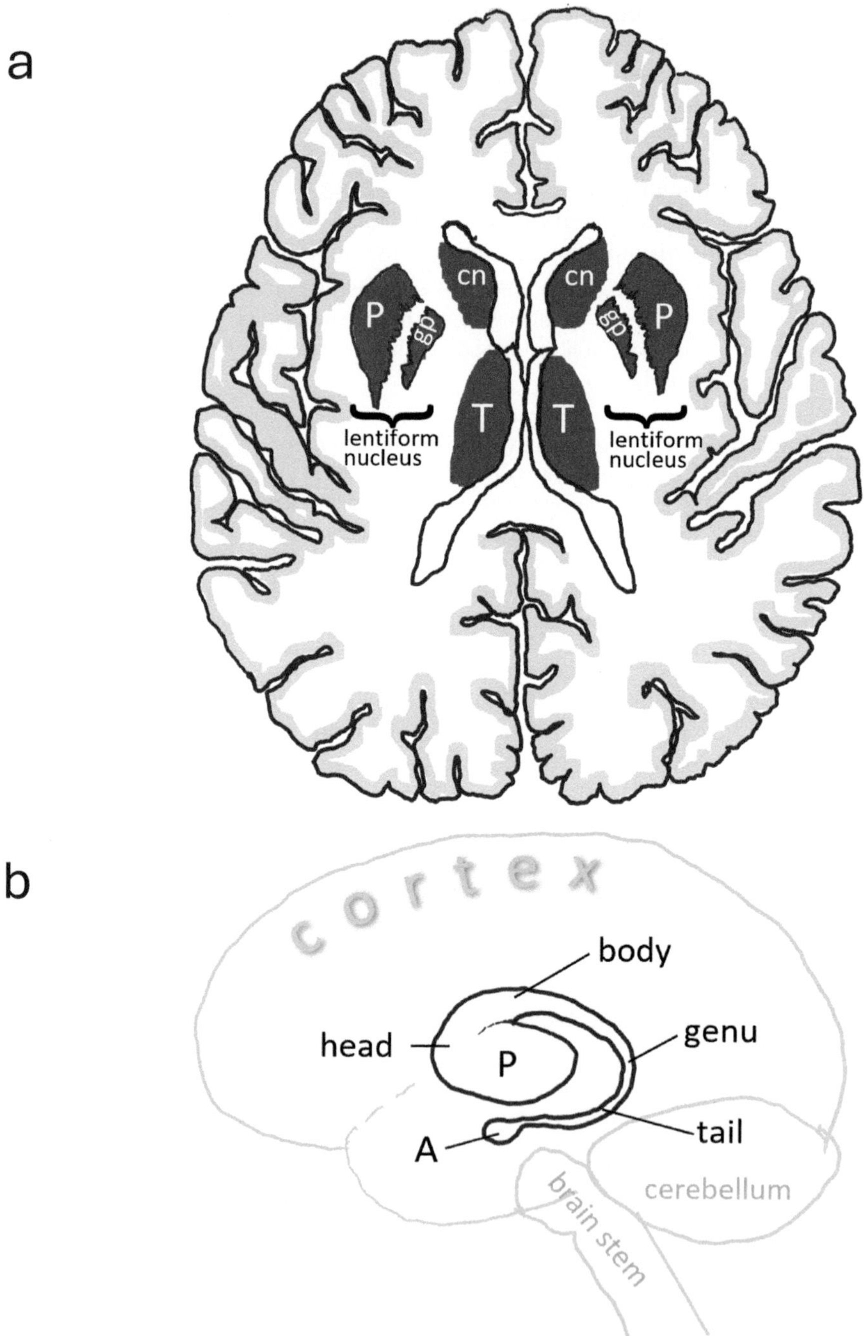

Figure 3: a: Transverse section of the human brain at the height of the thalamus (T) and basal ganglia (dark grey). The latter structures consist of the putamen (P) and the globus pallidus (gp), which form together the lentiform nucleus. Also part of the basal ganglia is the caudate nucleus (cn), whose head is visible in this transverse section. b: Lateral view of one caudate nucleus. A: amygdala, P: putamen.

Anatomy and Functions of Hippocampus, Amygdala, and the Cingulate Gyrus

These three structures are part of what is commonly referred to as the "limbic system"—a term that is less precise than limbic lobe, but perhaps more appropriate, given the functional diversity and systemic interdependence of its components. The archaic nature of the limbic system justifies a conceptual separation from the evolutionarily younger cortical lobes. While the location of limbic structures can be clearly illustrated (see Figure 4), their system function cannot easily be captured by listing the roles of individual parts.

> The hippocampus is most commonly associated with memory formation. Informal tests conducted with over 100 neuropsychology students (unpublished observation) revealed that, in a 1-minute category fluency task for associations to the word "hippocampus," terms like "memory," "learning," or "amnesia" appeared among the first three responses in 100% of participants. However, terms such as "navigation," "hypoxia," or "seizure" were rarely mentioned.
>
> For those readers engaged in teaching, I recommend administering controlled association tasks like this early in a course. The results may significantly inform the content of future lectures, allowing instructors to highlight neglected aspects and correct common stereotypes, many of which may stem from popular media coverage.

Some hippocampal neurons fire selectively in specific spatial locations (place fields) when an animal moves through an environment. Shapiro (2001) described this functional domain as "hippocampal neuropsychology." Disorders of spatial learning and topographagnosia (see Chapter 7) may be rooted in low-level associative processing within the hippocampus.

Temporal lobe epilepsy (TLE) due to hippocampal sclerosis is the most common form of epilepsy in humans. A key aspect of TLE's neuropsychological semiology has already been summarized above (see p. 25-26).

The hippocampi work in close concert with the amygdalae, which add emotional salience to what is learned and familiarity to what is explored in space. The amygdala is central to the fight-or-flight response and is often viewed as specialized for processing negative emotions such as fear and anger. It modulates aggressive impulses and contributes more broadly to an affective tone to experiences and memories. (See Chapter 11 for a discussion of the lateralization of emotion, with the left and right amygdala preferentially involved in positive and negative emotions, respectively.)

The amygdala's functions extend from autonomic and endocrine regulation to decision-making and empathetic responses in social interactions (Penagos-Corzo et al. 2022). Clinical neuropsychologists assessing patients after a selective amygdala–hippocampectomy (Hoyt and Smith 2016) must look beyond standard tests of learning and memory. Emotional, social, and behavioral changes are often equally significant.

The cingulate gyrus is another core component of the limbic system. Its neural architecture supports attentional control and emotional awareness. It includes the cingulum bundle, one of the brain's longest white matter fasciculi, which projects to all cerebral lobes. Modern connectivity studies have greatly expanded our understanding of this structure. Maldonado et al. (2020) demonstrate how the earlier view of the cingulum as a single fiber tract was first revised into a two-, then a three-part division, and today, more than five meaningful subcomponents are recognized.

Due to its broad connectivity, both within the limbic system and beyond, the cingulate gyrus plays a central role in processing emotion, reward sensitivity, action planning, and social communication.

Figure 4 sketches the location of the limbic structures discussed in this section.

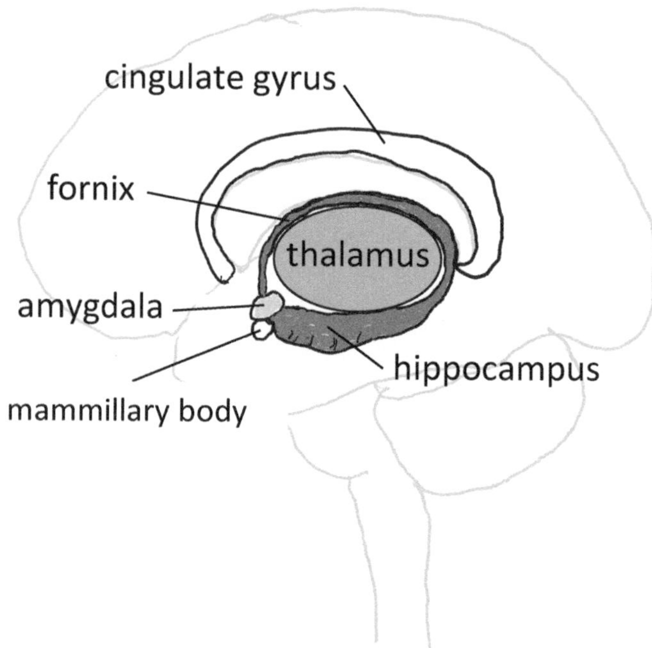

Figure 4: The major components of the limbic system: cingulate gyrus, just overlaying the corpus callosum, hippocampus, and amygdala. Hippocampus and fornix surround the thalamus—the former fuses with the latter in its posterior part. Together with the mammillary bodies, these structures play a crucial role in the consolidation of memories (see Figure 4, Chapter 6).

Anatomy and Functions of the Brain Stem

The brainstem connects the spinal cord, cerebellum, and cerebrum. It is composed of three sections, from inferior to superior: the medulla oblongata, the pons, and the midbrain. Brainstem functions are vital, regulating breathing, blood pressure, and the sleep-wake cycle. The modulation of wakefulness and attentiveness is discussed in Chapter 3, where Figure 1 also depicts the localization of brainstem structures within the human brain.

A rare but striking neuropsychological condition is peduncular hallucinosis, a form of visual disturbance that follows midbrain lesions (Lhermitte 1922), despite intact eyes and visual cortex. Neuropsychologists should routinely inquire about such phenomena in patients presenting with brainstem signs. These hallucinations are typically visual, but can also be multimodal (Taylor et al. 2005).

Anatomy and Functions of the Cerebellum

The cerebellum—literally, the "little brain"—is attached to the brainstem and consists of left and right hemispheres and a midline structure called the vermis. Its surface features much finer convolutions than those of the cerebral hemispheres. Importantly, in contrast to the cerebrum, the cerebellar hemispheres control the ipsilateral side of the body.

The cerebellum plays a central role in motor coordination, receiving indirect input not only from motor areas of the cerebral cortex but also from sensory areas and from cortical regions involved in language and emotion. Core symptoms following cerebellar damage include ataxia, vertigo, balance problems, and deficits in eye movement control.

? **What do neuropsychologists need to know about the cerebellum?**

You basically need to know two things.

First, the cerebellum is no longer thought to play a purely motor or balance-related role. Damage to cerebello-cerebral and cerebello-limbic circuits can result in a constellation of deficits across multiple cognitive domains, including spatial cognition, language, executive functions, and emotion regulation. This constellation is known as the Cerebellar Cognitive Affective Syndrome (CCAS) or Schmahmann's syndrome (Schmahmann 1998). The idea was already anticipated by the Russian neuropsychologist Alexander R. Luria (1902–1977); see the review by Budisavljevic and Ramnani (2012).

The nature and variety of CCAS symptoms depend on a patient's age and the location of the cerebellar lesion. Posterior lesions tend to lead to more severe deficits than anterior ones. While some neuropsychologists question whether CCAS truly qualifies as a distinct syndrome, others argue that it is underdiagnosed due to a general lack of familiarity within the scientific and medical communities.

Second, neuropsychologists should be aware of a phenomenon called cerebro-cerebellar diaschisis. This term refers to the finding that a unilateral cerebellar lesion—commonly an infarct—may result in a contralateral deficit in the frontal lobe. For example, a right cerebellar stroke may lead not only to right-limb ataxia (due to ipsilateral motor control) but also to anomia, reduced verbal fluency, or impaired verbal source memory (Tamagni et al. 2010). Conversely, left cerebellar lesions are frequently associated with reduced nonverbal fluency or other signs of right frontal dysfunction.

Similar crossed relationships exist in the other direction: left or right frontal cortical lesions may lead to right or left cerebellar symptoms, respectively. The most likely mechanism underlying these cortico-cerebellar symptom constellations is the disruption of cortico-ponto-cerebellar white matter tracts, resulting in hypometabolism of the contralateral cerebellar hemisphere (Infeld et al. 1995). Sokolov et al. (2014) even identified a structural pathway between the left cerebellum and the right superior temporal sulcus.

Not to be Forgotten: Anatomy and Function of the Ventricles

The ventricles are cavities within the brain that produce and contain cerebrospinal fluid (CSF), which circulates through the cerebrum and spinal cord. These cavities are interconnected and together contain about 150 ml of CSF in the human central nervous system. By absorbing and redistributing force, the ventricles act as a kind of "shock absorber," helping to reduce mechanical damage from impacts to the head or spine.

Beyond its protective function, the CSF plays an important role in cerebral metabolism. It delivers glucose to brain tissue and helps remove waste products. It also maintains a homeostatic balance of ions, which is essential for proper neuronal functioning.

The two lateral ventricles, symmetrically arranged, are the largest and produce the bulk of CSF (approximately 500 ml per day, meaning the total CSF volume is renewed about four times daily). CSF production occurs in a specialized tissue known as the choroid plexus.

The third ventricle lies along the midline, just beneath the corpus callosum and between the two thalami (see Figure 3). The fourth ventricle connects to the subarachnoid space surrounding the brain and spinal cord and is located between the brainstem and the cerebellum. The cerebral aqueduct is a narrow channel that links the third and fourth ventricles.

The location of the ventricular system is displayed in Figure 5.

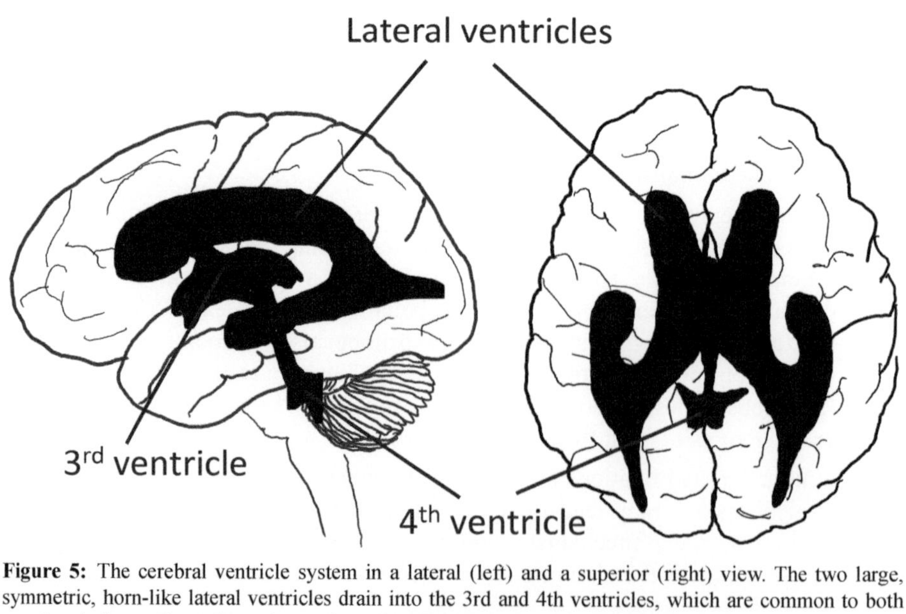

Figure 5: The cerebral ventricle system in a lateral (left) and a superior (right) view. The two large, symmetric, horn-like lateral ventricles drain into the 3rd and 4th ventricles, which are common to both hemispheres. From there, the cerebrospinal fluid runs down the central canal of the spinal cord.

Why should a neuropsychologist care about the brain's ventricles?

There are several conditions involving dysfunctional CSF circulation that affect cognitive processing. A common example is acquired hydrocephalus, in which cerebrospinal fluid accumulates either due to excessive production or a failure to reabsorb it (communicating hydrocephalus), or due to a blockage in normal flow (non-communicating or obstructive hydrocephalus). The resulting increased intracranial pressure can lead to severe neurological problems, including ataxia, motor or sensory deficits, or even death.

The primary neurosurgical intervention is the placement of a ventricular shunt. Neuropsychological functions, such as learning, memory, attention, fluency, and cognitive flexibility, can also be affected. However, it remains debated whether these deficits are domain-specific or simply the consequence of a general slowing of processing speed and reactivity.

Before placing a shunt, neurologists may attempt a diagnostic CSF tap, reducing intracranial pressure by removing a small amount of CSF via syringe. They then monitor for neurological improvement, which should be immediate and observable. Neuropsychologists may be asked to assess improvement in cognitive functions that were previously impaired.

This is not a simple task. It requires rapid differentiation between genuine domain-specific gains and non-specific improvements linked to arousal and processing speed. Standardized tests are of limited value in this context. What is needed are brief, repeat-resistant tasks that emphasize mental chronometry and allow for nuanced behavioral observation. In this sense, the ventricles serve as a reminder that neuropsychology can be an art (see Chapter 12): their dysfunction calls for clinical flexibility and highlights the limitations of overly standardized assessment.

Once we are hooked on to neuroanatomy, what else should not be forgotten, but is beyond the scope of this book?

☺ Among the basics not even touched here are:

(1) The functional anatomy of single neurons
(2) The peripheral nervous system and the spinal cord
(3) The cytoarchitecture of the cortex and other structures
(4) Cerebral vascularization (but see Figure 6, below)
(5) The neurotransmitter and neuroendocrine systems

These topics are covered in depth in Schoenberg and Scott (2011)—particularly in Chapters 3 and 13.

Figure 6 provides a rough sketch of the brain's blood supply by the three largest cerebral arteries, the arteria cerebri anterior, media, and posterior, respectively. In the bottom part of the figure (the medial view), please note one detail: while most of the corpus callosum is supplied by the ACM, its most posterior part, the splenium, receives blood from the ACP. One important implication of his fact for

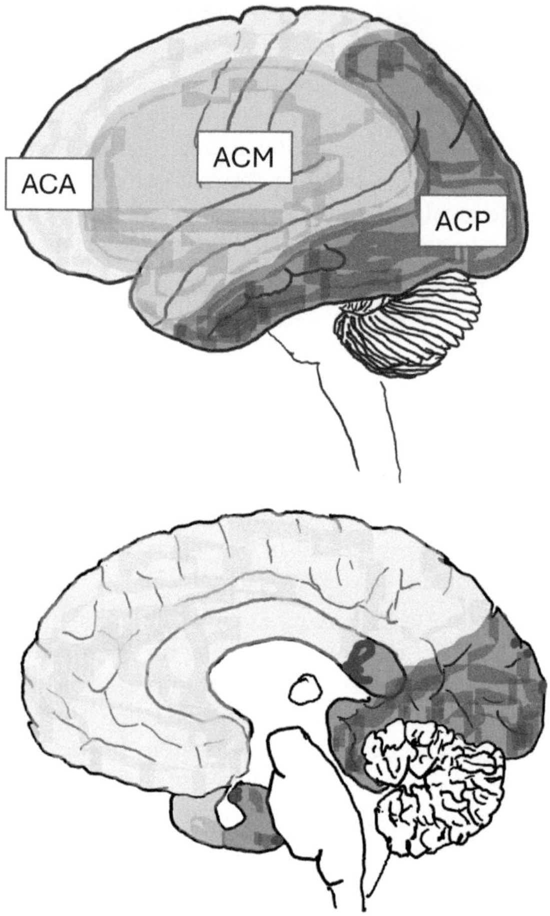

Figure 6: Lateral (top) and medial (bottom) view of one hemisphere showing the rough distribution of blood supply by three large cerebral arteries: ACA (arteria cerebri anterior, light grey), ACM (arteria cerebri media, grey), and arteria cerebri posterior (ACP, dark grey). The brain stem is supplied by the vertebral and basilar arteries, and the cerebellum is provided by three distinct cerebellar arteries.

neuropsychology was already recognized around the turn of the 19th to the 20th century based on autopsied cases. It is the conceptualization of pure alexia (i.e., the acquired inability to read while the ability to write is spared; see Chapter 4, p. 71). For the contributions of Carl Wernicke and Joseph Jule Dejerine, in the context of modern views of the disorder, see Bub et al. (1993).

Table 5 lists some neuropsychological impairments according to the vascular territories involved. All these impairments are discussed in the present book, and reference to this Table may be provided. The term "watershed zones" refers to areas of brain tissue bordering the territories of different cerebral arteries. Microinfarctions in these zones significantly contribute to age-related cognitive decline (Kapasi et al. 2018).

Table 5: Vascular territories and a selection of associated cognitive syndromes.

Arterial Territory	Typical Cognitive Syndromes
Anterior Cerebral Artery (ACA)	- Akinetic mutism - Abulia - Executive dysfunction - Alien limb syndrome - Hemisomatognosia (less often)
Middle Cerebral Artery (MCA)	- Various types of aphasia, depending on the arterial branch (mostly left hemisphere) - Apraxia (mostly left hemisphere) - Hemispatial neglect (more often right hemisphere) - Amnesia - Body schema disorders (depending on type, either left or right hemisphere) - Gerstmann syndrome (if left angular gyrus involved) - Visual field defects (optic radiations)
Posterior Cerebral Artery (PCA)	- Alexia without agraphia (left occipital + splenium) - Cortical blindness (with or without Anton's syndrome (bilateral occipito-temporal) - Prosopagnosia and topographagnosia - Memory impairment based on thalamic or hippocampal contributions
Watershed Zones	- Transcortical aphasias (especially between MCA and ACA or MCA and PCA) - Hypoxic encephalopathy - Disconnection syndromes

Connectomics

We live in an era that values connections—in technology, in the social sciences, and in neuroscience. The step from localization to the study of connections marks a significant leap forward in the brain sciences in general—and in neuropsychology in particular.

? **What is the difference between connectomics and hodology?**

Brain connectomics investigates the general wiring diagram of the nervous system. It has a much broader scope than hodology, which focuses on specific neural tracts and examines their functional neuroanatomy in detail. For instance, you would not typically refer to a "connectomics of memory" or a "connectomics of hallucinations," but you might well speak of the hodology of these functions or perceptual experiences (Ross 2008; Ffytche 2008).

On a methodological note, both connectometry and tractography are neuroimaging techniques used to study white matter pathways. These methods are complementary: the former highlights areas of interest, while the latter offers more detailed structural analyses of the relevant tracts.

Connectomics has become a far more common term—PubMed lists over 12,000 publications using it since the year 2000, compared to fewer than 60 for hodology (last accessed January 1, 2025). Its rise must be seen in the context of the expanding number of "-omes" across biological disciplines. The term "genome," for example, dates back to the 1920s and refers to the complete set of an organism's DNA, including all its genes (Sporns 2013). The microbiome comprises the full range of microorganisms living in a specific part of an organism, such as the gut. The Human Connectome Project (van Essen et al. 2012) played a major role in popularizing the term connectomics, contributing to its explosive adoption in recent years.

Brief History of the Study of White Matter

Precursors of connectomics can be traced back to the 19th century. Several of today's leading contributors to the field provide an outstanding and beautifully illustrated historical overview (Catani et al. 2013). Among the figures they discuss is René Descartes (17th century), whose drawings hinted at the existence of connecting fibers, though they lacked anatomical precision. In the 18th century, Galvani's concept of "animal electricity" introduced the idea that nervous activity could be understood in terms of electrical currents—a concept supported by early experiments that evoked peripheral muscle movement via central electrical stimulation.

The 19th century witnessed a significant leap in microscopic techniques, aided by the invention of the precision microtome and the development of advanced staining methods, which enabled researchers to track individual white matter tracts in great anatomical detail.

By contrast, 20th-century neuroimaging introduced novel tools for studying connections, but—as Catani et al. (2013) warn—also introduced new pitfalls. Contemporary brain mapmaking often relies on mathematical or conceptual networks rather than physiological ones. As they note, "we tend to see in our connectomes patterns of neuronal organization that may just represent a biased view of the real anatomy."

From a ranger's point of view, you can only value the forest once you've learned to value the trees. To paraphrase Brodmann (see p. 19): in our quest for a comprehensive connectome, we must never lose sight of anatomical grounding.

Types of White Matter Tracts

White matter tracts are classified into three major categories, depending on the anatomical regions they connect:

1. Commissural Tracts
   These interhemispheric pathways connect homologous regions across the two hemispheres. The most prominent is the corpus callosum, the largest white matter tract in the human brain (see Figure 6a). Other commissural fibers include the anterior and posterior commissures. In "split-brain operations" (commissurotomies; see Chapter 11), not all these connections are necessarily severed. A particularly important link is the hippocampal commissure, which

connects the two hippocampi and supports the synchronization of memory formation.

2. Association Tracts
   These fibers link different regions within a single hemisphere. Notable examples include the superior and inferior fronto-occipital fasciculi, which connect the frontal lobes to the posterior regions (parietal, temporal, and occipital) and are involved in semantic processing. The inferior longitudinal fasciculus (ILF) links the temporal and occipital lobes to support visual recognition and memory. The uncinate fasciculus connects orbitofrontal regions to limbic structures, likely contributing to emotional–cognitive integration (see Figure 7). Additionally, U-fibers connect neighboring gyri in a short-range arc, forming the shape characteristic of the letter U.
3. Projection Tracts
   These fibers connect cortical regions with subcortical structures, including the thalamus, basal ganglia, brainstem, or spinal cord. Most prominent is the corticospinal tract (also called the pyramidal tract), which is responsible for voluntary movement and extends from the motor cortex to spinal motor neurons. In the pyramids of the medulla, these fibers cross, such that the left motor cortex controls the right side of the body, and vice versa.

In addition to efferent motor fibers, projection tracts include afferent sensory fibers, which carry signals from the body to the somatosensory cortex. A less often mentioned example is the geniculo-calcarine tract, or optic radiation, through which visual information passes from the thalamus via the optic chiasm to the calcarine fissure in the occipital lobe (see Figure 2 of Chapter 11).

A richly illustrated and highly accessible introduction to white matter tracts is the article by Bullock et al. (2022). Figure 7 shows a selection of the "21 major tracts for the 21st century" discussed by these authors.

A phylogenetic perspective on the mammalian white matter connectome, with particular emphasis on hemispheric asymmetries, is provided by Güntürkün et al. (2020). These authors illustrate how comparative studies may help address foundational questions, such as whether certain major white matter tracts exert predominantly excitatory or inhibitory influences.

A critical issue for clinical neuropsychology is the ontogenetic development of the functional connectome. The prenatal organization of white matter pathways may offer biomarkers for identifying developmental disorders. Turk et al. (2019) demonstrated that some of the major features of adult human brain wiring are already evident in the fetal brain during the second and third trimesters of pregnancy.

A recent technical study by Voigt et al. (2023) highlights the advantages of combining hemodynamic (MRI-based) and metabolic (PET-based) approaches to connectivity analysis. These multimodal methods are crucial for understanding the intricate links between brain structure, brain function, and the emergence of cognition and emotion.

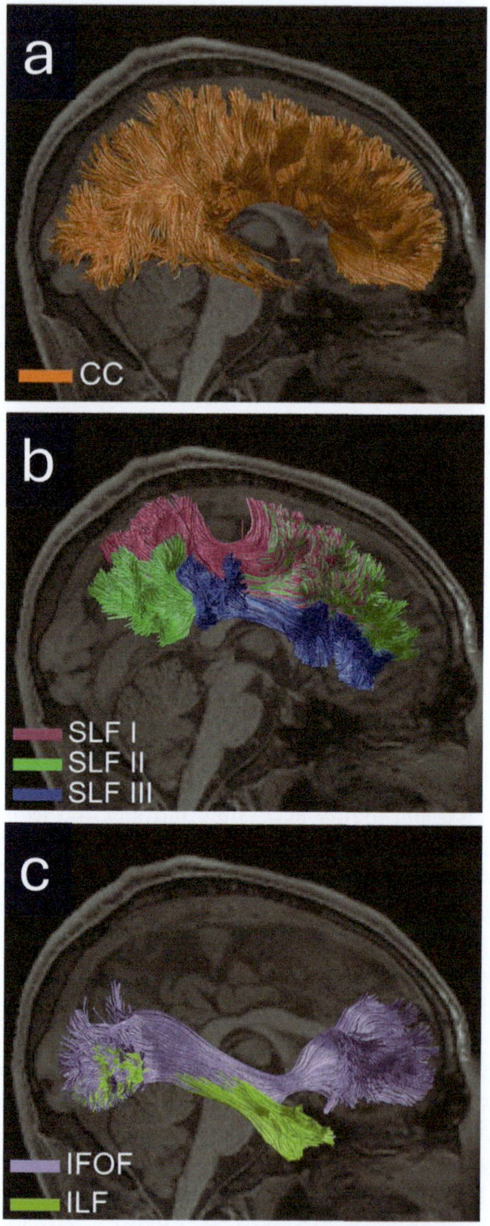

Figure 7: Sagittal views of selected white matter tracts of the human brain:
(a) Corpus callosum, the largest commissural fibre bundle.
(b) Three components of an association fiber tract, the superior longitudinal fasciculus (SLF): SLF I (pink), SLF II (green), SLF III (blue).
(c) Two more white matter association tracts, the inferior fronto-occipital fasciculus (IFOF; pink), which connects the frontal and occipital lobes, and the inferior longitudinal fasciculus (ILF; green), a long tract, whose precise endings are disputed.
Reprinted, with permission from Oxford University Press, from Bullock et al. (2022; Figures 2, 3, and 6, respectively).

? Speaking of connections: does not neuroscience miss something important when only concentrating on the neural connections within the central organ? Are not the connections beyond the CNS, between the brain and other organ systems, somewhat neglected?

The question is justified. While connections between the brain and other organ systems may not be primary topics in most neuropsychology textbooks, they are by no means neglected. In fact, substantial literatures exist on several key interactions, particularly between:

(1) the brain and heart,

(2) the brain and reproductive system, and

(3) the brain and gut.

(1) In the field of cardio-neurology, brain–heart connections are bidirectional, mediated primarily by the vagus nerve and the autonomic nervous system. Disruption of these links is associated with emotional disorders and vascular-cognitive decline. A striking example is Takotsubo syndrome, which can result in circumscribed neuropsychological deficits. It should not be reduced to a mere "broken heart" phenomenon—positive emotional states can also trigger it (Ghadri et al. 2016).

For neuropsychological rehabilitation, heart rate variability (HRV) is of special interest. HRV is widely used as a proxy for autonomic regulation of cognition and emotion, and biofeedback techniques may be useful in therapy. Increasingly important in experimental settings is the observation that the precise cardiac phase (systole vs. diastole) during which a stimulus is presented significantly influences multisensory integration (Saltafossi et al. 2023).

(2) The nervous and reproductive systems are connected primarily through hormonal pathways, especially those regulated by the hypothalamic–pituitary–gonadal axis. These connections influence cognition, mood, and social behavior. Sex hormones play vital roles in neuroplasticity, cognitive aging, and psychiatric vulnerability. They also have implications for understanding sex differences in neuropsychological function, which are too often attributed solely to sociocultural factors (Holden 2005). Tailored approaches—e.g., for individuals undergoing hormone therapy or for postmenopausal women—could optimize cognitive rehabilitation outcomes.

(3) The gut-brain axis is now recognized as a critical interface regulating affective states, cognition, and potentially neurodegenerative diseases (Morais et al. 2021). The enteric nervous system, also known as the "second brain," is part of the autonomic nervous system. Alterations in the gut microbiome have been linked to anxiety, depression, autism spectrum disorders, and Parkinson's disease. Gut-targeted therapies may soon complement traditional neuropsychological interventions, as gut flora significantly influence neurotransmitter activity. With reference to the brain-mind connection, Harrington (2007) considered the possibility that essential parts of the mind are not strictly housed in the brain, but move freely across a CNS-gut-immune system entity.

Even more revolutionary is the concept of "holobiont biology" (Bordenstein et al. 2024). A holobiont is the assemblage of a host organism and all the microorganisms living in or on it. There is growing evidence that certain parasites can affect their host's neuropsychological performance and even personality—effects that go beyond mere infection (Flegr et al. 2023).

Just as we once believed that planet Earth was the center of the universe, we are still tempted to think of the human individual as a unitary being. But in reality, we are "superorganisms" (Kramer and Bressan 2015), carrying more bacteria in our mouths than there are people on the planet. Acknowledging this reality may pave the way for a striking convergence of microbiological and cognitive sciences.

Where Can We Read More?

Fix, J.D. Neuroanatomy (3rd edition). Lippincott, Williams & Wilkins: Philadelphia, 2002.

Henley, C. Foundations of Neuroscience. Michigan State University Libraries: East Lansing, 2021.

Vanderah, T.W. and Gould, D.J. Nolte's The Human Brain: An Introduction to Its Functional Anatomy. (9th edition). Elsevier: New York, 2025.

Schoenberg, M.R., Marsh, P.J. and Lerner, A.J. Neuroanatomy primer: Structure and function of the human nervous system. *In*: Schoenberg, M.R. and Scott, J.G. (eds.). The Little Black Book of Neuropsychology, Springer: New York, 2011, 59–126.

See also Table 1.

CHAPTER 3

Awake, Alert, Attentive

What's the Difference?

Wakefulness is a prerequisite for any neuropsychological examination, but it is not an all-or-nothing state. Patients may drift in and out of sleep, require constant stimulation to remain awake, or need to be kept engaged through motor activity. Mild drowsiness is common and often tolerated to some extent by the examiner. Here, however, we are concerned with more profound disturbances of wakefulness. The Glasgow Coma Scale (Teasdale and Jennett 1974) provides a standardized method for quantifying deviations from normal waking consciousness. It comprises three subscores: eye opening, motor response, and verbal output (Table 1).

Although neuropsychologists do not typically administer the Glasgow Coma Scale (GCS), understanding its scores is important when reviewing patient records. A total score of 8 or below is generally considered indicative of a comatose state—the lower the score, the deeper the coma. In the context of traumatic brain injury, GCS scores are commonly used to classify severity: scores of 13 to 15 are labeled "mild," scores of 9 to 12 are labeled "moderate," and scores below 9 are labeled "severe."

? **At the end of a coma is death. Who determines that a person is dead, and by what criteria? What do "near-death experiences" teach us about a potential life after death?**

There was a time when holding a mirror to the nostrils or mouth of an unconscious person was considered the ultimate test of life or death. Today, the cessation of respiration and cardiac activity is no longer sufficient. Instead, the focus has shifted to the central nervous system, and the concept of "brain death" has become a central issue. A flat surface EEG is not in itself diagnostic, as deep brain activity may persist. To declare brain death, neurologists must confirm the complete absence of all brainstem reflexes. This includes the pupillary, oculovestibular, corneal, gag, and cough reflexes—each tested at the bedside. These findings must be confirmed in a second evaluation conducted 24 hours later. Confounding factors, such as hypothermia or the use of muscle relaxants, must be excluded. Despite brain death, artificial support can maintain respiration and circulation. Nevertheless, medical consensus holds that without such support, (1) the

DOI: 10.1201/9781003384991-3

individual would be unable to sustain cardiopulmonary function, and (2) even if that function were maintained, the person would no longer be capable of being who they once were. The irreversible loss of "personhood" ultimately becomes a matter of trust in the medical process, as well as in broader cultural and ethical beliefs. "Brain death," also known as "death by neurological criteria," is thus not only a medical but also a social and cultural phenomenon. It plays a crucial role in decisions about organ donation, underscoring the importance of advance directives and living wills. One particularly sensitive issue concerns the use of anesthesia during organ retrieval in brain-dead patients (see Brown et al. (2023) for the US; Opdam and Radford (2021) for the UK). Greer et al. (2024) provide an in-depth account of the medical criteria used to determine death. As for your second question: science and belief collide here, and a full treatment of that collision lies outside the scope of this book. Organizations such as the Near-Death Experience Research Foundation (NDERF) present themselves as a bridge between science and spirituality. They catalogue thousands of self-reported near-death experiences (NDEs) in more than 25 languages, often elevating belief over empirical scrutiny. Despite the emotional power of these accounts, there is no scientific evidence for any form of existence after death. It is therefore striking—and most troubling—that, according to various recent polls, about 75% of adults in the United States believe in an afterlife.

Table 1: The Glasgow Coma Scale (GCS) has a maximal score of 15, indicating normality.

Response Category and Grading	**Observed Behavior**
Eye Response (Score)	
4	Eyes can be opened and kept open normally
3	Eyes only opened and kept open when told to do so
2	Eyes only opened when feeling pressure under the eyebrow (on supraorbital notch)
1	Eyes remain closed
Verbal Response (Score)	
5	Fully oriented to person, place, and time
4	Confused and not oriented to all qualities
3	Responses to questions do not make sense (while speaking and understanding commands is preserved)
2	Cannot talk but can make intelligible noises
1	Cannot talk but can make sounds
Motor Response (Score)	
6	Follows instructions on how to move and act
5	Intentionally moves away from pressure*
4	Moves away from pressure* only in a reflex action
3	Flexes muscles in response to pressure*
2	Stretches muscles in response to pressure*
1	Does not move in response to pressure*

(* "pressure" here on nailbeds and on trapezius, respectively, i.e., the muscle connecting the shoulder and neck)

Between coma and full wakefulness lies a state known as stupor. Stuporous patients can be roused by verbal or tactile stimulation but lapse back into an unresponsive, sleep-like state once stimulation ceases. Three additional conditions—akinetic mutism, the persistent vegetative state, and the locked-in syndrome—occupy the terrain between coma and stupor.

Akinetic mutism arises from extensive bilateral frontal lobe damage (see Chapter 8). Patients neither speak nor initiate movement. A persistent vegetative state may follow hypertensive crisis, anoxia, or toxic-metabolic encephalopathy. Patients in this condition do not respond meaningfully to their environment, although spontaneous eye opening and occasional visual tracking may occur. Whether such behaviors reflect consciousness or mere reflex is an intensely debated topic. Neuroimaging has not resolved this controversy but has instead intensified it (Owen et al. 2006; Nachev and Husain 2007). In locked-in syndrome, patients may initially appear comatose and unresponsive, but close examination reveals the ability to communicate—often through vertical eye movements. The condition results from a lesion in the pons that disrupts descending corticospinal and corticobulbar tracts while sparing the ascending pathways. The bestselling memoir *The Diving Bell and the Butterfly* (Bauby 1997), later adapted into a film, poignantly captures the internal world of a locked-in patient.

? What other scales used by neurologists should neuropsychologists know?

One important tool is the NIHSS—the National Institutes of Health Stroke Scale. Neurologists and other healthcare professionals widely use it to quantify stroke severity across 11 dimensions (the first of which includes three subcomponents). Several of these dimensions are directly relevant to cognitive functioning. As a neuropsychologist, you should consult a patient's NIHSS score before planning your examination—it will typically be available in the medical chart. The scale's individual dimensions and their scoring ranges (with zero indicating normal performance) are listed here:

Dimension:		**Scoring Range:**
(1) Levels of Consciousnes	a) Alertness	0–3
	b) Orientation	0–2
	c) Commands	0–2
(2) Best Gaze		0–2
(3) Visual Fields		0–3
(4) Facial Palsy		0–3
(5) Motor Upper Limbs		0–4
(6) Motor Lower Limbs		0–4
(7) Limb Ataxia		0–2
(8) Somatosensory Functions		0–2
(9) Best Language		0–3
(10) Best Speech: Dysarthria		0–2
(11) Extinction, Hemi-Inattention		0–2

While the total NIHSS score (maximum impairment: 34) offers no more detail than a general screening tool such as the MMSE (see Chapter 12, p. 272), the subscores provide valuable information. The NIHSS includes pictorial material to assess visual recognition, naming, comprehension, and even neglect-related dyslexia (see Chapter 9, p. 196). The full scale, along with supplemental materials and scoring instructions, is available as a downloadable PDF from the NIH Stroke Scale website.

Another scale worth knowing is the Unified Parkinson's Disease Rating Scale (UPDRS), which remains the gold standard for quantifying symptoms of Parkinson's disease, including their fluctuation over time. Familiarizing yourself with its various subscales will help you tailor your examination to the patient's condition.

Finally, when rapid screening for delirium is indicated (a condition not further addressed in this chapter), I recommend the 4AT (Bellelli et al. 2014). This is a brief and effective instrument that you may choose to administer yourself.

Arousal and Its Modifications

While arousal in everyday language often refers to emotional or sexual excitement, in neuropsychology, the term denotes a person's degree of wakefulness and their ability to sustain vigilance and alertness appropriate to the demands of a task.

? I am a bit confused after reading the few paragraphs above: terms like "arousal," "alertness," and "vigilance" appear alongside each other. Are these terms interchangeable?

No. These terms refer to related but distinct aspects of activation and attention. Arousal reflects a general, long-term activation level—essentially, a person's overall responsiveness and capacity for action. It spans the sleep–wakefulness continuum.

Vigilance refers to a state of sustained attention, characterized by the ability to monitor a stream of monotonous stimuli for rare targets over an extended period.

Alertness, by contrast, is a more short-term readiness to respond rapidly to a specific signal. As tonic alertness, it overlaps with vigilance; as phasic alertness, it describes enhanced detection following a warning cue.

We will return to the anatomy of the arousal system next, and examine finer distinctions within attentional function in the section "Forms of Attention" below.

Anatomy of the Arousal System

Disturbances in consciousness and arousal can result from either focal lesions in the brainstem or a broad range of medical conditions, including metabolic imbalances, toxic exposure, inflammation, cardiovascular events, or seizures. The neural substrate of arousal is the ascending reticular activating system (ARAS; see Figure 1), which originates in the lower brainstem (medulla and pons), projects through the midbrain, and relays to the thalamus, hypothalamus, and ultimately the cerebral cortex. These projections include dopaminergic, serotonergic, and noradrenergic fibers, among others. Cortical areas in turn exert top-down control—both on thalamic relay nuclei (see Chapter 8, p. 166) and more directly on the ARAS itself.

Not shown in Figure 1 is the lateral hypothalamus, a key region involved not only in wake-sleep regulation but also in motivated behavior more broadly. It houses neurons that express the neurotransmitters hypocretin-1 and -2 (also called orexin-A and -B). This region has been implicated in hyperattention to detail in autism spectrum conditions and may contribute to certain autism-related talents (Baron-Cohen et al. 2009).

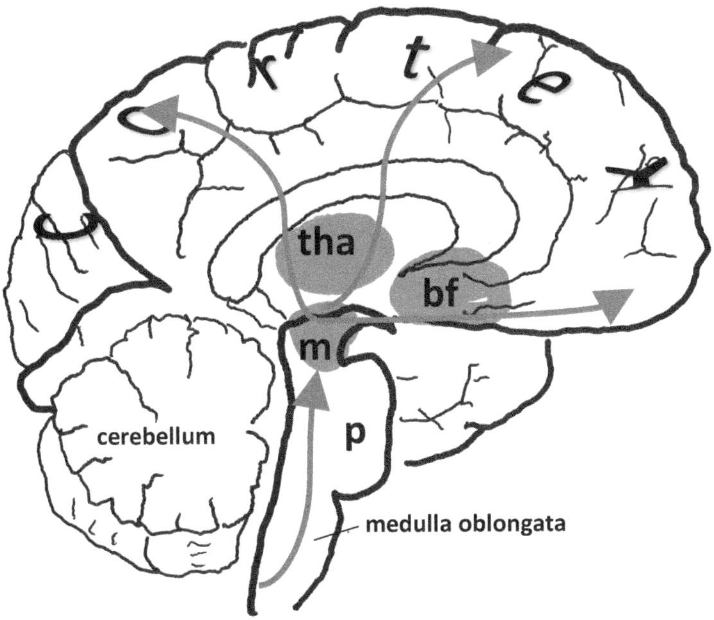

Figure 1: The ARAS (ascending reticular activating system) originates in the lowest regions of the brainstem and projects upward through the midbrain to the thalamus and basal forebrain, ultimately reaching diverse cortical areas.
bf = basal forebrain; m = midbrain (mesencephalon); p = pons; tha = thalamus

? What about "fatigue"? Does it exist? What differentiates it from sleepiness? Who suffers from it, and how can I "objectively" assess it?

Although some dismiss fatigue—particularly conditions such as chronic fatigue syndrome—as a modern medical fad, the symptom itself has long been documented in the neurobehavioral literature (Poore 1875; Carrieu 1878). Unlike everyday drowsiness, fatigue is marked by a persistent, often debilitating sense of exhaustion that is not alleviated by rest or sleep. The causes of fatigue are usually nonspecific and varied, including poor sleep hygiene, chronic stress, depression, nutritional deficiencies, and thyroid dysfunction, among others. In contrast to disorders of the ascending reticular activating system (ARAS), fatigue is more plausibly linked to disruptions in circuits connecting the thalamus, basal ganglia, and frontal cortex.

Fatigue is widespread in neurological conditions such as multiple sclerosis and in post-viral states like Long COVID (also referred to as "post-COVID-19 syndrome"). In oncology, patients frequently report fatigue as one of the most distressing consequences of cancer and its treatment. This cancer-related fatigue is not necessarily part of "chemobrain" (i.e., chemotherapy-induced neurocognitive deficits; see Chapter 6, p. 122), but arises from a complex interplay of biological, genetic, psychological, and behavioral factors (see Bower 2014 for review).

Efforts to identify biomarkers or "objective" indicators of fatigue are ongoing. Some studies have linked fatigue to inflammatory markers, though findings remain inconclusive. Several rating scales exist to assess subjective fatigue. One useful instrument is the scale by Penner et al. (2009), which distinguishes between motor and cognitive components of fatigue.

Forms of Attention

"Attention" is an umbrella term that refers to the capacity to react—either reflexively or deliberately—to environmental stimuli. It includes the ability to select relevant signals from noise, to maintain focus, and to shift attention when needed. A related term, concentration, refers more specifically to the sustained focus on a task while ignoring distractions. It overlaps with interference control (see Chapter 8, p. 168).

Attention and concentration are supported by widely distributed neural networks, which cannot be depicted as straightforwardly as the ascending reticular activating system (ARAS; see Figure 1). Nevertheless, all forms of attention depend on this foundational arousal system and are orchestrated by cortical and subcortical structures in dynamic coordination. Because attention is so widely represented in the brain, attentional deficits may arise from lesions in nearly any region.

What follows is a typology of significant forms of attention. These categories, however, are somewhat artificial. In everyday functioning, attentional processes interact constantly—sometimes hierarchically, sometimes in parallel. Kolb and Whishaw (2008) provide a helpful overview of attentional types. The late US neuropsychologist Muriel Lezak (1927–2021) also offers detailed descriptions and test references for each category in her foundational work, *Neuropsychological Assessment* (Lezak et al. 2004).

Reflex Attention

Two closely linked structures in the midbrain, the superior and inferior colliculi, mediate reflexive shifts of attention. The superior colliculus responds to visual stimuli, while the inferior colliculus responds to auditory input. These structures coordinate rapid eye movements, gaze shifts, and head turns in response to salient signals. Reflex attention is a prerequisite for stimulus detection and for subsequent voluntary attentional engagement.

The superior colliculus also plays a crucial role in "blindsight"—the ability to respond to visual stimuli without conscious awareness, typically following damage to the primary visual cortex. It achieves this by routing visual information along an alternative pathway, bypassing the occipital cortex (see Ajina and Bridge 2016).

Span of Attention

The so-called "magical number seven (plus or minus two)" (Miller 1956) refers to the typical number of digits a person can recall in order after hearing them spoken aloud—usually between 5 and 9. This digit span depends on age, concurrent cognitive load (e.g., reciting a rhyme), and the randomness of the digit sequence.

Digit span reflects basic attentional capacity and influences performance in many tasks that do not overtly appear attention-based.

A high digit span is not necessarily a sign of high intelligence. In simple span tasks, chimpanzees have outperformed medical students (Inoue and Matsuzawa 2007), and it is debated whether some individuals historically labeled as "idiot savants" may have shown similar advantages.

Patients with Transient Global Amnesia (TGA)—who lack memory for recent events (see Chapter 12, p. 280)—may demonstrate unusually large attention spans, possibly as a compensation for the rapid forgetting that often accompanies this condition. Remarkably, training can substantially expand one's span of control over time. Ericsson et al. (2017) describe individuals who achieved digit spans between 80 and 200 through extended mnemonic practice. Similar findings are reported by Yoon et al. (2018), emphasizing the role of strategy over raw memory capacity.

Activating Attention: Two Types of Alertness

Neuropsychologists distinguish two forms of attention activation:

* Phasic alertness refers to a brief enhancement of attention following a warning signal. For example, in computerized tasks, a tone may precede a visual target. The difference in reaction times—with and without this cue—serves as a measure of phasic alertness. Patients with lesions to the right inferior parietal lobule, right prefrontal cortex, thalamus, or anterior insula often show impairments in this domain.
* Tonic alertness refers to the ability to maintain a general state of readiness to respond over time. External stimuli do not trigger a tonic alertness response. Tonic alertness rather reflects a baseline, stimulus-independent attentional tone. This form of alertness is particularly vulnerable to damage in the anterior cingulate cortex or disruptions to the locus coeruleus–norepinephrine system. Bilateral prefrontal cortex lesions and thalamic damage can also impair tonic alertness.

Sustained Attention

Often used interchangeably with "concentration," sustained attention is the capacity to maintain goal-directed focus on a task or stimulus over an extended period.

It is typically assessed using continuous performance tests (CPTs), which are often computerized, lengthy, and deliberately monotonous—thus, they are often unpopular with patients.

Neurally, sustained attention relies on distributed activation in the prefrontal cortex, parietal regions, and anterior cingulate cortex. Neuroimaging studies have shown that sustained attention is associated with the suppression of the default mode network (DMN), a system typically active during rest or self-referential thought.

? **I am a bit confused (or was I inattentive? 😉): what, then, is the difference between sustained attention and tonic alertness?**

You were attentive enough to notice that there's an overlap between the two concepts—well done! Still, there are key distinctions:

Feature	Tonic Alertness	Sustained Attention
Time span	Minutes to hours	Task-determined
Dependency	Not task-dependent	Depends on engagement with a given task
Focus	General readiness	Task-specific focus
Effort	Passive	Requires active effort
Neural system	Locus Coeruleus for arousal	Parietal, prefrontal, and anterior cingulate cortex for control

Attentional Switching and Selective (Directed or Focused) Attention

The American philosopher and psychologist William James (1842–1910) once remarked: "Everyone knows what attention is. […] It implies withdrawal from some things in order to deal effectively with others" (James 1890).

In that statement, James alluded to two interrelated attentional capacities: attentional switching and the ability to select and maintain a novel focus—what we now refer to as "selective" or "directed attention." The latter refers to the ability to concentrate on specific stimuli or tasks while ignoring others.

Attentional switching involves shifting focus between different stimuli or tasks in rapid succession. It requires cognitive flexibility and the ability to reallocate attentional resources. Directed attention, in contrast, involves maintaining focus on a single stimulus or task for a sustained period. It depends on persistence and the inhibition of switching impulses.

These two forms of attention operate in tandem. Together, they support seamless transitions between subgoals without losing track of the overarching objective. In the spatial domain, unilateral neglect offers a striking example of impaired attentional switching: patients often fail to disengage from the non-neglected side in order to attend to the neglected field. The typical lesion site is the (right) inferior parietal lobe.

Divided Attention

Divided attention is traditionally defined as the ability to allocate attentional resources to more than one task simultaneously—commonly referred to as multitasking.

More recent interpretations suggest that what appears to be simultaneous attention is often a matter of very rapid attentional switching between tasks. Regardless of theoretical framing, divided attention is central to everyday functioning.

We rely on it constantly: when cooking several dishes while half-watching TV, listening to a child's story, and mentally planning a birthday gift. Even simpler combinations can engage the same basic capacity to divide or shift attentional focus, such as talking while walking or listening while writing.

> "Talking while walking" was the focus of an influential dementia study by Camicioli et al. (1997), who found that patients with Alzheimer's disease exhibited greater walking impairments than healthy controls when asked to perform a simultaneous verbal fluency task. Since then, numerous dual-task experiments have confirmed the disruptive effects of speaking—i.e., engaging motor speech regions—on locomotion. Importantly, the degree of motor performance decrement under dual-task conditions has been shown to predict fall risk in older adults.

Dividing attention is generally more difficult for patients with brain damage. Individuals with traumatic brain injury (TBI) have been shown to perform more slowly than healthy controls when completing two tasks simultaneously (dual-task paradigm), despite showing no impairments when performing each task in isolation.

However, the label "TBI" is of limited use when attempting to identify specific functional circuits compromised by trauma. The effects of TBI vary widely depending on lesion site and extent. Lesions to the prefrontal cortex, for example, are particularly associated with impairments in divided attention, while *selective* attention may remain relatively intact.

? Is there anything like "hyper-attention"?

Hyperfocusing refers to the inclination—and in some cases, the ability—to intensely focus on a specific task or detail for a prolonged period. Some authors describe it as the opposite of distractibility. While this capacity may benefit high-functioning individuals in certain contexts, it can also interfere with broader, goal-directed behavior: an intense focus on details may hinder completion of a larger task.

Hyperfocusing is commonly observed in attention deficit hyperactivity disorder (ADHD), schizophrenia, and autism spectrum disorders (Baron-Cohen et al. 2009). It is also a hallmark of patients with frontal lobe damage, who may be unable to maintain focus on a task because their attention is hijacked by irrelevant stimuli—a sound in the hallway, an odor in the room, or the specks left by a copier on the Rey Figure.

Patients with hemispatial neglect (see Chapter 9) may exhibit a form of spatial "hyper-attention": an inability to disengage from the non-neglected field due to an abnormal attraction toward the ipsilesional space. This attentional "magnetism" prevents orientation toward the contralateral (neglected) side.

Of particular relevance to neuropsychology is Hans J. Eysenck's theory of personality, which links differences in extraversion and introversion to underlying brainstem activation and cortical arousal levels. Though primarily psychological in origin, this theory has occasionally been supported by neuropsychological evidence—for example, in an extensive electrophysiological study by Jawinski et al. (2021) involving healthy participants.

Where Can We Read More?

Cohen, R.A. The Neuropsychology of Attention. 2nd edition. Springer: Berlin, 2013.

Hopfinger, J.B. and Slotnick, S.C. (eds.). The cognitive neuroscience of attention. Current Debates and Research. Routledge: New York, 2022.

Posner, M.I. (ed.). Cognitive Neuroscience of Attention. 2nd edition. Taylor and Francis: New York, 2011.

CHAPTER 4

What Are Aphasias?

Aphasia is an acquired impairment of language due to cerebral damage. There are many different types of aphasia, broadly and conveniently divided into productive and receptive language disorders (Table 1).

Language production is not restricted to speech. Think of writing, of communication through smoke signals, and especially of sign language. Likewise, language reception is not limited to the understanding of spoken words. Consider reading text, "reading" sign language, or even the interpretation of co-speech gestures.

Before we turn to "visual and hand-motor language" (mainly reading and writing), we begin with the disorders of spoken language.

? This chapter is strongly focused on deficits. Where can we learn about the organization of language in the healthy brain?

This note is more than justified, and I must apologize. Language is such a fundamental function, with relevance to communication, evolution, and consciousness, that I have decided not to cover the topic in its full breadth here, but to focus instead on clinical neuropsychological patterns encountered in everyday practice. However, I can point to some introductory monographs and overviews (in addition to the references listed under "Where Can We Read More?"). Steven Pinker's *The Language Instinct* (1994/2007) offers a broad perspective on language, from evolution to individual development to varieties of language disorders. The book sparked an ongoing debate about the innateness of language. Brennan (2022) provides a more focused introduction to neurolinguistics, while Pulvermüller (2003) is a classic in that field—though it requires at least some graduate-level background. A useful overview with a special focus on bilingualism can be found in Wong et al. (2016). Last but not least, the late Michael Corballis' *From Hand to Mouth* (2002) is a must-read for anyone interested in the evolution of language and hemispheric asymmetries. See also the concluding section of this chapter.

? Deficient or healthy: how much does language rely on genes?

Around the turn of the millennium, a point mutation was discovered in a gene from a family with a dominantly inherited language disorder (Lai et

DOI: 10.1201/9781003384991-4

al. 2000). It was found exclusively in those family members affected by the disorder. The gene, dubbed *FOXP2*, regulates the activity of a large number of other genes—many of which are not directly related to language. Far from being "the language gene," as it was often portrayed in the media, FOXP2 appears to be involved in controlling developmental processes necessary for normal language acquisition. Notably, FOXP2 is found in both the human and mouse genomes. It is the task of evolutionary developmental biology and biolinguistics to determine the genetic contributions to the complex behavior we call "language." Any simple nature-versus-nurture dichotomy should be avoided (see Bishop 2002 for critical reflections, and Bates 1997 for a nuanced discussion of innate versus epigenetic modularity in language).

Disorders of Spoken Language

What is the Difference Between "Speech" and "Language"?

Speaking involves the well-orchestrated activation of various organs in the mouth and throat region. It is essentially a motor activity, dependent on both subcortical and cortical motor centers. Language, by contrast, encompasses non-motor, linguistic cognition and is mediated by widespread neural networks, predominantly, but not exclusively, across the cortical areas of the left hemisphere.

Speaking involves the vocal production of sounds that can be categorized into vowels and consonants, fricatives and sibilants, as well as other hissing, humming, murmuring, or crackling sounds. When used for communicative purposes, these sounds follow rules of articulation, intonation, and prosody. Articulation refers to the correct formation of speech sounds (for example, lisping or the failure to produce an "r" sound represents articulation deficits). Intonation conveys meaning—for instance, by raising the voice at the end of a sentence. Prosody refers to modulations in pitch, loudness, and rhythm across a spoken utterance. Together, these elements of spoken language constitute what we call "speech." Speech disorders resulting from cerebral damage are not considered aphasias. Speech therapists treat them.

Spoken language is not a unitary function. It involves selecting the correct word to convey a thought, constructing word sequences that conform to the grammar of a given language, monitoring one's verbal output, and adapting it to the conversational partner's responses. Productive and receptive language functions are thus tightly interwoven. Together, they engage a widespread neural network that extends well beyond the "perisylvian language network" traditionally described as the core substrate of verbal communication.

This chapter focuses on the types of aphasia most commonly encountered by neuropsychologists. Non-aphasic speech disorders, as well as language-related impairments in reading and writing, are also addressed. Table 1 lists the major disorders, some of which are discussed in more detail in the following sections.

Table 1: Some major disorders of speech and language.

DISORDERS OF SPEECH	**Prominent Signs**
Mutism (in milder form also "abulia")	Abolished verbal production.
Aphemia (apraxia of speech)	Almost mutistic. Effortful articulation.
Dysarthria (in the most severe form, "anarthria")	Deficit in articulation. Slurred or mumbling speech. Often slow and too soft (hypophonia).
Aprosodia	Loss of melodic aspects of speech, no pitch variation.
DISORDERS OF SPOKEN LANGUAGE	**Prominent Signs**
Global aphasia	Nonfluent, pressed speech is often reduced to repetitive sequences of syllables (e.g., "ta-ta-ta") or words (e.g., can't do) or expletives. Auditory comprehension, reading, and writing are comparably impaired.
Broca aphasia	Nonfluent spontaneous word production with impaired repetition and markedly better comprehension. Often "telegram style," absent grammar.
Wernicke aphasia	Fluent verbal production with spared articulation and intonation, but barely understandable content. Often with logorrhoea, empty phrases, and neologisms. Abolished comprehension and impaired repetition. Anosognosia.
Anomic aphasia (also "amnestic aphasia")	Fluent speech, but reduced output due to marked word-finding difficulties. Often, circumlocutions or the use of vague terms and semantic paraphasias are used. Comprehension spared. Typical signs of embarrassment are facing the deficits.
Conduction aphasia	Impaired repetition of auditorily presented sentences and words (especially pseudowords such as "hethnep" or "ropilan"). Spontaneous speech and comprehension are comparably spared.
Transcortical aphasias	* Motor: Essentially a Broca aphasia with markedly better repetition. * Sensory: Essentially a Wernicke aphasia with markedly better repetition. * Mixed: Essentially a global aphasia with spared repetition. Echolalia may be present.
Subcortical aphasias	Typically nonfluent, occasionally dysarthric speech with relatively spared repetition and fairly well preserved comprehension.
DISORDERS OF READING AND WRITING	**Prominent Signs**
Agraphia	Depending on the lesion site, signs of apraxia can be accompanied by impaired reading or occur without it. It can also be a spatial deficit.
Dyslexia/Alexia	Developmental dyslexia: Struggles in acquiring proficiency in reading and spelling. Deep dyslexia: Acquired reading disorder, typically with semantic paralexias.
Alexia with agraphia	Writing and reading are comparably impaired.
Alexia without agraphia ("pure alexia")	Inability to read, even text just written by the patient a minute before. Often, "letter-by-letter reading" is possible. Auditory and tactile reading are spared (understanding words spelled by the examiner or "written" on the skin).

? A look at this table makes me hesitate: is it not dated to speak of "Broca" and "Wernicke" when we talk about aphasias?

Dated terms may no longer be considered modern enough to use. Outdated terms, on the other hand, are not simply old-fashioned—they no longer represent what they originally referred to and are therefore obsolete. There is an ongoing debate about whether terms like "Broca area" or "Broca aphasia" are outdated or merely dated. This book argues that while these terms may indeed be dated, we currently lack a modern terminology for aphasias that is more clinically helpful than the traditional one. Tremblay and Dick (2016) present an excellent review that boldly declares, "Broca and Wernicke are dead." However, they fail to offer clinical neuropsychologists a viable alternative. That is why this book prefers old labels over bold fables.

An entirely different concern regarding terms like "Broca" or "Wernicke aphasia" is that they are eponyms—that is, they contain a person's name. Neuropsychology, along with neurology and psychiatry, is rich in such eponyms (see the table in the appendix to this chapter). Across all scientific disciplines, there are both pros and cons to the use of eponyms (see Whitworth 2007 vs. Woywodt and Matteson 2007). Possibly the strongest argument against them is found in the so-called "law of eponymy" (Stigler 1980), which states that no scientific discovery is named after its original discoverer. While this may not always be true, it certainly applies to the case of "Broca aphasia" (see next question).

? I have read that the "Broca" area and "Broca" aphasia should instead be labeled "Dax" area and aphasia, respectively. How come?

You are pointing here to perhaps the most embarrassing priority dispute in the history of neuropsychology. Marc Dax (1770–1837), a French country physician, discovered left hemisphere dominance for speech in a cohort of 40 aphasic patients—the first of whom he examined in 1800, well before Paul Broca (1824–1880) was even born. He presented his findings at a medical conference in 1836, one year before his death, and roughly 30 years prior to Broca's famous publication. As he passed away shortly thereafter, he was unable to publish his work.

His son, Gustave Dax (1815–1893), also a physician, read his father's manuscript—enriched with his own observations—at a meeting of a prestigious medical academy in Paris in 1863. Broca attended that meeting but did not acknowledge the Daxes' work in his own 1863 paper, which has since become the landmark publication on language lateralization. Broca's case series included fewer patients than Dax's, but he was more precise in localizing the speech-relevant area to the third frontal convolution. For sources, biographical context, and reflections on the sociology of scientific discovery, see the comprehensive essay by Finger and Roe (1996).

Disorders of Speech

Mutism

Mutism refers to the absence of verbal utterances in a wakeful and attentive patient. In its milder, less apathetic form, it may be referred to as abulia. The causes of mutism or abulia are heterogeneous—ranging from cerebral lesions to psychological trauma—and their manifestations are diverse.

When the onset is acute, mutism may represent the initial stage of aphasia or aphemia (see below). If mutism develops gradually and progressively worsens, a degenerative process should be suspected, such as cortical degeneration or advanced-stage extrapyramidal disorders like Parkinson's disease or Huntington's disease. In patients with severe concussion, transient mutism is often observed during the recovery phase from coma.

In akinetic mutism, the lack of initiation extends to limb motor activity; patients barely initiate eating or drinking. Bilateral lesions of the cingulate gyrus or medial forebrain bundle can result in akinetic mutism by disrupting ascending input from the brainstem's reticular formation. (S)elective mutism, by contrast, is psychogenic in origin and may fall under the broader category of anxiety disorders.

? How can I differentiate a psychogenic mutism from a comparably severe mutism of neurogenic origin?

History of onset, neurological, and general physical examinations may reveal signs associated with mutism of cerebral origin. A special case involves bilateral lesions to the frontal opercula (i.e., the inferior part of the precentral gyrus). These often produce a dissociation between voluntary and involuntary activation of facial and pharyngeal muscles. Patients show a loss of speech and are unable to perform facial movements on demand (oral apraxia; see Chapter 6). However, reflexive facial movements are spared—for instance, during spontaneous laughing or crying. Oral apraxia is not compatible with psychogenic mutism. Moreover, patients with psychogenic mutism typically lack the sometimes desperate attempts to communicate through alternative means, especially writing.

Aphemia

Aphemia is an apraxia of speech (see Chapter 6 for apraxia). Broca coined the term, and the disorder was initially considered a "little aphasia." Although aphasic signs may be present and patients may exhibit bucco-facial apraxia, pure cases of aphemia—isolated deficits in articulation—have been described (Fox et al. 2001). Writing is entirely preserved. Comprehension, including written statements, remains intact, and even complex sentence repetition is possible.

Various subtypes of speech apraxia have been identified (Mailend and Maas 2020). One particularly intriguing and debated variant is ANFAS (acquired neurogenic foreign accent syndrome; Priftis et al. 2020). Patients, typically after a stroke, suddenly begin to speak with a specific accent, including accents of languages they were never exposed to.

Dysarthria (and Anarthria)

Dysarthric patients struggle with articulation. Their speech does not show signs of aphasia but is typically slow and noticeably altered. It may be slurred or mumbled. A common characteristic—as in Parkinson's disease—is that speech is too soft to be understood (hypophonic) and poorly modulated (monotonous). Lesions are typically found in the lateral motor cortex—more often in the left hemisphere—or in the underlying white matter. Dysarthria may also result from brainstem lesions affecting the vagus, hypoglossal, or glossopharyngeal nerves.

Dysarthria is frequently seen in Parkinson's disease and other neurodegenerative conditions such as Multiple System Atrophy (MSA). Cerebellar degeneration may also produce dysarthria. Anarthria refers to dysarthria so severe that no vocalization is possible—not even whispered speech.

Aprosodia

Have you ever heard the robotic voices in old science fiction movies? They sound tinny and lack tonal variation—it is surprisingly difficult to imitate such flat and choppy speech! The loss of expressive prosody results from anterior, pre-Rolandic lesions in the right hemisphere. A corresponding deficit in the appreciation of prosody points to more posterior lesions in the same hemisphere (Ross (2021) provides an overview).

Both expressive and receptive prosodic deficits can cause serious social difficulties. Misunderstanding or misusing prosodic features may lead to misinterpretations of irony, sarcasm, or even tender or seductive tones embedded in spoken language.

? Are the words of a parrot reflecting the bird's capacity for language or just for speech? And why is it primarily parrots that speak? What about linguistic expertise and numerical cognition in dogs, horses, and chicks?

Parrots' vocal mimicry is grounded in the properties of their speech apparatus: unlike most other birds, they have a single sound-producing larynx rather than two syrinxes, and they possess a particularly strong, muscular tongue. Whether their utterances reflect genuine language capabilities is a matter of ongoing debate. American professor Irene Pepperberg, a chemist by training, devoted much of her life to the study of what she would call parrots' language and cognition. A fascinating account of her collaboration with her first African Grey parrot, Alex, is found in Pepperberg (2008).

As for dogs' capacity to process human language, studies on so-called Gifted Word Learner Dogs (Dror et al. 2023) offer little to (animal) neuropsychology beyond the trivial insight that dogs can form and retain auditory–visual associations. A particularly instructive case of scientific self-deception is the story of the German horse Clever Hans, which allegedly performed complex arithmetic. Meticulous experimentation in the early 20th century revealed that the horse was responding to subtle, involuntary visual cues from his trainer. Such pseudo-communication phenomena have since been labeled the Clever Hans effect.

The Clever Hands effect (Wegner et al. 2003) is especially relevant to neuropsychology. It demonstrates how unconscious hand movements by therapists—particularly in interactions with autistic patients—can give rise to striking but misleading impressions of communicative intent (see Chapter 5, p. 86).

Finally, a word about chicks: research with this avian species has contributed significantly to our understanding of human neuropsychology, including topics such as numerical cognition (Rugani et al. 2015), hemispheric lateralization, and even the phenomenology of spatial neglect (see Brugger 2015).

Disorders of Spoken Language

Global Aphasia

Patients with this most severe type of aphasia are severely impaired in both productive and receptive language functions (including reading and writing). Production is often limited to single words or stereotyped, nonsensical syllables (e.g., "No!", "My God!", and "Tan-tan").

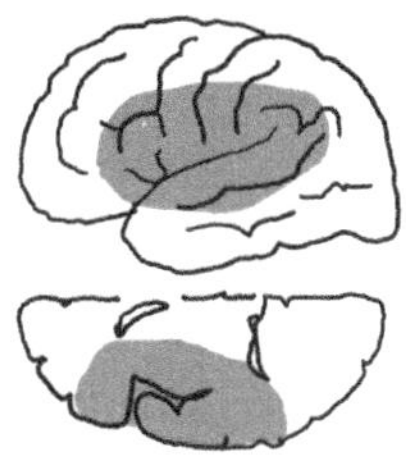

An initially global aphasia may evolve spontaneously into a Broca aphasia; typically, comprehension recovers earlier and more fully than expressive abilities.

For the Global, Broca, and Wernicke types of aphasia, lesion locations are shown here on a lateral surface view (top) and a transverse section of the left hemisphere (templates redrawn after Schnider (1997)).

Broca Aphasia

This type of aphasia is characterized by nonfluent speech, typically containing few content words, verbs, and adjectives (*agrammatism*). Comprehension is largely preserved, but passive sentence structures are often misunderstood. For example: "A lion was eaten up by a tiger. Which animal is dead?" Patients will typically assume that the first-named word (lion) is the agent.

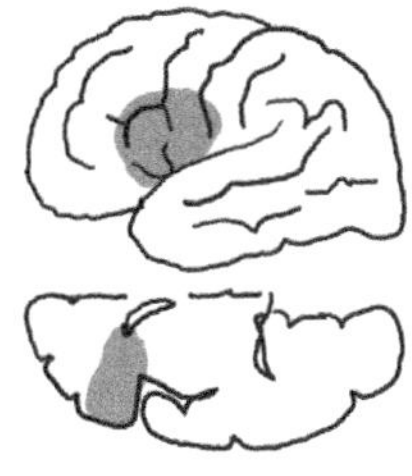

Writing is also nonfluent. The underlying lesion involves the inferior frontal gyrus of the left hemisphere, specifically Brodmann areas 44 and 45. Even within Broca's region, there is notable specialization: some subregions are more involved in phonological aspects of language, others in semantic processing (Heim et al. 2008).

? Can patients with global or Broca aphasia speak during sleep?

Embarrassingly, this question cannot be answered based on empirical data. Sleep talk remains a neglected topic in neuropsychological research. It can occur as a consequence of seizure activity in the right hemisphere or emerge in the course of certain degenerative disorders (Kimura et al. 1997). However, to my knowledge, no systematic group study has ever examined Broca aphasia patients with a known pre-stroke habit of sleep talking.

Even a single case study—featuring a patient with a minimal active vocabulary during wakefulness and carefully monitored sleep (using EEG and acoustic recordings)—could offer valuable theoretical insight into the neuropsychology of language. There is anecdotal evidence (unpublished personal observation) of one patient with a five-word vocabulary who was able to produce fluent speech during sleep, beyond fixed or formulaic phrases. This was a classical symptom constellation following a classical lesion, and the patient's sleep partner reported the observation.

Wernicke Aphasia

Comprehension is abolished, usually also for written language. Despite fluent and often disinhibited verbal output, a Wernicke patient's speech is unintelligible. It is sometimes referred to as "word salad": a mixture of phrases, paraphasias, and neologisms (see Table 2). Writing is similarly incomprehensible. Intonation and prosody are typically intact and often situationally appropriate. A nearly defining feature of this disorder is anosognosia: it seems as though the patient's own sentences do not reach their ears. What they perceive themselves to be saying are orderly, coherent thoughts, not the gibberish that others hear.

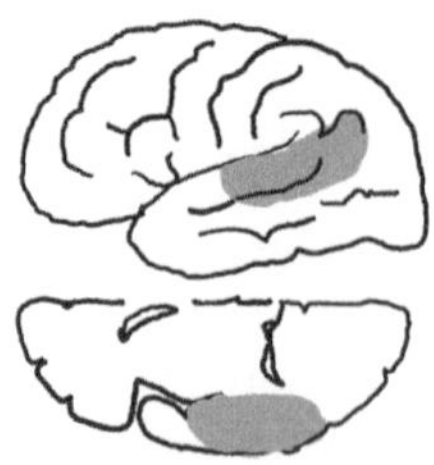

? What differential diagnoses do we have to consider once a patient shows abolished language comprehension?

First of all, the patient could be deaf and simply unresponsive to auditory input. If the cause were a peripheral lesion (affecting the acoustic nerve or brainstem), auditory evoked potentials (AEPs) would be pathological.

Alternatively, the patient might suffer from:

* Cortical deafness, caused by bilateral lesions to the primary auditory cortex (Heschl's gyrus). In such cases, AEPs are normal. Auditory reflexes are preserved, and patients may orient spontaneously to sudden loud sounds.
* Auditory verbal agnosia (also known as pure word deafness) refers to the inability to understand verbal sounds, while recognition of nonverbal auditory stimuli is spared, such as howling wind, crying babies, ringing phones, or the sounds of machines and vehicles. The lesion site typically involves the left Heschl's gyrus and underlying white matter connections between the right auditory cortex and left hemispheric language areas. (Nonverbal auditory agnosia presents the opposite clinical picture: speech sounds are understood, but environmental sounds and music cannot be recognized).

Anomic Aphasia

As the term suggests, this aphasia subtype is characterized by anomia—that is, a marked difficulty in retrieving words, primarily nouns and verbs. A synonymous label is amnestic aphasia, as patients often feel as though they have "forgotten" the words they are trying to produce.

Anomic Aphasia: Optic Aphasia Type. This variant is also known as visual anomia, as the naming deficit is limited to visually presented objects. Spontaneous speech is normal and fluent unless the name of an object must be retrieved from its visual appearance (including line drawings). Visual recognition remains intact. Patients can name the object when it is heard (e.g., the ringing of a bicycle bell) or described verbally (e.g., "What is the name of the bent yellow fruit on whose peel one can easily slip?").

Optic aphasia is fundamentally different from associative agnosia, which is a perceptual disorder marked by visual similarity errors. In contrast, patients with optic aphasia tend to make semantic errors. Their naming deficit is equally severe for photographs, detailed drawings, and simple sketches, whereas patients with associative agnosia often perform better with more realistic images. Perseverative naming errors are frequent in optic aphasia but absent in associative agnosia (see Iorio et al. 1992).

Subcortical Aphasias

These aphasias most commonly result from lesions in the anterolateral thalamus, the internal capsule, or the anterior striatum (typically on the left side). Clinical signs include nonfluent, hypophonic, and dysarthric speech, impaired naming, and relatively spared comprehension and repetition.

Patients with thalamic lesions may also show impaired naming and comprehension and produce many paraphasic errors. Subcortical aphasias often have a good prognosis and may ultimately be caused not by the lesion itself, but by a secondary cortical hypometabolism (Olsen et al. 1986).

Conduction Aphasia

The hallmark of conduction aphasia can seem puzzling to those unfamiliar with neuropsychology: patients have preserved spontaneous speech, comprehension, and expressive language, yet they are unable to repeat even a single sentence—often not even a single word—and especially not pseudowords or pseudophrases (e.g., *sandicule, ulphemize tin lorthy gamblate!*).

Conduction aphasia results from a disconnection between the parietotemporal and frontal (motor/premotor) regions in the left hemisphere. Functional imaging studies have identified area SPT (superior-posterior temporal) as a key gray matter contributor to this condition. These findings have refined the earlier notion that conduction aphasia is merely due to a disconnection between Broca and Wernicke areas via the arcuate fasciculus (Acharya et al. 2025).

Three Types of "Transcortical Aphasia"

Transcortical Motor Aphasia. This form arises when the frontal perisylvian language areas are disconnected from the supplementary motor area of the left hemisphere. Patients can repeat words and sentences but produce very little spontaneous speech. Reading, writing, and auditory comprehension are impaired. Clinical signs resemble those of a mild Broca aphasia.

Transcortical Sensory Aphasia. This condition resembles a mild form of Wernicke aphasia. The lesion lies slightly posterior to Wernicke's area.

Transcortical Mixed Aphasia. Signs correspond to a mild global aphasia. The lesion combines features of both the transcortical motor and transcortical sensory types.

? **This section is on oral language production and its reception. Yet it is restricted to talking. What should a neuropsychologist know about singing and whistling?**

Humming and singing are important in neuropsychology and aphasiology because they may be preserved in patients with productive aphasia (see Johnson and Graziano (2015) for a historical review). That humming might be spared is perhaps not surprising, since it contains no words; but singing does involve words, and yet may remain relatively intact. Every aphasia assessment should include a test of the patient's ability to sing. This can be revealing. Many patients who cannot say "Bah, bah, black sheep, have you any wool?" without intonation can sing the entire tune with perfectly intelligible words.

Studies of the neural commonalities and differences between speaking, singing, and humming may provide valuable insights into the co-evolution of language and music.

Whistling, in contrast, tells an entirely different story. In some regions of the world where long-distance communication is necessary, whistled languages have developed (Meyer et al. 2021). In proficient users, these activate receptive language areas, whereas control subjects—who perceive the signals merely as whistles—do not show such activation (Carreiras et al. 2005).

Primary Progressive Aphasia (PPA)

A type of frontotemporal dementia, PPA is characterized by the progressive deterioration of language abilities. Learning and memory may remain intact for a long time. Initially, spontaneous speech becomes impoverished, while comprehension is relatively preserved. Naming becomes increasingly difficult.

Many patients shift to reading and writing as alternative modes of communication with their partners. Eventually, however, even these visual-motor forms of verbal communication break down. Cognitive flexibility and conceptual thinking are progressively impaired.

Crossed Aphasia

Aphasia is called crossed when it results from a strictly unilateral lesion in the hemisphere *ipsilateral* to the dominant hand, most often, a right-hemispheric lesion in a right-handed individual. This may reflect a "mirrored" or bilateral representation of language functions.

Crossed aphasia is more frequently observed in multilingual individuals. The specific aphasic symptoms vary and are not distinctive. However, they are often accompanied by signs typical of right-hemisphere damage, such as left-sided neglect.

? The words "lick", "pick," and "kick" all sound similar. I assume, they engage an almost identical part of the network for language?

Wrong assumption! Think of the meanings each of these verbs conveys, and the different body parts involved in performing the respective actions. Elegant research has shown that hearing or reading the words "lick," "pick," and "kick" activates not only temporal lobe regions involved in word meaning but also neuronal assemblies associated with tongue, finger, and leg movements, respectively (summarized in Pulvermüller 2003).

Phonematic and semantic similarities between words depend on distinct neural substrates, and these are differentially affected by various neuropsychological disorders. The lick–pick–kick example illustrates that beyond phonological and semantic dimensions, language is also embodied—grounded in sensorimotor systems. This offers a sense of how widespread the so-called "cerebral language network" really is.

? What do we know about self-talk and "hearing voices" in aphasic patients with impaired language?

Self-talk takes many forms. It can occur as silent inner speech, closely tied to the relationship between language and thought (Vygotsky 1943/1962). Interestingly, inner speech is reportedly just as frequent in aphasic patients as in healthy individuals (Alexander et al. 2024). Self-talk can also manifest as audible speech directed at oneself—known as soliloquy. One might expect soliloquy to be preserved in Wernicke aphasia but reduced in Broca or Global aphasia. However, no systematic studies have examined the incidence or characteristics of aphasic soliloquy.

Table 2: Some signs of aphasic or otherwise impaired language (auditory or related to reading/writing).

Designation (Alphabetic Order)	**Description and Context**
Agrammatism	Speech lacks grammatical structure. Typical in Broca-type aphasia. "Paragrammatism" refers to incomplete or faulty use of grammatical rules, mostly seen in fluent types of aphasia.
Echolalia	Repetition of another person's speech, especially the examiner's questions, is a sign of frontal lobe damage. It has also been described in normotypical contexts (Pruccoli et al. 2021) and is just one of many echo phenomena, including echopraxia, the involuntary mirroring of observed hand actions.
Environmentally dependent speech	A variant of the environmental dependency syndrome (Lhermitte 1986) in which patients with frontal lobe damage adopt the occupation or personality traits of their conversational partners (also Zelig symptom, after Woody Allen's 1983 film *Zelig*; see Appendix to this Chapter).
Graphomimia	Writing-like activity consisting of poorly formed graphemes or simply mimicking the act of writing.
Hypergraphia	Disinhibited writing. A typical sign of temporal lobe epilepsy (TLE). It may involve a compulsive urge to write repetitive phrases, but it can also lead to literary creativity, as in Dostoevsky. A fascinating case involving TLE-related hypergraphia in a classic murder trial is a must-read for students of forensic neuropsychology (Poston 2024).
Hyperphasia (forced)	A variant of environmentally dependent speech in which the patient shows compulsive, involuntary commenting. For example, when the examiner waves a hand, the patient says, "The doctor is waving his hand" (Tanaka et al. 2000).
Hyperverbalisation	Not merely excessive talking (logorrhoea), but a compulsive tendency to verbalize everything.
Logorrhoea	Disinhibited speech—"oral diarrhoea." The patient's speech is difficult to interrupt. A sign of intra- or interhemispheric disinhibition. It can be seen as one extreme on a speech-rate continuum from mutism to hyperlalia (Braun et al. 2004).
Macro-/ Megalographia	Excessively large handwriting. Seen after cerebellar lesions, in Essential Tremor, Huntington's disease, autism spectrum disorders, or mania.
Micrographia	Excessively small handwriting. Typical in Parkinson's disease, where handwriting often shows progressive diminution of letter size across a line.
Paraphasia (Paragraphias, paralexias)	Substitution of spoken words with unintended, related words. Common in fluent aphasias. Types include phonemic paraphasia and semantic paraphasia. * Phonemic paraphasia: e.g., "cat" for "hat" * Semantic paraphasia: e.g., "cat" for "dog" Paragraphias (in writing) and paralexias (in reading) are analogous. Semantic paralexias are reading errors characteristic of deep dyslexia—often vivid and surprising (e.g., "Russia" for "Tsar") and may reflect the release of right-hemisphere associations from left-hemisphere control (Landis et al. 1983).

Table 2 contd. ...

...Table 2 contd.

Designation (Alphabetic Order)	Description and Context
Schizophasia	Disorganized or idiosyncratic language typical of psychotic thought disorders or schizophrenia-like episodes, including those seen in TLE. There are several variants of schizophasia.
Catatonic speech	Overproductive, hyperkinetic speech featuring stereotypies and echophenomena. Mostly seen in psychiatric diseases.
Pressed speech	It can be a sign of thought disorder. Sentences are delivered too quickly for clarity.
Clanging	Speaking in rhymes. Patients may produce compulsive rhyming strings. In some cases, this becomes "compulsive versifying," with spontaneous poems, puns, or doggerels—often after temporal lobe damage (Woollacott et al. 2015).
Tangentiality/ Circumlocutory speech	Answering questions by circling the point. The message may eventually be conveyed through successive approximations, but this requires considerable patience from the listener.
Word salad	Speech consisting of a random mixture of words—gibberish. Seen in patients with thought disorder, but also a characteristic feature of Wernicke aphasia.

Notably, soliloquy as visible self-talk is also seen in individuals who use sign language (Zimmermann and Brugger 2013), though nothing is known about its modification in sign language aphasia.

As for auditory verbal hallucinations in aphasia, an early review by Anastasopoulos (1966) addresses this rare but striking phenomenon. One case involved a patient with fluent aphasia who experienced transient verbal hallucinations—remarkably, confined to the right ear—following an infarction of the left superior temporal gyrus (Tanabe et al. 1986).

I feel compelled to add one thing, although you did not ask about it: Hearing voices is not necessarily a pathological thing! On healthy voice hearing, see Hugdahl (2009) and Luhrmann et al. (2023).

Disorders of Writing and Reading

Writing is, beyond its motor component, an *expressive* language function. Likewise, reading is, beyond its reliance on visual input and eye movement control, a *receptive* language function. Acquired deficits of writing and reading may affect both functions equally, or may impact one while sparing the other.

Agraphia

Agraphia is the acquired inability to write, not due to motor dysfunction or apraxia of writing movements. Lesion sites are typically frontal (Brodmann area 6, or Exner's area) or around the angular gyrus (both left-hemispheric). Linguistic classifications

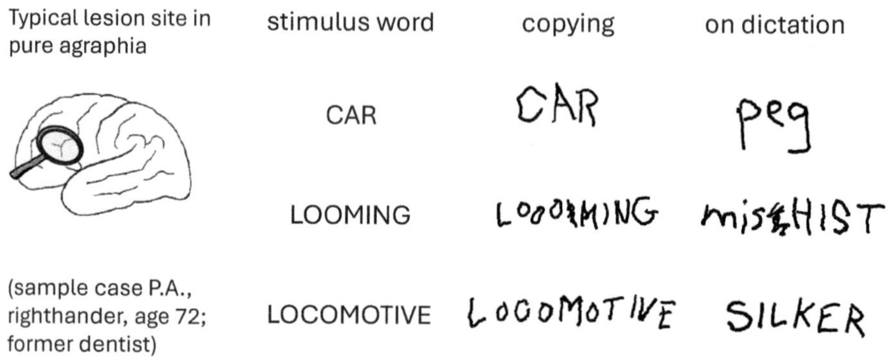

Figure 1: Pure agraphia (without aphasia, no alexia, no apraxia) in a 72-year-old right-hander after an ischaemic stroke targeting BA 6 and the underlying white matter. The patient (a native English speaker) was almost flawless in copying visually presented words but failed to write spontaneously or transpose heard words to writing (dictation).

differentiate between a lexical and a phonematic form of agraphia. In the latter, especially pseudowords and uncommon words cannot be written.

Patients with pure agraphia (without alexia and without apraxia) may show a striking dissociation: they are unable to write to dictation, but can copy text without difficulty (see Figure 1). The typical lesion site was once assumed to be at the foot of the second frontal gyrus, but parietal locations have also been reported. Some authors describe non-apraxic agraphia as a disconnection syndrome between the parietotemporal and frontal cortex (Croisile et al. 1990).

In deep agraphia, patients produce semantic paragraphias (e.g., writing "plane" when dictated "helicopter"). The lesion site is the left supramarginal gyrus and the underlying insular cortex. Agraphia may affect only the left hand (a "split-brain" symptom), indicating a lesion in the medial parts of the corpus callosum (see Chapter 11).

Alexia

Alexia—which is called dyslexia in a milder form—presents in several distinct types. Developmental dyslexia is essentially a learning disorder, specific to the acquisition of text-based language encoding. There are multiple subtypes of both developmental and acquired alexia (Hancock et al. 2017). One particularly instructive dissociation is between *alexia with* and *alexia without* agraphia. This contrast is of great instructional value to students of neuropsychology because it illustrates the underlying anatomy and allows for surprisingly straightforward differential diagnosis.

Developmental Dyslexia

This learning disorder, which has a genetic basis, is usually diagnosed in early childhood. It is specific to the acquisition of visual reading skills, independent of IQ. Reading difficulties tend to persist throughout life, though compensatory strategies may allow affected individuals to succeed academically.

In languages that rely more on logographic than phonological encoding (such as Chinese or Japanese kanji), right-hemispheric involvement is more pronounced. This underscores the culture-bound nature of dyslexia networks. *ICONO*, a novel picture-based language, may help children with developmental dyslexia (Kramer 2023).

Another developmental condition, hyperlexia, is characterized by an obsessive interest in reading from an early age—often regardless of content (Smith and Tsimpli 1995). Williams syndrome is frequently associated with above-average, even savant-like, reading ability (Levy 1996 for review).

Deep Dyslexia

An acquired reading disorder, especially marked for pseudowords and function words (e.g., "and," "but," and "however"). Errors are more frequent with verbs, adverbs, and adjectives than with nouns, and with abstract words more than concrete words. Reading errors typically involve semantic paralexias (e.g., "grave" for "tombstone"; "Russia" for "Tsar"). Visual errors also occur (e.g., "think" for "thing").

Among other hypotheses, the right hemisphere hypothesis suggests that left-hemisphere damage not only impairs normal reading but also disinhibits right-hemisphere contributions, leading to aberrant reading patterns (Coltheart 2000; Landis et al. 1983).

Alexia With Agraphia (Figure 2, Left)

This combined reading-and-writing impairment results from left hemisphere lesions near the angular gyrus and is often part of Gerstmann syndrome (see Chapter 7, p. 163). Often, even single letters or digits cannot be read, and there is no visual-based understanding of text.

Frontal Alexia (Also "Third Alexia"; Figure 2, Middle)

Reading aloud is marked by phonematic paralexias (e.g., "purse" for "nurse"). Text comprehension is significantly better than reading performance suggests. Nouns, in particular, may be understood despite misreading and are often associated with Broca-type aphasia.

Pure Alexia (Alexia Without Agraphia; Figure 2, Right)

This striking condition involves an inability to read despite intact writing ability. Patients may show letter-by-letter reading, although in some cases, even individual letters cannot be recognized. A key diagnostic clue is spared auditory reading—when a word is spelled aloud (e.g., S–P–O–O–N), it is easily recognized, providing attention span is sufficient.

Kinesthetic reading is also preserved, and many patients spontaneously trace letters with their fingers to help decipher meaning. Pure alexia is caused by a lesion to the left occipital cortex and the splenium of the corpus callosum (typically after a left PCA stroke). The splenial lesion disconnects visual input from the right occipital cortex and language processing in the left hemisphere.

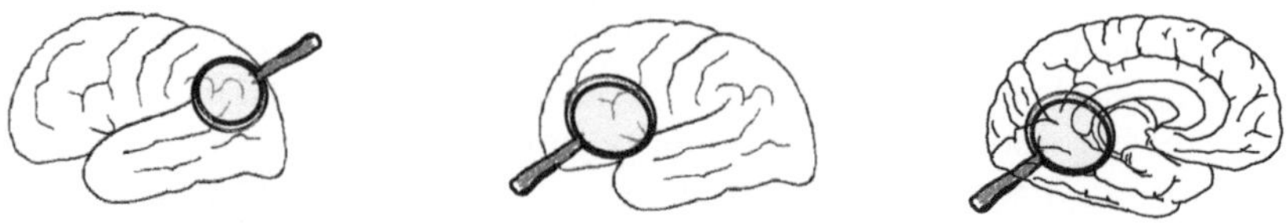

Figure 2: Three types of alexia. See text for details.

Hemianopic Reading Disorders

Pure alexia is often accompanied by right-sided hemianopia or upper quadrantanopia. The neuropsychology of alexia sheds light on the differential contribution of visual field defects versus genuine language-related reading deficits. Efficient reading requires perception of approximately three to four letters to the left and about ten to the right of fixation (in left-to-right reading cultures; reversed for right-to-left scripts). Pflugshaupt et al. (2009) discuss how simple eye movement analyses can help distinguish these conditions and guide treatment.

Neglect Dyslexia

A non-aphasic reading disorder caused by failure to attend to the left side of words or lines of text. This symptom of hemispatial neglect is discussed in Chapter 9, p. 196.

? **Reading is an exceedingly important function. The brief comments on alexia we read here do not appear to tell the whole story that neuropsychologists should know. Where could we find a more complete picture?**

The reading of sign language will be discussed immediately below. Tactile “reading” via the Braille system—either embossed on paper or displayed on a refreshable electronic surface—will also be addressed.

Another form of tactile reading, graphesthesia, is used in the neurological assessment of somatosensory function. A neuropsychologist may “write” a letter on a patient’s skin and ask her to “read” it (with eyes closed). Beyond assessing stereognosis, the identification of the letter *E*—as regular or mirror-reversed—offers insight into cutaneous spatial reference frames (Lenggenhager et al. 2016).

Healthy volunteers almost always perceive the letter *E* as correctly oriented when drawn on the back of the head. But when written on the front, it may appear either mirror-reversed or regular, depending on whether the participant imagines the letter from a first-person or third-person viewpoint (Corcoran 1977). Letter identification on the back of the hand strongly depends on hand posture relative to body orientation (Sekiyama 1991; see also Stracke (1947) and Arnold and Auvrey (2017), who recommend using *d*, *b*, *p*, and *q* for testing).

Graphesthesia is also used in assessing callosal disconnection following commissurotomy—especially when testing the left hand.

Finally, "reading" can be abstractly extended to include social cognition, such as the ability to "read" another person's mind from subtle facial cues—especially the eyes (Pavlova and Sokolov 2022 for review). Evolutionary neuropsychology has asked what the visual word form area might have been used for before the invention of writing. One proposal: it helped decipher animal tracks and other meaningful patterns in nature.

Aphasia for Sign Language

A stroke affecting the left middle cerebral artery (MCA) may lead to an inability to produce sign language (in the case of anterior lesions near Broca's area) or to understand it (in the case of more posterior lesions). See Goldberg and Hillis (2022) for an overview. Higher-order auditory association cortex around Wernicke's area is thus involved in processing the meaning of visually perceived signs.

Interestingly, "reading" manual language may be easier than reading written script for some dyslexic children, as it involves less reliance on spelling. After damage to language-sensitive brain regions, signing deaf individuals may also show paraphasic errors or lose grammatical and syntactic structure in their signing, depending on the specific location of the lesion.

? Does a deaf patient with "anomia" in sign language also have difficulties in gesturing (e.g., waving goodbye, signaling thumb-up)?

Just like spoken language, sign language is particularly vulnerable to left hemisphere damage. Patients with Wernicke's area damage typically exhibit fluent jargon signing, while damage to Broca's area leads to nonfluent, grammar-impaired signing.

But signing is not the same as gesturing. Sign retrieval is not equivalent to gesture retrieval. Marshall et al. (2004) described a patient with sign language aphasia following a left-sided stroke. The patient showed prominent difficulties retrieving signs, with both phonological and semantic errors. Strikingly, she could still perform everyday gestures, such as miming how to smoke a cigarette, indicating that the anomia was not due to motor or apraxic deficits, but rather reflected a breakdown in linguistic sign access.

? When we try to retrieve a word but fail, we have what we call a tip of the tongue experience. What about deaf people who communicate with sign language—do they report analogous "tip of the fingers" experiences when they fall short of retrieving a particular sign?

Although they may not use that exact phrase, this is precisely what they report. Thompson et al. (2005) elicited such tip-of-the-fingers (TOFs) experiences using methods analogous to studies on tip-of-the-tongue (TOT) states. Deaf participants using American Sign Language were shown pictures of faces, printed words for cities and countries, and rare vocabulary. They were asked to retrieve proper names or signs, and to indicate when they felt a sign was known but just beyond reach.

Signs are composed of four basic parameters—analogous to phonemes in spoken language: handshape, location, movement, and orientation. TOFs occurred at rates comparable to those in TOT studies. As with TOTs, TOFs were most common when attempting to retrieve proper names. Participants often guessed individual components of the sign's form, but these guesses were unrelated to resolving the TOF.

In sum, Thompson et al. (2005) provided strong support for a separation between semantics and phonology-like components in sign language—similar to what is observed in spoken language. This separation had previously been questioned due to the iconic nature of signed words.

Having a *tip-of-the-tongue* (TOT) experience can be highly revealing. Many instructive questions for neuropsychology are collected in Brown (1991). An entertaining and informative website on illusions and mental teasers—*Mind Bluff: Illusions, Brain Teasers, Mental Tricks & More!*—refers to a method for eliciting TOT states. It presents a list of object or action descriptions, each requiring the participant to retrieve a specific target word (examples originally from Schacter 2002). Here are a few examples (with solutions in the footnote):

(1) Inscription on a tomb;

(2) Material once used for fireproofing, later recognized as toxic;

(3) Tough, elastic tissue forming part of the skeleton;

(4) Heavy, broad-bladed knife or hatchet used especially by butchers.[1]

These and similar items can be useful for testing patients' naming ability, especially when they fail to name visually presented objects and you are uncertain whether the failure is due to visual perception problems. In these cases, target objects should be presented visually (e.g., a drawing of a banana) and on other occasions auditorily ("the yellow, curved fruit").

Alexia for Braille

This disorder is, of course, extremely rare—mainly because the number of individuals who read Braille is small, and the subset who then suffer a stroke or space-occupying

[1] (1) epitaph, (2) asbestos, (3) cartilage, and (4) cleaver.

lesion is even smaller. Strategically, a lesion causing Braille alexia must damage higher-order tactile processing of tactemes (the tactile analogs of graphemes in print) in the posterior parietal lobes (Liu et al. 2023).

An early case report by Gloning et al. (1955) is especially noteworthy. The authors describe not only Braille alexia but also agraphia in a patient who learned Braille as their only writing system. Remarkably, the agraphia extended to writing with a typewriter, which requires coordinated bimanual activity, not a trivial finding.

Another case, involving a blind organist, was translated by Fisher and Larner (2008). The patient had suffered a left MCA stroke, resulting in Wernicke aphasia and alexia, yet retained the ability to hum melodies from standard musical notation, suggesting no amusia.

Since reading Braille also requires spatial-tactile processing, contributions from the right hemisphere would be expected. Supporting this, Karavatos et al. (1984) found a shift from the typical right ear advantage in dichotic listening (see Chapter 11, Table 1) to a left ear advantage during the acquisition of Braille reading in congenitally blind children.

The broader issue of cross-modal plasticity in blind Braille readers is addressed in the PET study by Sadato et al. (1996), which demonstrated activation of the visual cortex during somatosensory processing. The interpretation of this cortical "reorganization" remains complex (see Chapter 10, p. 214, in relation to he functional reorganization within the somatosensory system).

? **I have recently assessed a left-handed man in his 70s, who was born in Portugal. A tiny lesion in his right thalamus was neurologically silent. To my amazement, his neuropsychological functions were almost normal—except for severe alexia (sparing specific brand names and newspaper titles) and a nearly complete agraphia (his own name and some family names were spared). How can this be?**

The chances are high that this gentleman never learned to read and write. In certain parts of Europe, especially in the mid-20th century, this was not uncommon. Illiteracy is rarely volunteered by patients, who may feel embarrassed or stigmatized.

The brains of individuals who have never acquired literacy show reduced functional hemispheric asymmetries in certain language-related tasks, as well as generally diminished large-scale network connectivity (Petersson et al. 2007).

For a valid neuropsychological assessment of illiterate individuals, it is imperative to use tests with high ecological validity. Assessments linked directly to activities of daily living help avoid underestimating cognitive functioning in this special patient group (Petri et al. 2023).

Concluding Remarks: On the Meaning of "Language"

The classical model of aphasia—the so-called Broca–Wernicke–Lichtheim–Geschwind model—is named after the pioneering informed researchers who first formulated it (for a critical review, see Fridriksson et al. 2018). More than a century later, clinical neuropsychology still relies heavily on its core assumptions.

A more recent and nuanced model of language comprehension and production was proposed by Hickok and Poeppel (2007). It is based on the dual-stream concept of cortical brain organization. According to this model, the bilateral ventral stream contributes to language comprehension, while the left-hemisphere dorsal stream is primarily involved in language production. Although this dual-stream model has influenced theorizing in other areas of neuropsychology (see Chapter 7), its impact in aphasiology has so far been limited.

Nonetheless, the classical model has clear limitations. Its two-component distinction between receptive and productive language functions is an oversimplification (Tremblay and Dick (2016) offer a detailed critique). First, the exact functional boundaries of "Broca's area" and "Wernicke's area" remain unclear. Moreover, the connectionist view—focused primarily on white matter tracts linking these two regions—is also incomplete.

Language depends far more on subcortical contributions than is currently acknowledged. In addition, the brain's language network involves more connections than the traditionally emphasized *arcuate fasciculus,* which has long been portrayed as the main bridge between "Broca" and "Wernicke."

? Most textbooks treat "language" explicitly or silently as a function of the left hemisphere. What is your opinion on that?

The idea of absolute "dominance" of one hemisphere for a specific function is naïve. It perpetuates a narrow—and, tongue in cheek, left-hemisphere-dominated view of functional hemispheric asymmetries (FUHEMAS; see Chapter 11). This notion reflects less a biological reality than a culturally reinforced belief that language dominates life.

Statements such as "The hemisphere that controls language is referred to as the Dominant hemisphere" (taken from a leading neuropsychology textbook) do a disservice to the training of neuropsychologists. While it is true that classical Broca and Wernicke aphasias more frequently follow left-hemispheric lesions, any claim of dominance is at best relative.

The right hemisphere is "dominant" for several essential aspects of language, such as the understanding of humor, sarcasm, metaphors, poetry, and indirect associations. It also excels in comprehending and producing emotional prosody, planning and interpreting discourse, and many facets of social communication.

A dynamic view of FUHEMAS must also acknowledge that the right hemisphere's latent language capacities may only become fully apparent once the inhibitory influence of the left hemisphere is removed. This can occur after a left-hemispheric lesion (Regard et al. 1994), through transient disruption of left "language areas" (Hartwigsen and Siebner 2013), or even via vestibular stimulation (Schüeli et al. 1999).

For comprehensive reviews of right-hemisphere contributions to language, see Searleman (1977), Lindell (2006), Gainotti (2016), and Neophytou et al. (2023).

However, the critique of the classical Broca–Wernicke–Lichtheim–Geschwind model goes much further. Language is so much more than producing and understanding words and phrases. With Pinker and Jackendoff (2005), we must ask what kind of biological system language is—and how it relates to other cognitive systems, in humans and other species.

The concept of a narrowly defined "perisylvian language system" is insufficient for describing the true scope of the language connectome. Language encompasses emotion, poetry, and other forms of linguistic creativity. It enables both narrow and remote associations, supports causal inference, and underlies both rational and many irrational beliefs.

Language transcends verbal communication—it includes facial expressions, gestures, and the broader domain of embodied cognition. While clinical neuropsychology may not yet fully benefit from today's evolving theoretical visions (Tremblay and Dick 2016), it is clear that the traditional classification of linguistic impairments still lags behind contemporary conceptual frameworks.

Where Can We Read More?

Brennan, J.R. Language and the Brain. A Slim Guide to Neurolinguistics. Oxford University Press: Oxford, 2022.

Marshall, J. Aphasia. The Basics. Taylor & Francis: New York, 2024.

Pulvermüller, F. The Neuroscience of Language. On Brain Circuits of Words and Serial Order. Cambridge University Press: New York, 2007.

Poizner, H., Klima, E.S. and Bellugi, U. What the Hands Reveal about the Brain. MIT Press/Bradford Books: Cambridge, MA, 1987.

Appendix to Chapter 4: A Table of Eponyms

The routine use of eponyms in neuropsychology is a tradition worth questioning. While some offer convenient reference points, others carry problematic histories, obscure descriptive clarity, or perpetuate misleading associations. As an appendix to the chapter on aphasias, where the debate over "Broca's" and "Wernicke's" labels is particularly relevant (Tremblay and Dick 2016). This A–Z collection invites critical thought about the role of naming in science quite generally. Is it time to move beyond eponyms—or to at least use them with more awareness?

Eponym	**Meaning**	**Named After**
Anton's syndrome	Denial of (cortical) blindness.	Gabriel Anton (1858–1933), Austrian neurologist/psychiatrist. Due to his Nazi affiliations, alternative names have been suggested, e.g., "Bilateral Occipital Lobe Infarct Neglect Deficit (BLIND) Syndrome" (Shanmugam et al. 2021). Earlier descriptions exist (Das and Naqvi 2023).
Balint's syndrome	Triad of simultanagnosia, oculomotor apraxia, and optic ataxia following bilateral parieto-occipital lesions.	Rezső Bálint (1874–1929), Hungarian neurologist/psychiatrist.
Charles Bonnet syndrome	Visual hallucinations following peripheral (retinal) damage. Often interpreted as "phantom vision" after visual deafferentation.	Charles Bonnet (1720–1793) was a Swiss naturalist and philosopher.
Down syndrome	Neurocognitive disorder caused by triplication of chromosome 21 (trisomy 21); verbal abilities are more affected than nonverbal ones.	John Langdon Down (1828–1896) was a British physician. Earlier descriptions by Jean-Étienne Dominique Esquirol and Édouard Séguin.
Ekbom syndrome	(1) Restless legs syndrome (USA); (2) delusional parasitosis (UK), the belief that living or nonliving agents infest one's body.	Karl-Axel Ekbom (1907–1977) was a Swedish neurologist.
Fregoli-type delusional misidentification	The belief that familiar individuals appear in disguise as other people.	Leopoldo Fregoli (1867–1936) was an Italian actor and quick-change performer.
Gerstmann syndrome	Tetrad of dysgraphia, dyscalculia, finger agnosia, and left-right disorientation after damage to the left inferior parietal lobule.	Josef Gerstmann (1887–1969) was an Austrian-American neurologist.
Heschl's gyri	Transverse temporal gyri, mediolaterally oriented; site of the primary auditory cortices (Brodmann areas 41 and 42).	Richard L. Heschl (1824–1881) was an Austrian anatomist.

Eponym	Meaning	Named After
Island of Reil	The insular cortex, located deep within the lateral sulcus (BAs 13, 14, 16).	Johann Christian Reil (1759–1813) was a German physiologist and anatomist.
Jacksonian seizure/ march	Focal seizure beginning in one body part (e.g., a finger), progressing to adjacent areas (e.g., hand, arm). Consciousness is preserved.	John Hughlings Jackson (1835–1911) was a British neurologist.
Klüver–Bucy syndrome	Behavioral syndrome involving hyperorality, hyperphagia, hypersexuality, and hypo-emotionality due to bilateral medial temporal lobe damage.	Heinrich Klüver (1897–1979) was a German-American psychologist, and Paul Bucy (1904–1992) was an American neurosurgeon.
Lou Gehrig's disease	Popular name for amyotrophic lateral sclerosis (ALS), a progressive motor neuron disease.	Lou Gehrig (1903–1941) was an American baseball player who died of ALS.
Münchhausen syndrome	Fabrication or exaggeration of illness to assume the sick role. Now termed "factitious disorder imposed on self" (DSM-5).	Based on "Baron Münchhausen," a fictionalized version of Karl Friedrich von Münchhausen, known for his fantastical storytelling.
Nicotine replacement therapy (NRT)	Pharmacological aid to smoking cessation by stimulating nicotine receptors in the ventral tegmental area (dopamine release in the nucleus accumbens).	Jean Nicot (1530–1600), a French ambassador to Portugal, introduced tobacco to France.
Othello syndrome	Delusional jealousy concerning a romantic partner's fidelity.	Othello, the tragic protagonist in Shakespeare's play, is manipulated into falsely believing his wife has been unfaithful.
Pick's disease	Behavioral variant of frontotemporal dementia with 3R tau accumulation and "Pick bodies."	Arnold Pick (1851–1924) was an Austro-Hungarian psychiatrist.
Quetelet Index/BMI	Predecessor of the Body Mass Index: weight divided by height squared.	Adolphe Quetelet (1796–1874) was a Belgian statistician interested in "the average man" (Tafreshi et al. 2022).
Rey–Osterrieth Complex Figure (ROCF)	A "meaningless" figure used to assess visual construction, nonverbal memory, and executive function (copy and recall).	André Rey (1906–1965) was a Swiss psychologist (designed figure, 1941); Paul Osterrieth (1916–1980) was a Belgian psychologist (standardized procedures, 1944).
Schmahmann syndrome	Also called "cerebellar cognitive-affective syndrome," a constellation of symptoms after cerebellar lesions.	Jeremy D. Schmahmann was a South African-born American neurologist. See also Chapter 2, p. 37.

Eponym	Meaning	Named After
Tourette syndrome	A neurodevelopmental disorder with motor and vocal tics, usually beginning in childhood.	Georges Gilles de la Tourette (1857–1904) was a French neurologist.
Urbach-Wiethe disease	A rare genetic disorder affects the skin and medial temporal lobes, including the amygdala, and is associated with emotional processing deficits.	Erich Urbach (1893–1946) was an Austrian dermatologist, and Camillo Wiethe (1889–1949) was an Austrian ENT specialist. Described jointly in 1929.
von Economo's disease	Atypical encephalitis lethargica ("sleepy sickness"), also associated with "von Economo neurons."	Constantin von Economo (1876–1931) was a Romanian-Austrian neurologist and cytoarchitectonic pioneer.
Wernicke–Korsakoff syndrome	Combined effects of Wernicke's encephalopathy and Korsakoff's syndrome, often due to vitamin B1 deficiency after chronic alcohol use.	Karl Wernicke (1848–1905) was a German neuropsychiatrist, and Sergei Korsakoff (1854–1900) was a Russian neuropsychiatrist.
X Turner syndrome	A variant of Turner syndrome affecting only females, with severe intellectual disability.	Henry H. Turner (1892–1970) was an American endocrinologist.
Yentl syndrome[2]	Gender bias in medicine: women receive less aggressive treatment for conditions like coronary heart disease, which are considered "male diseases."	Yentl, the heroine of a story by Isaac Bashevis Singer. First introduced by Healy (1991), Bairey Merz (2011) showed the concept remains relevant.
Zelig phenomenon/ Zelig-like syndrome	Variant of delusional misidentification: patients assume the identities of others around them.	Inspired by Leonard Zelig, a fictional character in Woody Allen's 1983 film *Zelig*. See McGorry et al. (1987); Conchiglia et al. (2007).

[2] The eponym is included here because it is the only one in this table that refers to a woman.

CHAPTER 5

What Are Apraxias?

? What does "praxia" mean, and what brain function is disturbed in the apraxias?

"Praxia" is derived from Greek and means "action." "A-praxia" literally means "no action," but in the neuropsychological context, the term designates a deficit in coordinating the intention to perform a specific motor act, the planning of that act, and its proper execution. Classically, this act involves the pantomimed use of a tool. Gestures may also be impossible to perform on command—although they might still occur in reflex-like fashion when triggered by context or situation.

Apraxias involve a disturbance in the integration of intention, planning, and execution of motor behavior. They are not due to primary motor impairments (such as paralysis or paresis), nor are they the result of poor motor coordination. Apraxias also do not stem from a general dysexecutive syndrome in planning, nor from impaired language comprehension—though this has occasionally been claimed.

Apraxias are caused by acquired disruption of a large-scale fronto–parietal–basal ganglia network in the left cerebral hemisphere. They are paradigmatic examples of a disconnection syndrome.

? What are "disconnection syndromes," and which examples should neuropsychologists know?

Disconnection syndromes arise from the disruption of the neural pathways (white matter tracts) that link cortical regions, jointly contributing to a specific function. Paradigmatic examples include:

- Interhemispheric disconnection (split-brain; see Chapter 11, p. 249)
- Apraxias (this chapter)
- Conduction aphasia (see Chapter 4, p. 66)
- Pure alexia (alexia without agraphia; see Chapter 4, p. 71)
- Auditory agnosia (word "deafness;" see Chapter 7, p. 161)
- Alien hand syndrome (see Chapter 10, p. 223)

DOI: 10.1201/9781003384991-5

The fact that lesions producing apraxia primarily affect the left hemisphere aligns with this hemisphere's well-known superiority for a broad range of motor functions. However, this does not imply that right-hemispheric networks are uninvolved in the planning of goal-directed actions.

For instance, Wheaton et al. (2008), using connectivity analyses, demonstrated that right-hemisphere regions are substantially involved in what could be described as a praxia network.

Still, from early on, the left frontal and parietal cortex, the basal ganglia, and the white matter tracts linking these regions have been recognized as central to the genesis of apraxic disorders. The sketch of the left-hemispheric praxia network in Figure 1 highlights the heterogeneous loci whose disruption may lead to apraxia. It also helps explain the diversity of conditions and contexts in which different apraxia subtypes manifest.

? Which names should we associate with apraxic disorders (comparable to Broca and Wernicke in the context of language disorders)?

The German neurologist Hugo Liepmann (1863–1925) is the most influential figure in apraxia research. His foundational contributions to the neuroanatomy and neuropsychology of apraxias are reviewed in Goldenberg (2003).

Earlier figures also paved the way. German linguist Heymann Steinthal (1823–1899) and psychiatrist Carl Maria Finkelnburg (1832–1896) prepared the ground for Liepmann's more systematic analyses. Steinthal coined the term "apraxia" (*Apraxie*) and viewed apraxia as a disorder of the relationship between movement and objects. Finkelnburg focused on the failure to understand symbolic gestures, a deficit he termed asymbolia.

Figure 1 illustrates Liepmann's concept, which, in words, can be summarized as follows: The plan for a motor action originates in the left inferior parietal lobe. It is transferred anteriorly via sensorimotor cortical connections, notably the arcuate fasciculus, to the primary motor cortex (M1). In the healthy brain, this enables the action to be executed by the right hand. It can also be performed by the left hand, provided that the interhemispheric connection required to activate the *right* primary motor cortex is intact. If not, a selective apraxia of the left hand results (see also Chapter 11, p. 252).

In neuropsychological testing, the action plan is typically activated in response to a verbal command by the examiner (e.g., "Show how you use a hammer!"), which presupposes the integrity of connections from Wernicke's area to the angular gyrus.

Alternatively, the patient may be asked to imitate an action demonstrated by the examiner—this relies on the integrity of visual-parietal pathways.

Figure 1 may appear oversimplified and somewhat dated. Yet, it remains a clinically useful and conceptually valid representation of apraxia. Modern neuroimaging studies have refined this picture (see Figure 2). However, because even simple motor actions typically engage large and distributed networks, any attempt at greater accuracy inevitably adds complexity. See also the comments on Figure 2 on p. 87.

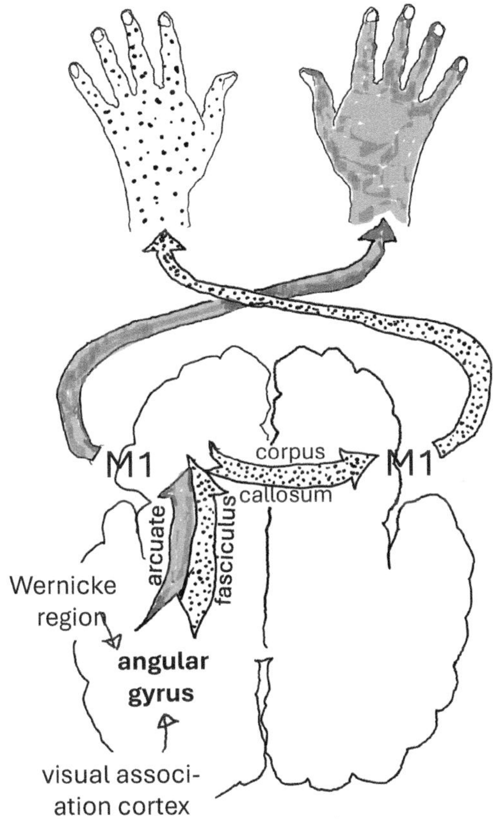

Figure 1: Apraxia as conceptualized by Hugo Liepmann around the turn of the 19th and 20th centuries. A lesion along any of the arrows (white matter paths) can lead to apraxia, which is usually manifested by both hands. Only the selective disruption of callosal pathways produces an isolated apraxia of the left hand. See the text for more details. Redrawn after a sketch by Liepmann and Maas (1907) and elaborations by Schnider (1997).

Apraxias are genuine neuropsychological deficits, but they do not constitute a unitary disorder. They can be classified according to the primarily affected body part or region:

* trunk apraxia
* limb apraxia (affecting upper or lower limbs)
* facial apraxia (also bucco(linguo)facial apraxia or oral apraxia)
* apraxia of the eyelid (closing or opening)

Apraxias are also sometimes named based on the specific *motor function* in which the deficit becomes manifest:

* apraxia of speech
* apraxia of writing
* apraxia of gait

In all these forms, the major network that links intention, planning, and execution—spanning fronto-parietal-subcortical regions of the left hemisphere—is at least partially compromised.

An important exception may be found in eyelid apraxias. While apraxia of eyelid closing (the inability to voluntarily close one's eyes, with reflexive closure spared) is due to a facial nerve paresis, apraxia of eyelid opening is most frequently observed after right hemisphere lesions (Nersesjan et al. 2020).

Terms like "dressing apraxia" or "construction apraxia" are also in use. However, these clinical presentations do not represent true apraxias. Although the disordered motor behavior appears mismatched with the intended goal, the underlying mechanisms differ and are *not* primarily due to the disruption of the typical left-hemispheric apraxia network.

Dressing apraxia—described evocatively by Perkin (1981) as "this complex and frightful problem of my clothes"—refers to the inability to put on clothes properly. Patients may put trousers on backward, push an arm through the top of a pullover, or struggle to orient a shoe correctly. The underlying deficit is typically a disturbance of body schema and is more frequently observed after right parietal lesions. Occasionally, frontal dysexecutive dysfunction—from left, right, or bilateral frontal lesions—can exacerbate dressing difficulties.

? Is there a valuable description of "dressing apraxia" in the belletristic literature?

A detailed first-hand account of dressing apraxia is given in Arnold Bennet's novel *Clayhanger* (Bennett 1910). It beautifully illustrates that the condition is more than an apraxic disorder in the narrow sense, and that dementia is involved. See Perkin (1981) for the literary and medical context.

"It was not that his father lacked the strength, but he would somehow lose himself in the maze of his garments, and apparently he could never remember the proper order of doffing or donning them. Sometimes he would ask, 'Am I dressing or undressing?'"

(cited after Perkin 1981)

"Construction apraxia" refers to the difficulty or complete inability to copy a two-dimensional drawing or a three-dimensional arrangement of cubes. Frontal and/or parietal lesions may cause it, ultimately reflecting a disturbance in planning and spatial representation (Gainotti and Trojano 2018).

Although stroke is the most frequent cause of apraxia, apraxia can also occur in the context of a neurodegenerative disorder, most notably corticobasal degeneration. In Primary Progressive Apraxia (PPA; note that the same acronym also stands for Primary Progressive Aphasia), apraxia is the first noticeable symptom (Fernandez and Frucht 2017). This condition allows for the gradual tracking of decline within the praxia network and its associations with impairments in spatial coordination and body schema.

In advanced stages, apraxia often co-occurs with autotopagnosia—a disorder of locating body parts. Patients may be unable to point to their nose, knees, or chin

(Sirigu et al. 1991). A case of probable Primary Progressive Apraxia is presented toward the end of this chapter (p. 91).

What follows is a brief description of the major apraxic disorders as typically differentiated in textbooks. It is important to emphasize that these are rarely seen as pure deficits; symptoms from one subtype are often intermixed with those of another.

Ideomotor Apraxia: A Paradigmatic Disorder of Pantomime Tool Use

Ideomotor apraxia is the inability to pantomime tool use on verbal command. Affected patients may be perfectly able to use a hammer, describe its function, imitate someone using it, or identify hammering in a photograph—but they are unable to perform a hammering movement when asked, "Show me how you use a hammer."

This is why such patients often remain unaware of their deficit in everyday life: pantomiming tool use on command is not a daily necessity. The problem typically becomes apparent only during clinical assessment, often to the patient's embarrassment.

More detailed analyses of tool-use pantomime reveal that ideomotor apraxia involves errors in both the spatial and temporal domains. Movements are inaccurate in terms of hand position in space, movement configuration, and amplitude, initiation timing, and sequencing of motion steps (Hermsdörfer et al. 1996).

Gross errors in apraxia testing include:

* Parapraxias: Substituting the required action with another, unrelated one (see Figure 4).
* Echopraxia: Involuntary mirroring of another person's actions (often the examiner). This belongs to the broader category of *echophenomena*, which includes echolalia and other "forced imitations" (see Chapter 4, Table 2).

A widespread phenomenon is the "body part as tool" (BPT) error. Here, the patient uses a part of their body to symbolically represent a tool, rather than pantomiming the tool's proper use. For example:

* spreading the index and middle fingers to represent scissors;
* holding the hand near the face with thumb and pinky extended to mimic a mobile phone;
* brushing teeth using the finger instead of pantomiming a toothbrush grip.

There is growing literature, including studies in healthy participants, on the neural basis of BPT errors (e.g., England and Nicoladis 2018). The left inferior parietal lobule appears particularly involved in translating symbolic representations into appropriate object-directed action.

One challenge in clinical testing is that BPT errors are not specific to apraxia. They can also occur in patients with neuropsychiatric disorders, after right hemisphere damage, or even in healthy individuals. Only when a patient continues to make BPT errors despite explicit instruction to form the hand as if holding the tool should the response be counted as apraxic.

Ideomotor apraxia should also be tested for leg actions, such as kicking a ball, mimicking bicycle pedaling (in a seated position), or stamping out a cigarette.

Indeed, claims that ideomotor apraxia affects primarily the upper limbs may simply reflect the fact that clinical tests typically focus on hand and arm actions. Lesions affecting leg-related ideomotor apraxia tend to be located more dorsally, along the arcuate fasciculus and within motor association cortex.

? The term "ideomotor apraxia" rings a bell: Is the concept of "ideomotor movements" not important in spiritualism and parapsychology?

Absolutely. The "ideo-" in ideomotor derives from the Greek word for "idea." In both contexts, a disturbed translation between idea and action is key. In ideomotor apraxia, the abstract idea behind an action cannot be expressed through pantomime, despite intact motor affordances. Ideomotor movements, on the other hand, are involuntary movements that seem decoupled from conscious intention. They are central to understanding phenomena such as dowsing rods, Ouija boards, and automatic writing.

Such phenomena should not be dismissed by neuropsychologists—or any clinician or scientist—as mere illustrations of mass delusion. They offer important lessons about unconscious muscle movements triggered by thoughts, expectations, and suggestion. Michael Faraday (1853) empirically demonstrated the ideomotor basis of table-turning, and analogous work was conducted for pendulum use (Chevreul 1854). Despite this early research, many people still attribute ideomotor movements to spirits or other external forces.

Of particular ethical concern are claims made in autism therapy, where patients' supposed communication "apraxia" is overcome by supporting their writing hands. Messages produced through Facilitated Communication (Biklen 1990) in fact originate from the facilitator, not the patient. The therapist's unconscious guidance, not the patient's will, is responsible (Wegner et al. 2003). For a detailed discussion of ideomotor phenomena and their relevance to neuropsychology, see Brugger (2001).

Apraxia for gestures is a subtype of ideomotor apraxia. No tools are involved; instead, the gestures convey social meaning (e.g., a military salute, hitchhiking gesture, or a threatening sign). Many patients with deficits in pantomime also struggle to perform symbolic gestures on command. However, gesture-related and tool-related ideomotor apraxias can also occur in isolation.

Gestures—unlike tool use—are likely to involve prefrontal networks associated with *theory of mind* and social cognition (Osiurak et al. 2021).

? Can patients with ideomotor apraxia imitate an observed action?

Imitating an action requires translating an observed movement into a motor command. Whether imitation is preserved or impaired depends on the integrity of pathways linking perception to motor planning. In mild apraxia, imitation is often spared. Similarly, patients with gestural apraxia may or may not imitate observed gestures correctly (Rumiati et al. 2009).

> Imitation of visually presented actions should be tested in all apraxic patients. A further distinction is crucial: imitation of meaningful gestures (e.g., a military salute) is conceptually different from imitation of meaningless gestures (e.g., touching the chin with an extended pinkie). Goldenberg (2009) emphasized that imitation of meaningless gestures is particularly dependent on the left inferior parietal lobe. Novel finger and hand postures require exact analysis of spatial relationships.
>
> There is also a functional distinction between dorsal- and ventral-stream processing: imitation of meaningless gestures has been associated with dorsal stream integrity, whereas imitation of meaningful gestures relies more on ventral-stream processing (Rumiati et al. 2005).

Patients with ideomotor apraxia may, in the context of a neurodegenerative process, progress toward ideational apraxia (described below) and may also exhibit signs of body image disturbances. As degeneration advances, the pantomime of tool use becomes increasingly difficult, and even actual tool use may become compromised.

Figure 2 shows the network involved in pantomime and tool use, based on multiple meta-analyses of neuroimaging and behavioral data (Osiurak et al. 2021). Compared to the early 20th-century model based on clinical observation and autopsy (Liepmann and Maas 1907; see Figure 1), the progress in understanding apraxia over the last 100 years is striking.

Liepmann's model of apraxia was deliberately narrow: he defined it as the failure, in the absence of paresis, to translate an action plan into a motor act, particularly in the form of pantomime. If we expand this concept to include knowledge of the body (corporeal awareness, see Chapter 10), theory of mind (as pantomiming is communicative), and technical reasoning,[1] then it is no surprise that the implicated neural network becomes broader.

Even so, it remains notable that the extended model presented by Osiurak et al. (2021)—despite its conceptual richness—still maps entirely onto the left hemisphere.

Ideational Apraxia

This form of apraxic disorder also involves a deficit in pantomime, but here the action to be imitated is more complex than the use of a single tool. Typical examples include brewing and serving coffee or using a hole puncher to prepare paper for filing. The impairment thus goes beyond isolated pantomime and reflects a failure to conceptualize, plan, and simulate an action sequence involving multiple tools or objects.

More or less isolated ideational apraxia (i.e., without accompanying ideomotor apraxia) is observed in cases of extensive left-hemispheric damage and in various forms of dementia. Some authors argue that ideational apraxia is not an actual apraxic

[1] Osiurak et al. (2021) discuss the term "atechnia" to refer to a specific failure in the use and understanding of technical instruments.

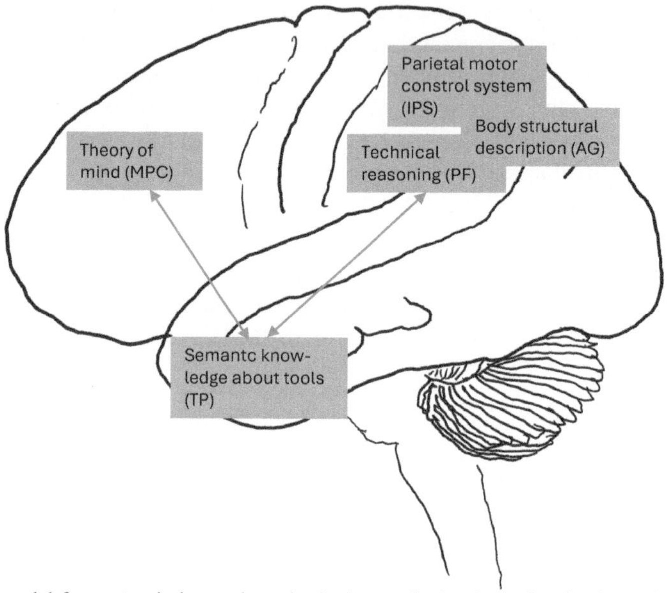

Figure 2: A model for pantomiming tool use in the human brain. As in the classic model of praxia by Liepmann (see Figure 1), a left hemisphere parieto-frontal network is involved. An inferior parietal system unites knowledge about body structure, technical reasoning, and motor control. In addition, semantic knowledge about tool use and theory-of-mind functions are necessary for successful pantomiming as a communicative action. See text and Osiurak et al. (2021) for details.
AG: angular gyrus; IPS: intraparietal sulcus; MPC: middle prefrontal cortex; PF: parietal F-complex; TP: temporal pole.

disorder, but rather a constellation of executive and memory deficits more accurately described as action disorganization syndrome (Humphreys and Forde 1998). This broader interpretation is supported by the observation that deficits are typically present regardless of whether the action sequence is performed spontaneously or on command.

Accordingly, lesions are more widespread than those typically implicated in ideomotor apraxia and often extend beyond the left inferior parietal lobe. Ideational apraxia is frequently encountered in neurodegenerative conditions.

Conceptual Apraxia

In contrast to ideomotor and ideational apraxia, conceptual apraxia reflects a more fundamental impairment in action knowledge. Patients may even lose understanding of the purpose of single tools. Misuse of objects is common—for example, using a spoon to cut meat. Patients may also fail to match tool images with appropriate usage scenarios (e.g., associating an axe with a bathroom rather than a woodcutting scene).

These difficulties extend to both execution and judgment of gestures: even highly familiar gestures are no longer reliably connected to their typical social or situational context.

Conceptual apraxia is statistically associated with posterior left hemisphere lesions and with damage to areas supporting semantic memory. In contrast to patients with ideomotor or ideational apraxia, those with conceptual apraxia are often profoundly disabled in everyday life (Goldman Gross and Grossman 2008).

Buccofacial (Orofacial) Apraxia

Buccofacial apraxia is the inability to perform skilled voluntary movements involving the face, mouth, tongue, larynx, or pharynx. The deficit becomes apparent when the patient is asked to perform specific actions on command—e.g., "Show me how you would kiss," "Wink with your eye," or "Drink with a straw." These same actions, however, may be performed flawlessly in spontaneous situations or when cued by the presence of the relevant object.

Buccofacial apraxia frequently co-occurs with ideational apraxia, but the two can also dissociate.

In accordance with the organization of the motor homunculus/femincula anterior to the central sulcus (see Chapter 10), buccofacial apraxia is most commonly caused by ventrally located lesions of the arcuate fasciculus and motor association cortex. Lesions may also involve the inferior frontal cortex, deep frontal white matter, insula, and basal ganglia—predominantly on the left side of the brain.

In various neurodegenerative disorders, the presence of buccofacial apraxia can aid differential diagnosis (Ozsancak et al. 2004).

? **Next week, I will see a deaf signer who has suffered an infarction of the left middle cerebral artery. Relatives say his gestures have become largely unintelligible. How can I differentiate whether his difficulties reflect an impaired language or an apraxic deficit?**

This is a very sensitive and instructive question, clearly demonstrating that neuropsychological assessments involve far more than the administration of standardized tests.

Deaf signers can indeed suffer from aphasia for sign language (see Chapter 4) and, independently, from apraxia of signing. Carefully designed assessments are required to distinguish between the two, which result from breakdowns in distinct but overlapping neural networks (Roby-Brami et al. 2012).

Aphasia for Sign Language

In receptive aphasia, the diagnosis is relatively straightforward: the patient has difficulty understanding sign language. In expressive aphasia (without coexisting apraxia), signing is impaired by omissions, morphological or semantic errors—essentially paraphasia in sign. However, the ability to mime simple actions (e.g., pantomiming toothbrushing) is typically spared.

Apraxia for Sign Language

In contrast, patients with apraxia understand what they want to communicate but fail to form the signs correctly. Difficulties may concern handshapes, trajectories, and spatial positioning. Crucially, the errors are not linguistic but motoric, and isolated miming of everyday actions is often equally impaired.

To reliably differentiate between the two, you must either be fluent in the patient's sign language or work closely with a professional interpreter experienced in clinical assessments.

Good luck with your case!

Limb-Kinetic Apraxia

The nature of limb-kinetic apraxia remains controversial—particularly whether it qualifies as a *genuine* apraxic disorder or whether it reflects a form of ataxia instead (Schnider 1997).

Clinically, it presents with clumsy, imprecise hand movements, most evident in the distal extremities. Patients struggle with fine tasks such as lifting a coin or unbuttoning a shirt. These difficulties are not due to primary motor, sensory, or coordination impairments.

Typically, limb-kinetic apraxia affects the limb (especially the fingers) contralateral to a central unilateral lesion and does not show the clear hemispheric asymmetry seen in other apraxias. Discrete impairments of the pyramidal tract have been suggested.

Unlike ideomotor apraxia, limb-kinetic apraxia tends to be independent of task demands—i.e., performance is similarly impaired whether the patient is imitating an action or responding to a verbal command. Also absent is the classic voluntary-automatic dissociation seen in most other apraxic disorders (Leiguarda and Marsden 2000).

? **Is there any case report illustrating the essence of what is contained in this chapter on apraxias?**

Yes. The following case, described in detail, is illustrative of multiple key points. The initial examination revealed:

1. The apraxia is "pure", meaning there is virtually no aphasia.
2. Action execution is *flawless with tools,* but pantomiming the same action is abolished (e.g., drinking coffee). A retest five months later shows that tool use has also deteriorated—even with the tool present.
3. Reflexive, non-intended pantomimes may still occur.
4. Apraxia affects not only the arms and hands—facial actions and lower extremities must also be assessed.
5. Over time, symptoms progress to include autotopagnosia and other body schema disturbances, particularly in hand-body actions.

On a general level, this case highlights the importance of ruling out alternative explanations for pantomime failure: receptive language deficits, lack of knowledge about tools, or motor impairments.

More broadly, the case underscores the value of video documentation. Recording patient interactions—of course, with written informed consent and a clear statement of intended use—can provide rich material for both clinical documentation and teaching purposes.

An Illustrative Case of Ideomotor Apraxia

IA (short for Ideomotor Apraxia) is a 53-year-old woman, left-handed, with an uneventful medical and psychological history. There is no family history of neurological or psychiatric illness.

After 14 years of schooling, IA began working as a secretary. For several years, she holds a full-time position with considerable responsibility. She then pauses her professional career for 15 years to raise her two children with her husband.

In her current role, she has been employed for about 10 years (the first three part-time) as the private secretary to the director of a large trading agency. Now divorced, she lives alone but is in a new relationship.

For about a year, her boss has observed increasing difficulties in IA's ability to perform calculations—long considered one of her strengths. His concerns led to her referral to a neurologist in private practice, who subsequently arranged for her inpatient evaluation at the neurology department of a university hospital.

> Neurological examination reveals discrete signs of an extrapyramidal syndrome, with pronounced rigidity on the left side. Upward gaze is slightly more impaired than downward gaze. Otherwise, the neurological status is unremarkable. EEG shows slight general slowing, with a frontotemporal decelerating focus over the left hemisphere. MRI reveals mild cortical atrophy bilaterally, most prominent over the left parietal cortex. FDG-PET indicates a bilaterally reduced metabolism in both parietal and temporal lobes, more pronounced on the right. CSF analysis and standard laboratory work-up are within normal limits.

In the neuropsychological evaluation, IA presents with a good, slightly elevated mood. She is fully oriented to time and place. However, her orientation to person and situation is conspicuous: she has difficulty articulating the reason for her admission. While she does not deny "occasional difficulties" at work, she clearly downplays her deficits—even when confronted with specific shortcomings during testing.

Compared to age- and education-matched normative data, her test results show:

* Mildly reduced performance ($z > -1.6$) in psychomotor speed, alertness, sustained and selective attention, and nonverbal learning and recall (with quantitatively and qualitatively normal recognition).
* Moderately reduced performance (z between –2.0 and –1.6) in divided attention, interference control (in both the Stroop test and a nonverbal analogue), concept finding (quantitatively normal but qualitatively perseverative), and verbal learning and recall (mildly impaired recognition with false positives).
* Moderate to severe deficits in calculation (based on oral and written tasks involving addition, subtraction, multiplication, and division; in-house norms).
* Normal performance in visual perception, visual cognition, visuoconstructive abilities, and body schema (including left-right discrimination).

Verbal and nonverbal fluency are quantitatively normal, though her verbal output shows a mildly elevated number of repetitions. Productive and receptive language functions are intact, aside from a mild naming deficit on clinical exam (despite normal performance on the Boston Naming Test) and a slight tendency toward logorrhea.

Apraxia is not assessed via formal testing, but its severity is striking and is videographically documented (see Figure 3).

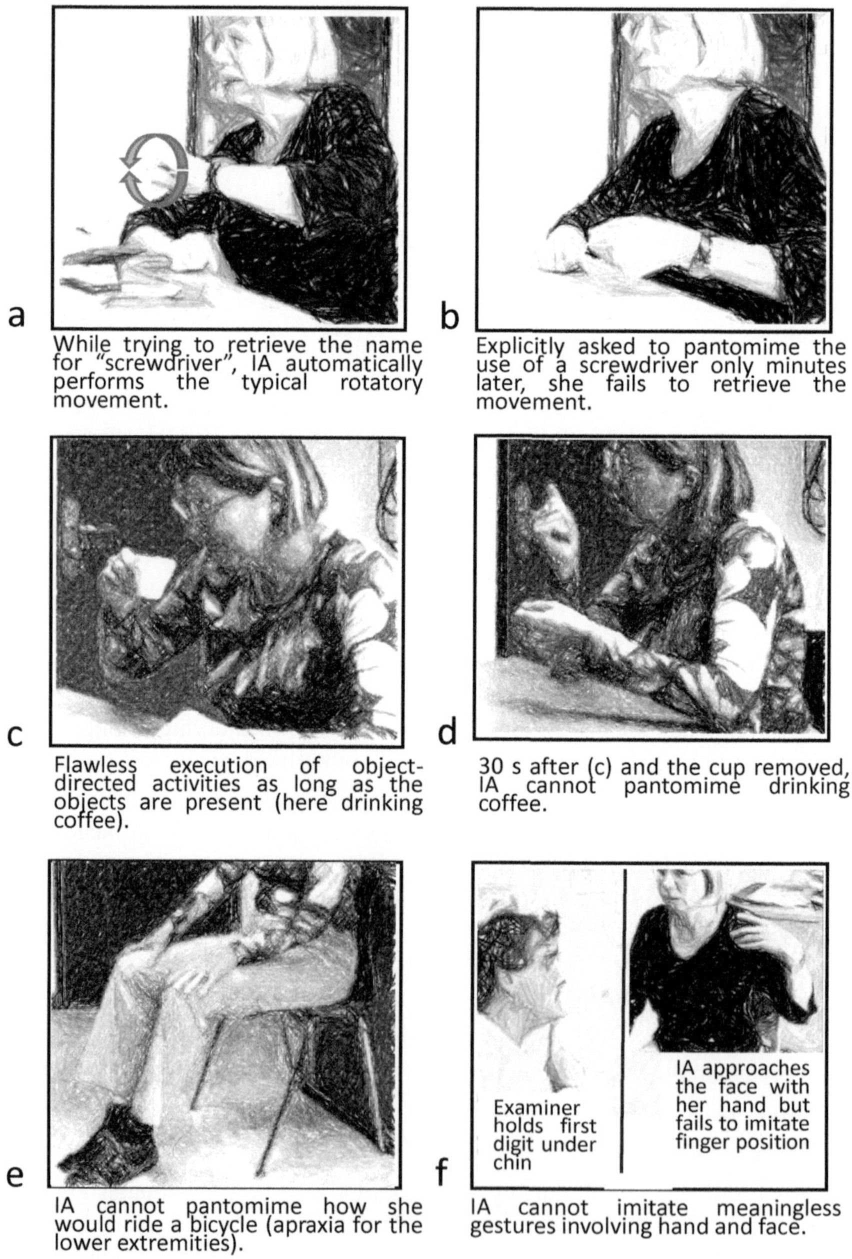

a — While trying to retrieve the name for "screwdriver", IA automatically performs the typical rotatory movement.

b — Explicitly asked to pantomime the use of a screwdriver only minutes later, she fails to retrieve the movement.

c — Flawless execution of object-directed activities as long as the objects are present (here drinking coffee).

d — 30 s after (c) and the cup removed, IA cannot pantomime drinking coffee.

e — IA cannot pantomime how she would ride a bicycle (apraxia for the lower extremities).

f — IA cannot imitate meaningless gestures involving hand and face.

Figure 3: Patient IA's behavior in 6 different situations. See text for more details.

First Assessment (Extended Legend to Figure 3)

IA is seated in front of an array of objects and tools (stapler, hole punch, hammer, drill, screwdriver, pen, paper clip, etc.). She is asked to:

1. Point to selected objects;
2. Name selected objects;
3. Demonstrate the use of selected objects.

(1) She correctly and without hesitation points to each object named by the examiner. This indicates that her visual recognition, understanding of object names, and comprehension of verbal instructions are intact.

(2) Her naming is largely intact—occasionally imprecise in articulation, but consistently followed by immediate self-correction. There is only a single instance of a naming block, involving the screwdriver (see below).

(3) IA easily demonstrates the use of all tools and objects. This includes both the basic movements associated with each tool and the correct sequencing of multi-step actions—e.g., punching holes in two sheets of paper and attaching them with a paper clip.

Figures 3a and 3b

A critical observation regarding IA's apraxia emerges during an incidental episode. When asked to name the screwdriver, she hesitates. During her word-finding attempt, she spontaneously pantomimes the correct use of the tool—with perfect form and fluidity (Figure 3a).

Later, after a few intervening pointing and naming trials, the examiner asks IA to pantomime the use of each tool on the table, one by one. She completely fails with all tools, including the screwdriver. When reminded that she had performed the correct pantomime just moments before, she acknowledges it, but is unable to reproduce any hand movement even remotely resembling the act of using a screwdriver (Figure 3b).

Figures 3c and 3d

Tool use remains *flawless* in real contexts. When casually invited to make coffee, IA performs a full action sequence: pouring milk, stirring with a spoon, sipping from the cup (Figure 3c). However, just 30 seconds after taking the last sip, she is asked to pantomime drinking coffee. Apart from bringing her hand repeatedly near her head, her movements are erratic and do not convey the intended action (Figure 3d).

Figure 3e

IA's apraxia extends to the *lower limbs*. She is unable to pantomime bicycling, extinguishing a cigarette on the floor, or kicking a ball toward a target. She also fails at coordinated hand–foot actions such as pantomiming the tying of shoelaces.

Figure 3f

IA shows apraxia for gestures. She cannot pantomime culturally familiar gestures such as a military salute or the hitchhiking sign. However, she readily imitates meaningful gestures when demonstrated. In contrast, imitation of meaningless gestures—such as novel bimanual finger-hand configurations or unusual hand–face postures—is unsuccessful.

Not Shown in Figure 3

IA also presents with buccofacial apraxia, though less marked than her limb apraxia. She is unable to pantomime drinking through a straw, kissing, or blowing out a candle. While she can wrinkle her nose, she cannot show how she would turn up her nose at a disgusting smell.

Follow-Up Assessment, 5 Months Later

IA arrives on time for the scheduled follow-up examination. She is still cheerful, though her mood is slightly less elevated, and she is a bit less talkative. She reports being on sick leave since the first neuropsychological evaluation, which she places "about three months ago." She remains oriented to time (± two days) and place. She acknowledges "some health problems," clarifying that they are "not physical, more about memory."

She continues to manage daily life largely independently (confirmed by her sister and general practitioner), though she now accepts assistance from her daughter two to three times a week. Her brother-in-law has been managing her finances for about a year. IA's anosodiaphoria is more pronounced than at the first assessment. She jokes about having forgotten an appointment with an old friend a few weeks earlier. When recounting her Christmas holidays, she offers only vague descriptions, with few details—though those she does recall are accurate (as confirmed by her sister). She expresses pleasure at the follow-up testing, remembering clearly how much she enjoyed it the first time. She also recalls being surprised that pantomiming the tool was harder than actually using it.

> Test scores (same tests, B-versions) are lower across the board compared to the initial assessment. The most preserved areas *are visual processing, nonverbal learning and retention,* and *spoken language function*. However, she now shows word-finding difficulties in spontaneous speech, and performance on the Boston Naming Test is mildly impaired.
>
> Agraphia is now present: even copying written words is reduced to one- or two-syllable items. *Reading* remains largely intact, but reading comprehension is impaired due to memory deficits. All calculation tasks yield floor-level performance.
>
> The most pronounced deterioration is seen in verbal memory (except for only mildly poorer attention and working memory span) and verbal and nonverbal fluency (phonetic, semantic, and design fluency, especially quantitatively). Measures of attention (focused and divided) and executive function also show significant decline, including those thought to be primarily mediated by the right prefrontal cortex.
>
> Tasks assessing body schema now reveal mild to moderate impairments: reduced speed in hand laterality decisions, difficulties with left-right discrimination of body parts, and occasional errors in pointing to one's own body or in interpersonal spatial actions.

A neurological follow-up examination produced findings largely identical to those 5 months before. Neither the vertical gaze palsy nor the left-sided rigor had increased. The repeat MRI showed a discrete symmetrical increase in the bilateral parietal atrophy. EEG and laboratory analyses were not repeated.

Apraxia—both ideomotor and ideational—remains among the most striking impairments. At first assessment, pantomiming even overlearned actions like brushing one's teeth or opening a door was already impossible. Now, many fewer automated actions can no longer be performed, even with the actual tool present.

For instance, punching holes in a sheet of paper is now clumsy and time-consuming. She cannot cut paper into two roughly equal pieces. She is at a loss when asked to use pincers, a file, or a mortar—despite being able to name the tools and verbally describe their function. All multi-step actions show marked hesitation and repetition.

She is still able to drink coffee with the examiner and shows no difficulty opening and closing a milk bottle.

A new sign appears: IA now exhibits autotopagnosia—she cannot localize body parts named by the examiner on her own body. (No attempts are made to test localization on the examiner's or a doll's body; see Sirigu et al. 1991). Had her failure to imitate meaningless gestures (Figure 3f) not already been documented at the first assessment—when autotopagnosia was absent—this imitation deficit might now be equally attributable to either apraxia or body schema disturbance.

Synopsis

IA presents with an exceptionally severe apraxia, in the absence of major impairments in other neuropsychological domains—at least initially. The first sign that prompted medical attention was a calculation deficit, noted by her employer. Indeed, calculation is severely impaired for all four basic operations (quantified via in-house tasks; more complex numerical reasoning, e.g., WAIS arithmetic, was not further assessed). This points toward left parietal dysfunction, aligning with the observed apraxia and MRI findings.

For an interim diagnosis, three tauopathies are considered: Alzheimer's disease, frontotemporal dementia, and corticobasal degeneration. After the first follow-up, IA moved to another region of Switzerland and was lost to follow-up. No long-term clinical data or pathology results are available to refine the diagnosis.

From a neuropsychological standpoint, IA appears to suffer from Primary Progressive Apraxia. Fernandez and Frucht (2017) describe two cases of isolated, profound ideomotor apraxia with preserved cognition, including intact language and motor functions. Their cases progress slowly (over five years) and show no correlative structural imaging findings. In contrast, IA's decline is more rapid. Within just five months, she exhibits measurable deterioration in memory and cognition, alongside cortical atrophy that corresponds to her deficits already at initial presentation.

? I do not understand a tiny detail in Figure 1a. Does not the automatic retrieval of the appropriate hand movement while searching for the label "screwdriver" prove that the "action plan" is somewhere stored? How, then, can the patient fail to pantomime the action?

The "tiny detail" you are referring to resists an easy explanation. We can only speculate that pantomiming while searching for a word resembles a natural situation. In contrast, executing a pantomime on verbal command in a clinical context is always somewhat artificial—it lacks the situational props that normally support action execution (Leiguarda and Marsden 2000).

One might say that the conscious attempt to retrieve an item from the "pantomime vocabulary" is comparable to the tip-of-the-fingers phenomenon in deaf signers who momentarily struggle to access a specific sign (see Chapter 4, p. 73). The key difference is that such retrieval is transiently blocked in healthy signers—but lastingly disrupted in apraxic patients. See also the earlier question on aphasia vs. apraxia in deaf signers (p. 89).

An Illustrative Case of a Selective Apraxia of the Left Hand

As this condition is rarely observed, I present here a case of "left-hand apraxia."[2]

Patient LHA (for "Left Hand Apraxia") is a 52-year-old, right-handed purchasing manager and recreational horse rider. During a forest ride, his horse suddenly shied, causing LHA to fall and strike his head against a tree. He sustained an intracerebral and intraventricular hemorrhage, with bleeding extending into the corpus callosum. Neurologically, he showed impaired motor coordination of the upper limbs, but no mirror movements. There was no hydrocephalus, and the patient did not experience seizures.

> A neuropsychological evaluation was conducted three weeks after the accident. LHA was fully oriented in all four dimensions (place, time, person, and situation). Nonverbal learning was mildly impaired with intact recall. In the verbal domain, learning was moderately impaired, recall was mildly impaired, and recognition was normal for both verbal and nonverbal material. Psychomotor speed was slow. However, error monitoring was relatively well preserved. Sustained attention was grossly normal, though divided attention tasks revealed moderate to severe deficits. Executive functions ranged from moderate impairment (color-word interference, design fluency) to severe impairment (quantitative aspects of verbal fluency). Language functions were intact (both expressive and receptive), and there were no signs of agnosia, constructional deficits, or apraxia, as long as testing involved the right hand. Buccofacial praxis was also normal.

2 To be precise: this was an ideomotor apraxia manifesting only during left-hand testing (hands themselves are not apraxic; see footnote 5 in Chapter 11, p. 252).

Figure 4 shows examples from the apraxia testing involving the left hand. It is essential in such examinations to avoid mirror imitation of actions previously demonstrated correctly with the right hand. Both unimanual and bimanual object use, as well as gesture production, should be assessed.

In Figure 4a, when asked to pantomime brushing his teeth with his left hand, LHA's right hand was gently restrained to prevent substitution. LHA murmured, "Ah, with the left… all right, let's see…," then bared his teeth for about 15 seconds without moving his hand.

Afterward, he was allowed to perform the pantomime with his right hand, which he did correctly (not shown). The examiner then rephrased the request: "Now, please show how you would hold a toothbrush with your left hand." LHA shaped his left hand into a fist and made repetitive up-and-down movements (Figure 4b), closely resembling the pantomime of hammering—a gesture he had executed correctly with both hands about four minutes earlier (the left simply imitating the right).

A further request to pantomime a bimanual action—"Hold a wine glass with your right hand and pour wine into it with your left"—resulted in appropriate right-hand action, but a complete absence of left-hand movement (Figures 4c and d).

Selective apraxia of the left hand was also evident during gesture production (Figure 4e). On some trials, parapraxias occurred (Figure 4f). For instance, when asked to pantomime sawing, LHA's left hand instead grasped the zipper of his jacket and attempted to open it. The involuntary assistance by the right hand remains puzzling, as LHA understood the request and could accurately execute the appropriate pantomime. This is reminiscent of intermanual conflict seen in split-brain patients. LHA reported that during the early weeks following his accident, he occasionally experienced such bimanual conflicts.

At a six-month follow-up, pantomime performance on verbal command was normal for both hands, and no further intermanual conflict was observed. Psychomotor speed and performance in all cognitive domains had returned to within normal limits, apart from slightly increased interference susceptibility on the Stroop test.

? Why are the chapters on apraxia in neuropsychology textbooks often very short?

They often are—and we can only speculate about the reasons. One possibility is that many neuropsychologists still regard apraxia as "too motor" to qualify as a true cognitive disorder (a misconception that Goldenberg (2013)—recommended below (p. 99)—carefully dismantles).

Another explanation may be historical. The concept of apraxia has deep European roots, and this might account for its more modest representation in certain Anglo-American neuropsychology handbooks. For instance, the nearly 1,000-page volume by Schoenberg and Scott (2011) allocates just over a page to what the authors refer to—rather redundantly—as a deficit of "motor movements" (p. 211 and elsewhere).

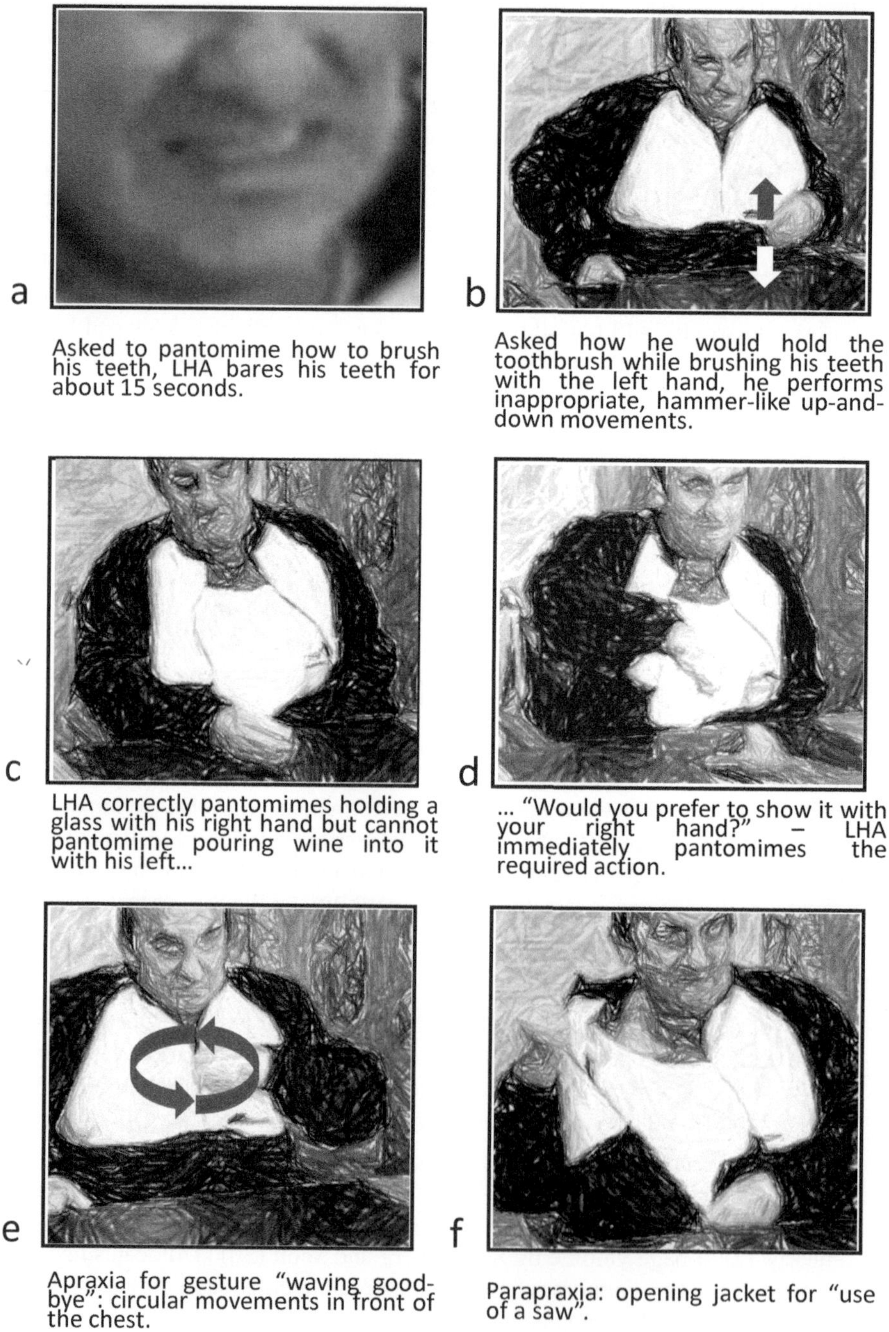

Figure 4: Patient LHA's apraxia is selective for left-hand use. See text for details.

A final explanation (also raised in Chapter 9 on spatial neglect, p. 182) is that diagnosing apraxia still requires clinical expertise more than psychometric instruments. While structured tools like those by Vanbellingen et al. (2011) exist, their role is supportive—not central.

Where Can We Read More?

Goldenberg, G. Apraxia. The Cognitive Side of Motor Control. Oxford Academic: Oxford, 2013. (also as a target article with the same title in a discussion forum in Cortex 2014; 57: 270–274 with comments by more than a dozen experts).

Gonzales Rothi, L.J. and Heilman, K M. (eds.). Apraxia. The Neuropsychology of Action. Imprint Psychology Press: London, 1997 (eBook 2014, DOI: https://doi.org/10.4324/9781315804545).

CHAPTER 6

What Are Amnesias?

Amnesia means the loss of memory. Different types of amnesia may have very different underlying causes and neural correlates. Is the storage of information impaired while learning is still intact? Is there a selective retrieval deficit? Is learning no longer possible, preventing both active recall and passive recognition? Does memory loss target verbal but not nonverbal material (or vice versa)? Is a patient's failure to learn new information a genuine memory impairment, or are attentional factors responsible for their mnestic deficits? Are there qualitative peculiarities in how a sequence of test items is recalled—such as repetitious recall, intrusions (items not originally part of the list but mentioned in another context), or confabulations (entirely "false" memories)?

? **When is a memory disorder amnesia?**

Every day forgetfulness is common, but when does a memory lapse qualify as amnesia? Memory impairments are typically considered amnestic when they occur in relative isolation, without significant deficits in other neurocognitive domains. Disturbances in vigilance, attention, executive functions, or one's mood often mimic a genuine amnesic disorder.

The Nature of Memory: Who Learns, Who Remembers, and Who Forgets?

As with all aspects of cognition, understanding impairments requires first grasping the nature of the function itself. Let's start out with some of the most general comments about "memory" and ask:

What is the Thing We Call "Memory"?

An informal "memory fluency" task given to undergraduates (approximately 200) asked them to generate as many associations as possible to the word "memory." Their responses revealed a striking anthropocentric and training-based bias. The mean number of associations provided for one minute was 10.8. The three most frequently generated items were "learning," "temporal lobes" (or one of its key structures, most

DOI: 10.1201/9781003384991-6

frequently the hippocampus), and "dementia". Absent were terms such as "Markov chains," "smart alloys," or "spontaneous alternation behavior"—ideas well-known in mathematics, materials science, and behavioral neuroscience, respectively, which illustrate the broader nature of memory.

"Memory" is more than the recall of personal experiences; it is an essential property of adaptive systems. In mathematics, for instance, Markov chains of different "memory lengths" define the dependency of an event on its past occurrences. In metallurgy, "memory alloys" can be shaped at a certain temperature, deformed at a different temperature, and then revert to their original shape when reheated, indicating a material form of memory that does not require a brain. Even single living cells can store traces of past mechanical forces, as recent research suggests (Jain and Ghosh 2024). The fundamental nature of memory extends beyond cognition, beyond individuals, perhaps even beyond neurons themselves. It is easy to be trapped in an anthropocentric view of memory, but in doing so, we risk missing a deeper understanding of what it actually is.

The following section will attempt to address these issues by taking a closer look at the biological roots of "remembering," focusing on the phenomenon of spontaneous alternation behavior.

The Mealworm's Memory Span

Spontaneous alternation behavior (SAB) refers to an animal's tendency to alternate consecutive binary directional decisions while exploring a maze. Although most frequently studied in rodents, this phenomenon has been demonstrated across a wide range of species, from primates to protozoa (Dember and Richman 1989). Some of the earliest studies quantifying memory capacity in a SAB paradigm were conducted with the mealworm (*Tenebrio molitor*). Grosslight and Ticknor (1953) had the insects crawl through mazes of varying lengths, comparing a forced turn and a free choice (between-subject design). They found that alternation rates steadily decreased as the distance increased. Even at a distance of about six times an individual's body length, there was still a small effect of remembering (58% alternation). A 50% alternation rate would indicate a complete loss of any influence from the previous choice, effectively defining the animal's memory span (Figure 1).

Later experiments in woodlice (*Porcellio scaber*) localized the memory effect to the animals' legs (Beale and Webster 1971). A turn in one direction induces asymmetric fatigue in a woodlouse's legs—right turns place more strain on the left legs (as the outer curve is longer than the inner curve), and vice versa. In an ingenious experiment, the researchers counteracted this asymmetry by implementing a transport belt covering one side of the runway between the forced and free choice. This intervention reduced the workload on the fatigued legs while relatively increasing the workload on the less fatigued ones, effectively balancing bilateral leg movements and "erasing" any memory effects.

The relevance of SAB for memory research cannot be overstated (Lalonde 2002). Studies on "lower" animals, particularly unicellular organisms, reveal that remembering is not necessarily tied to the hippocampus, the temporal lobes, or even

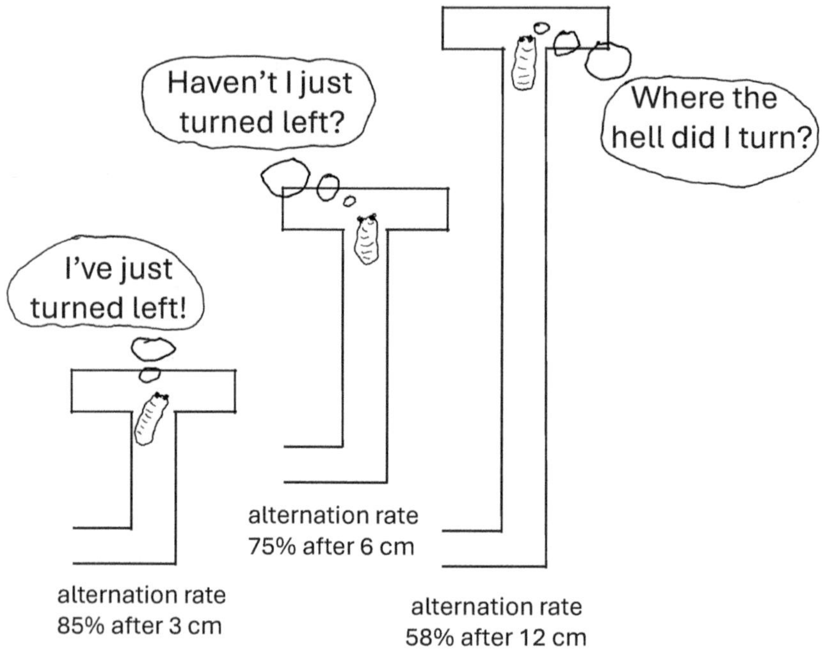

Figure 1: A mealworm's spatial memory can be inferred from the statistical dependency of directional choices at consecutive T-maze points. A high alternation rate (i.e., turning in the direction opposite to the previous forced turn) indicates memory of the earlier decision. As the distance between the forced and free choice increases, memory decays. Data from Grosslight and Ticknor (1953; Table 3) show alternation rates of ~85% for arm lengths up to 3 cm. At 6 cm, the rate drops to 75%, and at 12 cm, it declines to 58%—only slightly above chance level.

a proper brain. Rodent studies enable the testing of various substances that might enhance memory capacity—perhaps even in humans (Miedel et al. 2017). Humans also exhibit SAB (Rothacher et al. 2022), and for practical reasons, virtual mazes are recommended for research in this population.

? Do sperm cells remember?

This question is the title of a study on SAB in human spermatozoa (Brugger et al. 2002). We found that these forward-propagating cells also exhibit spontaneous alternation in left/right choices at consecutive decision points in microscopic mazes. In other words, their choice at time (t) is influenced by their choice at time (t–1). Consequently, the question must be answered in the affirmative: sperm cells—although merely haploid micro-machines *en route* to becoming a living organism—do retain information about past decisions. We hypothesize that the protein structures responsible for flagellar beating serve as the substrate for this memory. Specifically, asymmetric fatigue effects in the bind/relax cycling mechanism would lead to alternating directional turning at successive choice points.

A Taxonomy of Major Types of Human "Memory"

The dominant framework for classifying different forms of human learning and memory is based on two key dichotomies: short-term memory versus long-term memory, and within long-term memory, declarative (explicit) versus non-declarative (implicit) memory. This classification is outlined in Figure 2, incorporating refinements and a depiction of underlying neural systems.

A key concept in this taxonomy is consolidation, the two-step process that stabilizes newly acquired information. In the first phase, which occurs at the synaptic level, new memory traces are reinforced through long-term potentiation. The second phase, termed systems consolidation, takes place primarily in the hippocampus, rendering memory traces independent of hippocampal short-term storage and transferring them—presumably—to the temporal neocortex for long-term retention.

? I came across the term "re-consolidation." What does it mean?

Consolidation—the process of stabilizing newly acquired information for long-term storage—was once believed to be a one-time event. However, current research suggests that each reactivation of a memory trace triggers reconsolidation, a process that temporarily destabilizes the memory, making it susceptible to modification. This dynamic process allows for both vulnerability (intrusions) and adaptability (corrections). Sleep appears to play a role in strengthening episodic memories via both consolidation and reconsolidation mechanisms (Jones et al. 2022). For an overview of the concept, see Dudai and Eisenberg (2004); for molecular mechanisms, see Tronson and Taylor (2007); and for implications in psychopathology, see Bisaz et al. (2014).

? I find the terms "mnestic", "mnemonic", "mnemome", "mnemonist," and "mneme" confusing. What is the exact definition of these terms?

All these terms derive from the Ancient Greek word *mnēmonikós*, meaning "pertaining to memory."

Mnestic refers to memory-related functions, such as learning, storing, and retrieving information. Hence, "mnestic functions" means "memory functions."

Mnemonic describes anything designed to aid memory. For instance, the acronym FLAV is a mnemonic for four major types of dementia: Frontotemporal, Lewy body, Alzheimer's, and Vascular. Another mnemonic for insular functions is provided in Table 3 of Chapter 2.

Mnemonist denotes a person with extraordinary memory abilities. Some cognitively impaired individuals (formerly termed "idiot savants") can recall irrational numbers to 10,000 decimal places.

Mneme (pronounced as /'ni:mi/ with a silent *m*) refers to the faculty of memory. In Greek mythology, Mneme was the Muse of memory.

Mnemome follows the pattern of "-ome" words, signifying a totality of elements. Just as the genome represents an organism's complete set of genes, the mnemome refers to the full array of molecules involved in memory formation and maintenance (Frankland and Josselyn 2013). See Chapter 2 for further discussion of "omes" and "omics."

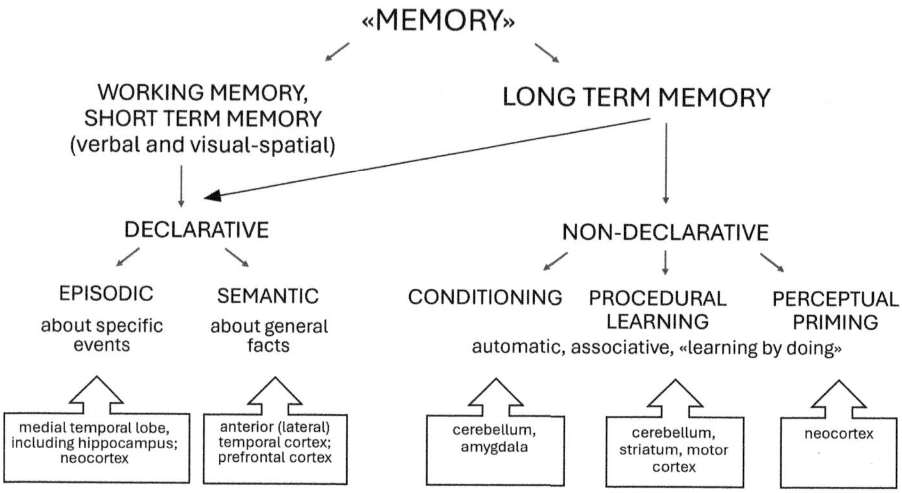

Figure 2: A rough classification of human memory systems and their primary underlying neural substrates.

Short-Term Memory and Working Memory

Short-term memory retains information for less than a minute and is limited to approximately seven discrete items—the "magical number 7 ± 2" (Miller 1956). We prefer the term working memory (probably first used by Miller et al. 1960), as it highlights one of the most influential concepts in memory research. Strictly speaking, working memory is not a memory function per se but rather an executive function (see Chapter 8, p. 176). It acts as a temporary buffer for information needed to complete cognitive operations.

There are distinct buffers for verbal and visuospatial information, which are associated with the left and right hemispheres, respectively. However, these buffers are not localized to a single brain region but instead emerge from interactions between the prefrontal cortex and other areas, including premotor and parietal cortices.

Long-Term Memory

Long-term memory includes both declarative (explicit) memory, which is accessible to conscious awareness, and non-declarative (implicit) memory, which operates without conscious control.

Declarative Memory

Declarative (explicit) memory concerns the storage and retrieval of factual knowledge, whether from personal experience (autobiographical memory) or general knowledge (semantic memory).

* *Episodic memory* consists of memories tied to a specific time and place. For example, the tie you wore at your wedding, your planned vacation destination, or a list of words you read 20 minutes ago.

* *Semantic memory* comprises general knowledge, such as the fact that a set of cutlery includes a fork, knife, and spoon; that Paris is the capital of France; or that a palindrome reads the same forward and backward.

Both episodic and semantic memory impairments typically result from bilateral temporal lobe damage. Episodic deficits are more pronounced with right-hemisphere involvement, whereas left-inferior temporal cortex damage is more critical for semantic impairments.

Non-Declarative Memory

Conditioning and priming are examples of non-declarative (implicit) memory functions.

* Conditioning is one of the most fundamental forms of associative learning. For example, an air puff to the eye causes an involuntary blink. If a tone consistently precedes the air puff, blinking occurs in response to the tone alone. This classical conditioning response is impaired in cerebellar degeneration. A striking dissociation exists between the hippocampus and amygdala in fear conditioning: lesions to the amygdala abolish the conditioned autonomic response (while sparing declarative memory for the stimulus pairing), whereas hippocampal lesions eliminate the declarative memory but preserve the autonomic response (Bechara et al. 1995).
* Priming facilitates learning by previous exposure to related stimuli. Perceptual priming (e.g., recognizing a word more quickly after prior exposure) is associated with plastic changes in the visual cortex, while semantic priming (e.g., responding faster to the word "doctor" after seeing "nurse") engages the prefrontal cortex. Even amnesic patients can be primed despite lacking conscious awareness of prior exposure. The mechanisms of priming are further explained in Chapter 11 (p. 261).

Non-declarative memory is rarely assessed clinically but is crucial for compensating for declarative deficits in amnesia. Its unconscious nature has led to the interchangeable use of the terms non-declarative and implicit memory. However, the processing mechanisms (e.g., single-item encoding vs. relational encoding) may be more critical than conscious awareness (Henke 2010).

? What makes implicit sequence learning special, and why is the paradigm important for neuropsychology?

Implicit sequence learning (ISL) refers to the ability to acquire the statistical structure of sequences without explicit awareness. Participants learn by predicting events through motor responses, yet they often cannot articulate the underlying rule (Batterink et al. 2019). Evidence for successful motor performance in the absence of verbal awareness is reminiscent of "unconscious perception" in patients with commissurotomy ("split-brain;" see Chapter 11). There is, in fact, evidence for a right hemisphere superiority for successful motor learning a sequence without being able to formulate the rule verbally (Anagnostopoulos et al. 2013). However, of more immediate clinical relevance are studies that support an impairment of ISL in Parkinson's disease and other subcortical degenerative disorders. Theories of learning consider the interactive contributions of medial temporal lobe structures and the basal forebrain during ISL as important. Some researchers suggest that ISL reflects a domain-general learning mechanism, potentially related to language acquisition (Conway and Pisoni 2008).

Other Dichotomies Within the Human Memory System

Several subtypes of declarative memory are often contrasted in the neuropsychological literature.

Collective vs. Autobiographical Memory

Collective memory refers to episodes and factual knowledge accessible to most members of a given culture. In contrast, autobiographical memory encompasses personal experiences that are not shared with others. Assessing autobiographical memory in neuropsychological examinations is crucial, as selective impairments may provide insights into the underlying neuroanatomy (Levine 2004).

A key distinction within autobiographical memory is that between field and observer memories. Events may be re-experienced from a first-person perspective (field perspective), as originally lived, or from a third-person perspective (observer perspective), as if viewed by an external observer. This differentiation is relevant to the emotional connotations of memories: Emotionally distressing events are more likely to be recalled from an observer's perspective—creating a detached, almost out-of-body experience—whereas emotionally neutral events are typically remembered from a field perspective (Nigro and Neisser 1983). An elegant experiment illustrating the effects of in-body vs. out-of-body perspectives on learning is described in Bergouignan et al. (2014). See also the case vignette on p. 125.

Past vs. Prospective Memory

Memory is commonly associated with past events, yet the ability to remember planned actions and execute them in the future—prospective memory—is equally essential for daily life. Prospective memory involves forming a mental representation of a future intention and retrieving it at the appropriate time. A review of its functional neuroanatomy can be found in Cona et al. (2015).

Item vs. Source Memory

This distinction corresponds to a what vs. when/where differentiation. Item memory (sometimes termed target memory) pertains to the content of stored information, whereas source memory includes details about the context in which the information was acquired. The neural basis of source memory and its role in false memory formation is reviewed in Mitchell and Johnson (2009).

Veridical vs. False Memories

Memories are not exact replicas of past experiences but reconstructions influenced by beliefs, expectations, desires, and fears. Experimental paradigms for inducing false memories have implications beyond forensic neuropsychology (Loftus 1979). A particularly influential approach is the Deese-Roediger-McDermott paradigm (Roediger and McDermott 1995), in which participants learn and later recognize words from lists containing lures—words that were never presented but are semantically related to the studied items. The likelihood of false memories increases with age and in the presence of dementia (Balota et al. 1999).

Retrograde vs. Anterograde Memory (and Amnesia)

This fundamental distinction is particularly relevant to this chapter, which primarily focuses on anterograde memory deficits.

* Retrograde memory encompasses all memories formed before the onset of an amnesia-inducing event.
* Anterograde memory refers to memories of events occurring after the critical event.

For completeness, a third term—congrade memory, primarily found in German-language literature—can be introduced (Figure 3). This refers to memories related to the time of the amnesia-inducing event itself. Including this concept is relevant, as some forms of amnesia (often functional or dissociative; see p. 121) selectively affect this critical time window.

? **I understand that anterograde memory deficits can have a retrograde component, yet are frequently "pure" in the sense of affecting exclusively episodes that occurred after a cerebral incident. What about "pure retrograde" amnesias? Do they exist?**

Reports of isolated retrograde amnesia do exist, but as you suspect, they are relatively rare. Such cases were already described in the 19th century (Ribot 1885), but the concept nearly vanished from the neurological literature in later decades. For a long time, pure retrograde amnesia was believed to occur only in cases of psychogenic origin (e.g., Schacter et al. 1982).

From a neuropsychological standpoint, it is indeed difficult to explain how a memory impairment could selectively affect only past events while sparing the encoding, storage, and retrieval of new material. One possibility is that plasticity mechanisms enable the formation and consolidation of new memories but fail to maintain retrospective access to older ones. However, the concept remains debated. Kopelman (2000) took an "exceptionally critical view" (as stated in his title), while Kapur (2000) defended the idea in "exceptionally benign" terms (as reflected in his title).

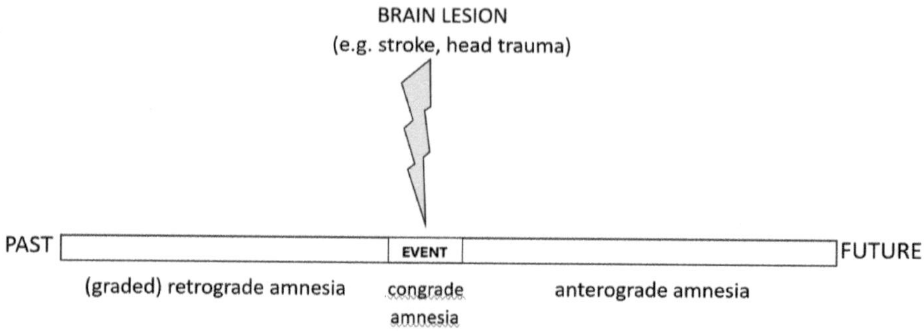

Figure 3: The meaning of retrograde, congrade, and anterograde amnesia.

Verbal vs. Nonverbal Memory

The broad distinction between verbal and nonverbal (or visual) memory reflects functional hemispheric asymmetries (FUHEMAS, see Chapter 11). The left hemisphere predominantly encodes verbal material, while the right hemisphere specializes in facial and other nonverbalizable information. This division applies to all aspects of memory, from working memory and learning capacity to long-term storage and retrieval. Memory tests should always be administered for verbal and nonverbal material. Ideally, the two separate tests should be identical in procedure and comparable in overall difficulty. This allows us to determine hemispheric differences in the level of learning, recall, and storage. See Table 1 for differential impairments of these three memory stages in the verbal modality.

Some of the most detailed insights into the laterality of memory functions come from studies of patients who have undergone selective unilateral amygdalohippocampectomy (AHE)—a surgical resection of the amygdala, hippocampus, and parahippocampal gyrus (for review, see Spencer and Burchiel, 2012). Presurgical evaluation of these patients may include the Wada test with amobarbital in a highly selective variant, targeting memory-relevant limbic structures without inducing general anesthesia in one hemisphere. Unfortunately, due to a predominant research focus on verbal memory, many neurological studies have examined only left-sided AHE cases, largely neglecting the verbal/nonverbal memory dichotomy. This highlights the crucial role of neuropsychologists in complementing neurosurgical teams.

Table 1: Performances of three fictitious patients in the learning, recall, and recognition phases of a standard auditory verbal learning test (Rey 1964). Recall is tested after 20 to 30 minutes, and recognition follows immediately. It consists of reading a list of 30 words to the patient, with the 15 critical words presented pseudorandomly along with 15 non-presented words.

Patients	**Learning Phase** Nb. of correctly recalled words in 5 consecutive runs	**Recall Phase** Nb. of correctly recalled words	**Recognition Phase** Nb. of correct words/number of false recognitions
Patient 1	4, 6, 9, 11, 11	2	14/2
Patient 2	6, 9, 9, 10, 13	3	7/7
Patient 3	5, 4, 5, 6, 6	5	6/0

Interpretation: Patient 1's learning curve is normal (age and education matched norms), and so is his passive memory (recognition). He suffers from a retrieval deficit. Patient 2's learning is normal as well, but recognition performance is at chance. He suffers from a storage deficit. Patient 3's recall and recognition are normal relative to the number of words learned. However, this number is below the norm, indicating that he suffers from a learning deficit.

Note that for this particular test, there is a comparably difficult nonverbal version with 15 designs, the Rey visual design learning test (Rey 1964). See Lezak et al. (2004) for procedure and sources to normative data.

Neuroanatomy of Learning and Memory and of Its Disorders

Figure 4 provides a schematic overview of the key structures and circuits involved in the encoding, storage, and retrieval of memories. Importantly, no single structure—whether gray or white matter—houses these processes in isolation; rather, memory functions emerge from interactions between distant brain regions. Nonetheless, lesions affecting specific areas and their connections, as depicted in Figure 4, can differentially impact distinct stages of information processing.

The three critical phases in memory formation are encoding, consolidation, and retrieval. These processes are mediated by large-scale, bilateral networks, including neocortical association areas, subcortical nuclei, and the medial temporal lobe (especially the hippocampus and parahippocampal cortex).

Medial Temporal Lobes and Memory Circuits

The hippocampi are central hubs in the brain's memory networks. Cortical association areas send information to the perirhinal, parahippocampal, and entorhinal cortices, which in turn relay it to the hippocampus.

* The perirhinal cortex, located in the anterior parahippocampal gyrus, is crucial for familiarity-based memory, allowing recognition of previously encountered objects. In the domain of semantic memory, it facilitates the integration of perceptual and conceptual knowledge about complex stimuli.
* The entorhinal cortex, positioned medially, acts as a gateway between the neocortex and hippocampus. It is essential for episodic memory formation and is an early target of neurodegeneration in Alzheimer's disease (Igarashi 2023). Notably, the right entorhinal cortex contains "place fields" (primarily known from the hippocampus), which are crucial for spatial navigation. Degeneration in this region contributes to the characteristic wayfinding difficulties in affected patients.

Three major memory circuits—highlighted in Figure 4—are the Papez circuit, the amygdala-orbitofrontal cortex (OFC) pathway, and the amygdala-basal forebrain loop.

* The Papez Circuit is historically the most well-known. Initially described by Papez (1937) in the context of emotion processing, it plays a fundamental role

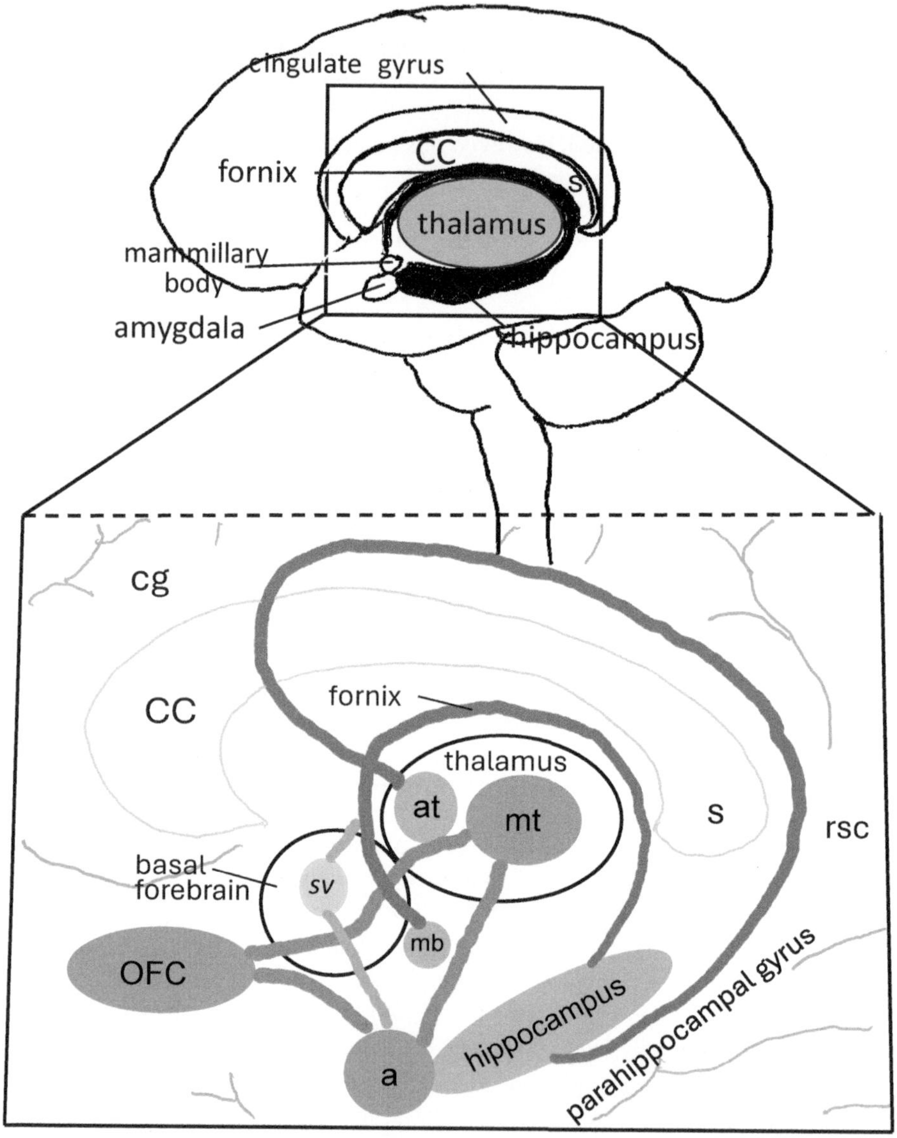

Figure 4: The Top part shows limbic system structures and cortex regions involved in the formation and retrieval of memories, and, if lesioned, leading to anterograde amnesia.

The bottom part shows an enlarged view of these structures to schematically depict the three most important circuits of memory formation: the Papez circuit in dark grey, the amygdala-OFC circuit in medium grey, and the amygdala-basal forebrain circuit in light grey. a: amygdala, at: anterior thalamus, CC: corpus callosum, cg: cingulate gyrus, mb: mammillary body, mt: medial thalamus, OFC: orbitofrontal cortex, rsc: retrospinal cortex, s: splenium, *sv*: *septum verum* (i.e., the grey matter part of the septum).

in declarative memory consolidation. This loop links the hippocampus (via the fornix) with the mammillary bodies (part of the hypothalamus), the anterior thalamus, the cingulate gyrus, and the entorhinal cortex. Damage to this circuit—whether due to a thalamic stroke or Korsakoff's syndrome—leads to profound anterograde amnesia. Notably, in cases of temporal lobe tumors or mass effects on the temporal lobes, declarative memory deficits may occur even without direct involvement of the Papez circuit. Neuropsychological assessments, which differentiate between storage failures, encoding strategies, and retrieval deficits, can help pinpoint dysfunctions within this and other memory-related loops (Carlesimo 2012). Aggleton et al. (2022) argue that understanding memory functions requires going beyond the classical Papez circuit.

* The Amygdala-OFC Circuit links the amygdala with the medial thalamus and OFC, playing a key role in reinforcement learning and fear conditioning. Dysfunction in this circuit can lead to exaggerated fear responses, as seen in post-traumatic stress disorder (PTSD). In social anxiety disorders, an imbalance between the amygdala and OFC is observed: while this connection forms a negative feedback loop in healthy individuals, pathological excitation dominates in affected patients, particularly in the right amygdala-OFC network (Sladky et al. 2015).
* The Amygdala-Basal Forebrain Loop, sometimes termed the basolateral limbic loop, is primarily responsible for encoding emotionally charged memories. It modulates arousal and attention via cholinergic activation. In Alzheimer's disease, early degeneration of cholinergic nuclei disrupts this process, impairing the beneficial effects of emotional salience on memory retention.

A fourth circuit—the hippocampus-striatum pathway—is not depicted in Figure 4 but is relevant for the integration of declarative and procedural memory. Its dysfunction contributes to procedural learning deficits in Parkinson's disease.

The Role of the Basal Ganglia in Memory Functions

The basal ganglia, a group of subcortical nuclei primarily involved in motor control, also play a critical role in procedural and habit learning. These structures—including the striatum (caudate nucleus and putamen), globus pallidus, subthalamic nucleus, and substantia nigra—support reinforcement learning by modulating behavior based on reward and punishment.

Dopaminergic projections from the substantia nigra and ventral tegmental area (frequently abbreviated as *VTA*) signal the value of actions, strengthening or weakening neural pathways based on outcomes. This system is essential not only for motor skill acquisition (e.g., learning to knit, ride a bike, or play a musical instrument) but also for implicit learning, where associations form outside of conscious awareness.

Basal ganglia circuits also interact with the prefrontal cortex, supporting habit formation and decision-making—especially in situations of uncertainty. Dysfunction in these pathways, as seen in Parkinson's and Huntington's disease, leads to deficits in motor learning and cognitive flexibility.

The Role of the Basal Forebrain in Relation to Memory

We briefly mentioned the basal forebrain in a figure depicting its location ventrally adjacent to the basal ganglia and expanding anterior to the ventral striatum (Figure 1 of Chapter 3). Beyond its crucial role in mediating wakefulness and attention, this system is important for memory consolidation by modulating cortical plasticity and facilitating synaptic changes necessary for learning. It is a major source of cholinergic projections to the cortex and hippocampus. Acetylcholine release enhances signal-to-noise ratios in cortical networks, promoting selective attention to relevant stimuli, thus strengthening memory encoding. In the hippocampus, cholinergic input supports not only spatial learning but also the transition of information from short- to long-term memory ("consolidation"). The septum, an area expanding in an anteroposterior direction between the corpus callosum and fornix, is a most relevant structure, as septo-hippocampal pathways directly contribute to the consolidation process (Lecourtier et al. 2011). Degeneration of basal forebrain cholinergic neurons disrupts this process and contributes to severe memory deficits. Basal forebrain structures also interact with the amygdala (Figure 4), linking learning to emotional and motivational states.

For a comprehensive overview of the mnestic functions, especially of the septal region of the basal forebrain, and of rehabilitative strategies of basal forebrain amnesia, see Tsanov (2022).

? What should neuropsychologists know about amnesia due to basal forebrain damage?

Amnesia resulting from basal forebrain damage most commonly occurs after the rupture of an anterior communicating artery aneurysm. Such damage typically affects the basal forebrain bilaterally, often including portions of the posterior orbitofrontal cortex. The resulting memory impairment is predominantly anterograde, affecting episodic memory for both verbal and nonverbal material. In some cases, a temporally graded retrograde amnesia is also observed (D'Esposito et al. 1996), and in certain instances, semantic memory may be compromised as well (Simard et al. 2003).

Degeneration of basal forebrain structures is also seen in Parkinson's disease, even when cognitive impairment is still mild. The extent of degeneration in this region can serve as an early predictor of future cognitive decline (Ray et al. 2018).

A distinctive qualitative feature of basal forebrain damage is the occurrence of spontaneous confabulations—fabricated memories composed of elements from the patient's past (Schnider and Ptak 1999). These confabulations arise due to a failure to suppress irrelevant information (for an overview, see Schnider 2003). It is crucial to differentiate them from confusional confabulations, which occur in patients with acute confusion but do not necessarily indicate severe amnesia. Basal forebrain confabulations are also different from provoked confabulations, which emerge during memory testing as intrusions from the test situation and lack anatomical specificity.

Spontaneous confabulations are highly characteristic of lesions to the basal forebrain and posterior orbitofrontal cortex. Schnider's 2017 monograph provides a comprehensive overview of confabulatory phenomena across different neurological conditions.

Different Types of Amnesia: A Selection

Difficulties with remembering are among the most common complaints patients present during a neuropsychological evaluation.

? **What is the etiology of amnesia?**

I have to caution you—this question is not quite properly framed. There is no single type of amnesia, and much less a single etiology responsible for amnestic deficits. Instead, amnesia can arise from a highly heterogeneous range of causes, including:

* Stroke
* Traumatic brain injury (concussion)
* Space-occupying lesions
* Seizure disorders
* Neurodegenerative diseases (dementia)
* Intoxications
* Hypoxia
* Viral infections
* Neuropsychiatric disorders

Even depression—whether endogenous or reactive—almost invariably hampers memory function.

Below, we explore some of these etiologies in greater detail.

Amnesias After Cerebrovascular Incidents

Ischemic stroke is a major cause of memory disorders, though the impact on memory depends on the affected vascular territory:

* Middle cerebral artery (MCA) infarctions typically do not produce isolated amnesia. Instead, memory impairment occurs in conjunction with deficits in attention and learning.
* Anterior cerebral artery (ACA) infarctions often result in reduced drive and attentional deficits, which can further compromise memory function.
* Anterior choroidal artery infarctions may produce a more isolated amnestic syndrome, as this artery supplies the head of the hippocampus.

* Posterior cerebral artery (PCA) infarctions are often bilateral and can lead to severe anterograde and retrograde amnesia, particularly if they damage the hippocampus.
* Bilateral thalamic strokes frequently cause profound memory impairment, especially when they disrupt the mammillothalamic tract—a crucial link between the thalamus and hippocampus. In such cases (paramedian thalamic stroke; Carlesimo 2012, 2022), amnesia is often long-lasting and relatively resistant to rehabilitation. Long-term recovery tends to be better following right-sided infarctions (Hermann et al. 2008).

Intracerebral hemorrhages—also referred to as hemorrhagic strokes—result from sudden bleeding into brain tissue (parenchyma), the ventricles, or both. They account for approximately 15% of all strokes and carry a higher risk of mortality than ischemic strokes. Both hemorrhagic and ischemic strokes can impair the ability to encode, store, and retrieve new information.

? What kinds of intracranial hemorrhages are there, and how do they affect memory and cognition?

Intracranial hemorrhages can be broadly classified into four types.

1. Intraparenchymal hemorrhage: Corresponds to a hemorrhagic stroke (as described above), with bleeding occurring within the brain tissue itself.
2. Subarachnoid hemorrhage: Bleeding into the subarachnoid space (between the brain and the surrounding membrane), typically caused by the rupture of an aneurysm. This can be either traumatic or spontaneous (i.e., without an external injury).
3. Epidural hemorrhage: Bleeding into the space between the skull and the dura mater (the brain's outermost protective membrane).
4. Subdural hemorrhage: Bleeding into the space between the dura mater and the arachnoid layer. Any of these hemorrhages can lead to the formation of a hematoma—a localized accumulation of clotted blood. Hematomas exert pressure on surrounding brain tissue, which can result in cognitive and memory impairments.

   A study by Ogden et al. (1993) assessed cognitive outcomes in patients with subarachnoid hemorrhage who had a good overall neurological prognosis. The degree of mnestic impairment 12 months post-incident was best predicted by neuropsychological test scores at hospital discharge rather than by the location of the aneurysm, the distribution of blood (as visualized via CT scan), or associated complications such as vasospasm and ischemia.

The type of stroke (ischemic vs. hemorrhagic) is not directly related to the severity of an observed memory disorder. An extensive US cohort study with more than 1,000 dementia-free individuals looked at cognitive outcomes 6 years after stroke (Levine et al. 2025). There were no differences in global cognition and, more specifically, memory scores (as standardized T-scores) between ischemic and hemorrhagic strokes nor among different types of the former (not enumerated here). Likewise, in a large rehabilitation population, the recovery and functioning 3 and 6 months post stroke did not differ between ischemic and hemorrhagic stroke patients (Oosterveer et al. 2022), regardless of the hospital treatment they had received.

Amnesia in Metabolic and Endocrine Disorders

Metabolic disorders disrupt essential biochemical processes, often affecting energy production, waste elimination, or enzyme synthesis. Endocrine disorders result from hormone imbalances due to dysfunction of hormone-producing glands.

While these conditions do not typically cause *pure* memory deficits, they frequently affect cognition more globally due to neurotransmitter imbalances, energy deficits, oxidative stress, and structural damage (for an overview, see Moon 2016). Below, we highlight specific metabolic and endocrine conditions where memory impairment is disproportionately pronounced.

Wernicke-Korsakoff Syndrome

Wernicke-Korsakoff syndrome is a two-stage disorder:

* Wernicke encephalopathy presents with gait ataxia, oculomotor dysfunction, and altered mental status.
* Korsakoff syndrome follows, characterized by severe anterograde amnesia, confabulations, and temporally graded retrograde amnesia. While semantic memory may also be affected, nondeclarative memory (priming, procedural learning, and implicit learning) remains intact (Kopelman et al. 2009). In later stages, executive dysfunction, emotional blunting, and apathy develop.

The term alcohol-related dementia (Sachdeva et al. 2016) is a misnomer when applied to Korsakoff syndrome, as it contradicts the definition of dementia as a non-reversible neurodegenerative process. Many patients show functional recovery following vitamin B1 (thiamine) supplementation.

? **What components of alcohol produce forgetfulness and confabulation?**

Wernicke-Korsakoff syndrome is not due to alcohol toxicity *per se* but rather to a thiamine (vitamin B1) deficiency resulting from chronic alcohol abuse. However, this deficiency can also arise from non-alcoholic causes, such as malnutrition, anorexia, chronic gastrointestinal disease, and certain types of chemotherapy.

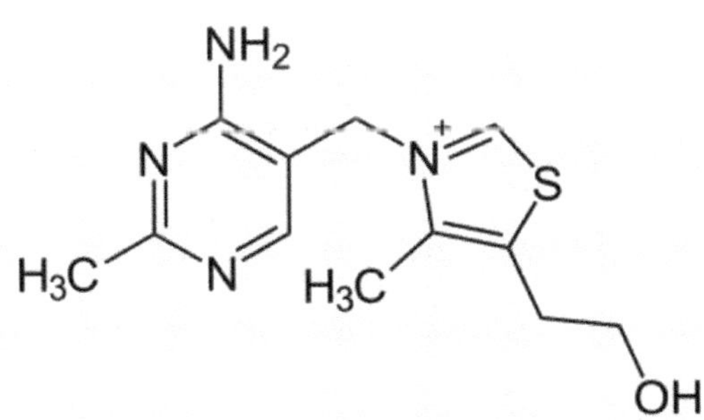

Above, you see the structural formula of thiamine (vitamin B1).

Chronic alcohol consumption impairs thiamine absorption in the gastrointestinal tract and disrupts its transport mechanisms. A relative thiamine deficiency primarily affects the cerebellar vermis, pons, medial thalamus, limbic system, and basal forebrain. This leads to the classic triad of Wernicke encephalopathy:

1. Ataxia (loss of coordinated movements);
2. Ophthalmoplegia (eye movement abnormalities);
3. Mental status alterations, including the core symptoms of Korsakoff syndrome, which are severe anterograde amnesia, a temporally graded retrograde amnesia, and confabulations.

If detected early, these symptoms can be reversed with thiamine supplementation. In contrast, ethanol toxicity itself plays a larger role in Fetal Alcohol Spectrum Disorders (FASD). Alcohol exposure during pregnancy does not cause selective memory deficits but instead leads to a range of cognitive and somatic abnormalities (for an overview, see Riley et al. 2011).

Liver and kidney diseases can lead to cognitive deficits due to metabolic imbalances. MELAS syndrome is a mitochondrial disorder that mimics ischemic stroke (for review, see Tetsuka et al. 2021). These conditions can cause metabolic amnesia, although motor, executive, and attentional deficits are often equally pronounced.

Endocrine Disorders

Among the endocrine disorders, hyperthyroidism is particularly associated with memory impairment. Excessive thyroid hormone production—stemming from various conditions—initially presents with psychomotor slowing, apathy, and depression. If left untreated, this can progress to cognitive decline and even dementia.

Kumar et al. (2024) examined structural brain alterations in 34 hyperthyroid patients compared to 35 controls. Hyperthyroid individuals exhibited increased cortical thickness in the bilateral superior temporal and superior frontal gyri, correlating with thyroid hormone levels. Memory performance (specifically visual recognition memory) showed a negative correlation with right superior temporal and superior frontal gyrus thickness. While the exact interpretation remains unclear, these findings suggest a need for further studies on structure-function relationships in memory deficits related to hyperthyroidism.

Other endocrine disorders, such as diabetes mellitus, Cushing's syndrome, and Addison's disease, are also associated with memory impairment. Chronic hyperglycemia and insulin resistance (diabetes) and excess cortisol levels (Cushing's) negatively impact hippocampal function, while cortisol deficiency (Addison's disease) has similar effects (Henry et al. 2014).

Toxic Amnesias and Hypoxia

Toxic amnesias result from exposure to external substances rather than intrinsic metabolic or endocrine dysfunction. Examples comprise heavy metals (e.g., lead and mercury), prescription drugs (e.g., benzodiazepines), and recreational drugs (e.g., fentanyl and cocaine).

Fentanyl abuse has been linked to an opioid-associated amnestic syndrome characterized by acute anterograde amnesia, often with hippocampal damage detectable on MRI (Walker et al. 2021). While chronic substance abuse generally leads to long-term cognitive deficits, acute and reversible amnesia has been reported following cocaine use (Ruiz Yanzi and Alessandro 2020).

Anoxia/Hypoxia

There are three main types of hypoxia-related amnesia that can be distinguished.

1. Toxic Hypoxia: Carbon monoxide (CO) poisoning causes hypoxia by binding to hemoglobin with ~ 200 times the affinity of oxygen, reducing oxygen transport despite normal oxygen availability. If untreated, coma and death occur within hours. After acute symptoms resolve, delayed neurological sequelae—including diffuse cerebral demyelination—can emerge, leading to severe memory deficits (Shprecher and Mehta 2010).

   I vividly remember the double case of a sister and a brother living together in a tiny two-story house. The sister, age 28, was a landscape gardener, her brother a polymechanic (age 34). One chilly March evening, they had a barbecue in their garden. They then brought the grill inside to profit from its warmth and watched TV on the first floor. He was lying on a sofa, she sitting upright in a chair. In this position, they were found the next morning around 8 a.m. by the brother's boss, who had come by after his employee failed to show up for work. Both appeared confused at first and were brought to a hospital, where she was able to relate what had happened (her brother remained in a confusional state). While she could recall 5 out of 10 words read to her after 5 minutes, he was not testable. Both received hypobaric oxygen therapy and a neurological examination. MRI and EEG were normal in her case, but his MRI showed bilateral subcortical diffusion restriction with abnormal T2 signal in both striata. The EEG signal was bilaterally altered over both frontal regions. Formal neuropsychological testing two days later was normal in her case (except for a borderline performance in figural and verbal recognition memory, with intact recall). Her brother's verbal and visual learning (15 items presented 5 times) was moderately impaired, as was his word recall (only slightly below norms for figures). Qualitatively, his learning and recall were highly repetitive for words and moderately so for figural learning. Visual functions were normal in both siblings, as was performance in all other neuropsychological domains—except, in the brother only, a moderate to marked slowing in psychomotor tests (e.g., TMT-A) and an elevated susceptibility to interference (Stroop test). Re-test four weeks later revealed entirely normal neuropsychological functioning in both siblings.

   The case is interesting because the differential severity of the cognitive deficits after carbon monoxide poisoning was, in all probability, due to the 50 cm difference in head position: CO had slowly filled the two-story house from below and reached the brother's nose and lungs before it reached the sister's.

2. Ischemic Hypoxia: Oxygen deprivation due to reduced blood flow, typically occurring during cardiac arrest, near-drowning, or near-hanging (McHugh and

Stout 1983). The severity of amnesia depends on the duration and completeness of ischemia. Neonatal hypoxia can lead to hippocampal atrophy, impairing memory later in life (Cooper et al. 2015).

3. Polonged or Chronic Hypoxia: Prolonged, mild oxygen deprivation occurs in heart and lung diseases. It compromises both memory and executive functions. High-altitude global amnesia has been reported in mountaineers climbing extreme altitudes without supplemental oxygen, independent of acute mountain sickness (Litch and Bishop 2000). Transient cognitive deficits after extreme altitude expeditions have been demonstrated (West 1986), and more extensive neuropsychological evaluations uncovered mild but persistent memory deficits (Regard et al. 1989). More recent work failed to show structural hippocampal alterations specific to habitually living at high altitudes (Zhang et al. 2022) or systematic structural grey or white matter alterations at the unselected whole-brain level (Luo et al. 2023). More importantly, however, both studies also failed to collect behavioral data alongside neuroimaging.

Amnesia After Infections of the CNS

Inflammation-related amnesia can result from various infections and autoimmune conditions affecting the CNS. The mechanisms often involve focal hippocampal damage or more diffuse immune-mediated neuronal dysfunction. Secondary effects of systemic inflammation must also be considered.

Limbic Encephalitis

Herpes Simplex Infection: Limbic system structures are particularly vulnerable to viral encephalitis. The most common cause is *Herpes Simplex Encephalitis (HSE)*, which typically leads to severe deficits in learning and episodic memory. Seizures and confusional states frequently accompany and exacerbate these neuropsychological impairments. Diagnosis is confirmed by detecting viral DNA in cerebrospinal fluid.

Human Herpesvirus 6: A related condition is caused by *Human Herpesvirus 6*, which primarily affects immunosuppressed individuals (e.g., post-transplant or HIV-infected patients, see below). Its radiological and neuropsychological profiles closely resemble those of herpes simplex encephalitis.

Autoimmune Limbic Encephalitis: This disorder presents with a *subacute anterograde amnesia* and MRI-detectable hyperintensities in the medial temporal lobes. It is often accompanied by confusion and, in many cases, responds to immunotherapy (e.g., corticosteroids).

HIV-Associated Mnestic Deficits

Two decades ago, the term HIV-dementia was common. Before the advent of highly active antiretroviral therapy, motor deficits were a hallmark of HIV-related neurocognitive decline. Today, the focus has shifted toward HIV-associated neurocognitive disorders (HAND).

HAND is categorized into three well-defined stages (Antinori et al. 2007):

* Asymptomatic Neurocognitive Impairment (ANI)
* Mild Neurocognitive Disorder (MND)
* HIV-Associated Dementia (HAD)

Memory deficits are just one component of HAND, affecting limbic and fronto-striatal circuits bilaterally. The historical focus on verbal memory assessments reflects practical testing constraints and neuropsychologists' verbal-system bias rather than clinical necessity.

Lyme Disease

Not every tick bite poses a risk for cognitive impairment. *Borrelia burgdorferi*, the spirochete responsible for Lyme disease, does not typically cause neuropsychological symptoms unless the CNS is involved (neuroborreliosis). Chronic borreliosis may lead to mild working memory deficits and episodic memory impairment, both of which can respond to antibiotic treatment.

Sporadic Creutzfeldt-Jakob Disease

One of the most rapidly fatal neurodegenerative conditions, Creutzfeldt-Jakob disease (a prion disorder), is characterized by progressive cognitive decline, with an average survival time of 6–12 months. While acquired prion diseases (e.g., infectious transmission) are rare, nearly half of affected individuals exhibit memory deficits as the first signs of dementia. The disease serves as a reminder to neuropsychologists that the trajectory of symptom progression can be diagnostically crucial.

Progressive Multifocal Leukoencephalopathy (PML)

PML is a rare demyelinating disease of the CNS caused by the JC virus, typically occurring in immunocompromised individuals. Cognitive deficits may accompany sensorimotor impairments. While PML rarely presents with isolated memory dysfunction, a recent case described predominant amnestic deficits with structural alterations in the left temporal lobe (Crowhurst et al. 2020).

COVID-19-Induced Amnestic Deficits[1]

COVID-19, caused by the SARS-CoV-2 virus, is associated with persistent cognitive deficits, with memory impairment being among the most frequently reported symptoms.

* Hampshire et al. (2024) assessed cognitive function in over 800,000 UK residents, confirming significant memory impairments post-infection.
* Hayward et al. (2024) demonstrated procedural learning deficits in long-COVID patients, with impaired initial learning and reduced overnight consolidation.

---

[1] Following the mock advice in Looi and Anderson (2021), the term COVID-19 is herewith included in this book. In fact, some empirical studies found limited support for long-term neuropsychological sequelae of COVID-19 (Priftis et al. 2022).

* Charles James et al. (2025) reported substantial impairments in verbal working and short-term memory in individuals with long-COVID.

Other CNS infections—including bacterial, fungal, and parasitic agents—rarely produce selective memory deficits. However, neurosyphilis, an advanced stage of syphilis (currently a growing health concern in high-income nations; Ramchandani et al. 2023), can target temporal lobe structures, leading to amnestic symptoms (Ahn et al. 2012).

Multiple sclerosis (MS) is sometimes categorized as a parainfectious autoimmune disease. MS affects cognition in ~ 50% of patients, with memory complaints being common. A useful introduction to its neuropsychological profile can be found in Bobholz and Gremley (2011).

Amnesia Associated With Traumatic Brain Injury

The term post-traumatic amnesia (PTA) refers to a state of altered consciousness following traumatic brain injury (TBI). PTA is a confusional state, often accompanied by agitation and a spectrum of cognitive impairments, with anterograde memory being particularly affected.

Structural damage can involve:

* Focal contusions, typically in frontopolar, orbitofrontal, and temporobasal regions.
* Diffuse axonal injuries, occurring at gray-white matter junctions, including memory-relevant temporal and frontal areas.

Post-Traumatic Stress Disorder (PTSD) and Amnesia

Memory impairments in PTSD straddle the boundary between organic[2] and functional disorders. While acute PTSD-related amnesia typically lacks structural correlates, persistent deficits may be accompanied by hippocampal and medial temporal lobe atrophy.

Amnesias of Transient Nature

Two forms of amnesia can be severe yet are temporally limited:

Transient Global Amnesia (TGA)

TGA is a sudden-onset anterograde amnesia lasting several hours, often triggered by:

* Cold showers
* Sexual activity
* Physical exertion
* Intense emotional reactions

[2] Use of the term "organic" is used here out of habit. It disappeared from DSM-IV onward and should be avoided (Benrimoh et al. 2019).

Key diagnostic criteria:

* Abrupt onset without trauma, seizures, or focal neurological signs
* Preserved consciousness and identity
* Profound impairment in immediate recall (e.g., inability to remember three words after one minute)
* Variable retrograde memory loss (ranging from hours to years)
* Resolution within 24 hours

MRI typically reveals small hippocampal lesions, detectable 2–3 days post-event but absent after several weeks. Standard neuropsychological assessment during an episode is impractical but, when feasible, cognitive testing can be an invaluable educational experience (see Chapter 12, pp. 280).

Transient Epileptic Amnesia (TEA)

TEA resembles TGA but differs in:

* Recurrence: While TGA is typically a one-time event, TEA recurs frequently.
* Duration: TEA episodes last 30–60 minutes, whereas TGA persists for hours.
* Additional seizure-like features: TEA may involve brief unresponsiveness or olfactory hallucinations.

For unknown reasons, TEA is twice as frequent in men as in women, whereas TGA occurs equally across genders. Subtle hippocampal atrophies may be detected in TEA, and surface EEG can reveal epilepsy-specific abnormalities during episodes. Unlike TGA patients, TEA patients often report persistent memory problems between episodes.

Psychogenic ("Functional") and Dissociative Amnesias

Psychogenic amnesia describes a memory impairment that is disproportionate to other neuropsychological deficits and lacks an identifiable neural correlate.

* Anterograde memory is relatively preserved, while retrograde and autobiographical memory are disproportionately affected (see the Question on retrograde amnesia on p. 107).
* Dissociative amnesia is situation-specific, affecting recall of a discrete traumatic event (Staniloiu and Markowitsch 2014).
* Global psychogenic amnesia (psychogenic fugue) involves sudden, transient loss of autobiographical memory and personal identity.

For a refined taxonomy of psychogenic amnesias, see Harrison et al. (2017). For a functional neuroimaging perspective, see Taïb et al. (2023).

? Are a patient's memory complaints after radiotherapy or chemotherapy not to be subsumed under the heading "functional amnesia"? Or is there something like a "chemobrain"?

"Chemobrain"—or more formally, cancer-related cognitive impairment—refers to cognitive deficits associated with chemotherapy for non-CNS cancers. These impairments can persist for years in some cases. Contrary to what the term suggests, these deficits are unlikely to be caused solely by chemotherapy; instead, they result from a complex interplay of biological, pharmacological, and psychological factors.

Cytostatics may impair memory through neurotoxic effects, leading to structural and functional alterations in the medial temporal lobes (Kesler et al. 2013).

Other mechanisms, including oxidative stress in non-targeted tissues, have been described as the collateral damage of chemotherapy (Chen et al. 2007).

Psychological factors, such as anxiety and depression in response to a life-threatening diagnosis, may further affect cognitive performance. Thus, memory deficits in chemotherapy patients should not be hastily classified as functional amnesia. That said, an additional functional component cannot be ruled out in individual cases (Pendergrass et al. 2018, for review).

? Can we learn during sleep, and if so, what are the consequences for clinical neuropsychology?

A substantial body of research supports the idea that learning can occur during sleep. Paller and Oudiette (2018) provide a useful introduction to the field. While acquiring entirely new information during sleep is limited, consolidation benefits significantly from sleep-dependent processes, involving hippocampal-neocortical networks.

Using a combination of EEG and fMRI, Züst et al. (2019) demonstrated that verbal learning during sleep is tied to peaks in slow-wave oscillations.

The hippocampus appears to play a role in implicit relational binding during both conscious and unconscious learning (Duss et al. 2014).

These findings have implications for amnesia models, particularly regarding memory stabilization (Grossberg 2017).

The role of slow-wave sleep in memory consolidation and stabilization is a clinically relevant topic (Atherton et al. 2016).

Wilkinson et al. (2017) even investigated memory retention in the fire salamander during brumation (a hibernation-like state), highlighting cross-species parallels.

Some Factors Influencing Learning and Memory

Both learning and memory are strongly influenced by factors related to wakefulness and motivation. The nucleus accumbens and striatum play key roles in dopamine-related motivated learning. These structures are tightly interconnected with broader networks supporting information acquisition.

While most textbooks emphasize the medial temporal lobes as the core structures for memory, it is important to consider their extensive connectivity with the rest of the brain.

The factors discussed below represent only a small selection of variables that can influence learning and recall. Many patients benefit from adjustments to standard neuropsychological testing procedures. Exploring ways to transform frustration into motivation through methodological adaptations can be rewarding for both patients and examiners. However, whether alternative testing approaches still allow for comparisons with normative data is another matter entirely.

Motor-Procedural vs. Content Learning

The dissociation between *motor learning* and *content learning* is well-documented, particularly in amnesic patients.

* Motor skills (procedural learning) remain intact even in severe amnesia (Corkin 1968). Previously learned motor programs (e.g., playing the piano, riding a bike, skiing) are retained despite profound deficits in episodic and declarative memory.
* Content learning depends on the medial temporal lobes, whereas motor learning relies on the basal ganglia and other subcortical structures that regulate motor programs.

A striking demonstration of this dissociation involves mirror reading. Amnesic patients can learn to read mirror-reversed words over repeated trials (showing improvement in procedural learning), yet they fail to recall the words themselves (declarative memory impairment; Cohen and Squire 1980).

Enactment effect

Interestingly, content learning can benefit from accompanying motor activity. The enactment effect suggests that memory for action-related words (e.g., hammering, walking, and laughing) is enhanced when the action is physically performed rather than merely read (Cohen 1989).

Phonemic or Semantic Relatedness

Most memory tests involve list learning, particularly in verbal tasks. Phonemic similarity (nurse purse) and semantic similarity (nurse–hospital) enhance recall. However, individual differences in the ability to benefit from these associations are large and may be disease-related. Qualifying this variability is diagnostically valuable, aiding in both differential diagnosis and individualized rehabilitation strategies.

It is recommended to use multiple memory tests that vary in their degree of between-item relatedness. However, this approach carries a risk; presenting multiple lists increases the likelihood of intrusions (i.e., interference between lists). This is especially problematic in patients with source memory impairments or frontal lobe dysfunction.

Redundancy and Interference

Item repetitions within a single learning trial enhance recall. However, standard list-learning tests typically lack within-trial repetitions. Instead, lists are repeated multiple times (usually 3–6 repetitions) to produce a learning curve. Some tests introduce an interference list—a second set of words presented once, followed by a surprise recall of the original list. The degree of recall reduction (post-interference decrement) may indicate frontal lobe dysfunction, as patients with executive deficits tend to show elevated distractibility.

Serial Position of List Item

The position of an item within a list affects recall probability. Murdock (1962) assumed the following laws:

* *Primacy effect:* The first few items are recalled better, likely due to long-term memory encoding.
* *Recency effect:* The last few items are also recalled well, as they remain in short-term memory.
* *Middle-position disadvantage*: Items from the middle are recalled least efficiently, as they have not yet been transferred to long-term memory and are no longer in short-term storage. These middle-list words suffer retroactive and proactive interference.

However, the interpretation of primacy reflecting long-term memory and recency reflecting short-term memory is not set in stone. Although serial position effects are widely studied, there is little research into whether list items are mentally represented in spatial terms. The idea of a "space to remember" in list learning paradigms remains an underexplored area of memory research.

? What exactly is meant by "the space to remember," and what is the evidence from clinical neuropsychology for such a space?

The term "space to remember" refers to the horizontal spatial organization of learned items across different time scales. These time points may span an individual's entire life or be separated by milliseconds. Within autobiographical memory, earlier experiences tend to be mentally represented on the left, while future episodes (i.e., prospectively imagined events) are represented on the right (Saj et al. 2014). A similar left-right spatial organization applies to sequences in short-term memory (Guida et al. 2020) and working memory (Rinaldi et al. 2015). This principle is known as the spatial-ordinal position effect, meaning that earlier items are encoded on the left, and later ones on the right.

A striking cross-cultural reversal of this phenomenon has been described: in left-to-right reading cultures, earlier information is mentally positioned on the left and later information on the right, while in right-to-left reading cultures, this organization is mirrored, meaning earlier items are represented on the right and later ones on the left (Fischer and Brugger 2011).

Support for the space to remember comes from patients with left-sided neglect (see Chapter 9). These patients show greater impairment in digit span backward (which requires recalling items in the leftward direction) compared to digit span forward (Weinberg et al. 1972; Rapport et al. 1994). In list-learning tasks, neglect patients exhibit greater deficits for primacy items (the leftmost part of the mental representation of a list) compared to recency items. The severity of this primacy-recency asymmetry correlates with neglect severity, as measured by a cancellation task (Loetscher et al. 2012). In contrast, patients from right-to-left reading cultures neglect the recency part of a word list more than the primacy part (Rasoulzadeh et al. 2024). These findings suggest that memory and spatial representation share common underlying mechanisms, influenced by both cultural factors (reading direction) and neurological conditions (e.g., hemispatial neglect).

Compatibility Between Learning Setting and Recall Setting

What is learned in a given environment is best recalled in that same environment. A frequently cited study supporting this context-dependent memory effect is that of Godden and Baddeley (1975). In their experiment, members of a university diving club learned and recalled matched word lists either 20 feet underwater or on shore. Each participant completed both a learning and a recall session in each location. There was no main effect of test site (i.e., underwater vs. shore did not differ in overall recall performance).

However, a highly significant interaction emerged: memory scores were vastly superior when learning and recall occurred in the same setting, compared to when they occurred in different settings. A similar compatibility effect extends beyond environmental context to body posture. Rand and Wapner (1967) demonstrated that recall performance is enhanced when the body position during retrieval (e.g., supine vs. upright) matches that of encoding.

Although standard neuropsychological test conditions inherently provide a stable setting for learning and recall, understanding an individual's susceptibility to context effects can improve the assessment of everyday mnestic performance.

A special kind of a mnestic compatibility effect was shown by one of our patients with an agnosia for heading direction, labelled "environmental allochiria" in the older literature (Binet 1894; see Chapter 7, p. 146-147). The otherwise healthy 39-year-old right-handed woman experienced seizure-like spontaneous reversals of front and back, i.e., felt the environment turned around by 180° while continuing to see things in front. She was almost entirely unable to recall anything learned during one directional state while feeling in the other. In school exams, she could only succeed once allowed to rearrange her seating direction during tests to match the one during information acquisition. She frequently experienced out-of-body experiences, but no overt seizures ever happened. Neurological status examination (including EEG) was normal. Apart from borderline performance in mental rotation, her neuropsychological profile was unremarkable. Bergouignan et al. (2014) provide a theoretical framework to account for the interaction between felt directional orientation and mnestic performance. These authors induced, in healthy volunteers, an out-of-body perspective during the recall of life events (evidently experienced during an "in-the-body" perspective). This mismatch of perspectives was associated with MRI-monitored activity changes in the posterior hippocampus.

Concluding Remarks

At the end of this chapter, we return to the question posed at its beginning (p. 100): What is this thing we call memory?

In stark contrast to our intricate understanding of the structures and circuits involved in acquiring, storing, retrieving, and recognizing information, we remain startlingly ignorant about the true nature of memory itself. Scientific specialization has amassed overwhelming evidence for the role of limbic structures—particularly the medial temporal lobe—in handling new information. Yet while entire books dissect the hippocampus and journals like *Hippocampus* analyze the microstructure of its layers, we are increasingly trapped in a narrow conceptual framework when discussing "memory."

Memory is more than recalling what you ate for breakfast—it is also recalling which things were not on your plate (Madeleines, for instance). It is more than the word "coffee" or its taste and smell; it is the unbidden associations the aroma evokes, the involuntary reveries that pull you into reflections on your identity—on the parts of you that act beyond your conscious self, perhaps even beyond your will. How many of these hidden influences will shape your actions today, long after they interfered with your fleeting joy of coffee drinking?

And still, at breakfast, an idea surfaces: What if memories are not stored in the hippocampus at all, but widely distributed across the cortex? What if the limbic system's role is merely to hold everything together, and our entire conception of its function has led us astray? What if the forces integrating an episodic memory—its emotional, sensory, and spatial aspects—emerge not *from* the limbic system but around it? (You might not realize that neither of these thoughts originated with you; you may just have skimmed Paller (2002) or Rolls (2019) at some point. And it wasn't even you who connected these two authors' thoughts—it was the smell of coffee!).

Those who perceive an underlying purpose in these "driving forces"—the spontaneous emergence of ideas, associations, and coincidences—often feel compelled to seek an overarching design, grasping at esoteric notions like "intelligent design" or, worse, the grand yet deceitful harmonizations of religion. As feeble as their reasoning may be, as ignorant as their faith demands them to remain, their convictions form part of the "dark matter" of cognitive science (Levin 2021). It is precisely this unexplored frontier that we must confront if we are ever to understand what we so casually call "memory."

Any questions? Yes, two of them may illustrate how important animal neuropsychology is for preparing answers to big questions on our way to climbing the mountain of memory:[3]

? **I read of an experiment with worms, which were trained on a task, "sacrificed" and fed to some conspecifics. Although these had never received any training on the task, they learned it much faster. Is memory edible, or is the story of the mashed worms just a myth?**

You are referring to the famous experiments by McConnell (1962) with flatworms (planaria). Planarians were conditioned to associate a light with an electric shock, making them contract in response to the light. They were then chopped up and fed to untrained ones (planaria are cannibals). Supposedly, the worms showed signs of having "inherited" the learned response. This led a number of authors to the wild-sounding idea that memory might be stored in RNA, since digestion would not preserve neural structures but could allow RNA molecules from trained worms to influence untrained ones. Among hundreds of follow-up studies, several could replicate the findings, while others failed or attributed the findings to artefacts (Walker 1966). Over the years, the hypothesis of "memory transfer," lost popularity. However, in recent years, there has been a resurgence of interest in non-neuronal memory storage, including epigenetics, "transgenerational memory" and RNA transfer. The basic paradigm goes back to McConnell et al. (1959) and takes advantage of the fact that planarians can easily regenerate two whole bodies after being cut in half. After having learned a conditioned response, individual worms are cut into a rostral half (containing the brain) and a caudal half (lacking a brain). After regeneration, both parts are tested again, and both show retention of the conditioned response. The question now is where the "memory" in the re-generated, but newly generated, brain might come from. Shomrat and Levin (2013) developed a computerized training and testing paradigm that avoids potential artefacts inherent to McConnell's paradigm. They could replicate the finding that regenerated tail parts of the decapitated planarians retain previously learned tasks and list several practical implications. See Levin (2021) for a number of similarly astounding experiments on extracerebrally mediated retention. I propose to follow this author's research agenda devoted to the study of cognitive processes in aneural systems. Its importance goes beyond the question of extra-cerebral memory; the topic has practical implications for regenerative medicine and neurorehabilitation (e.g., stem-cell-derived treatments of degenerative diseases).

[3] Eric Kandell, who received the Nobel Prize for his work on synaptic plasticity in the marine snail Aplysia, once stated that "*in the study of memory storage, we are now at the foothills of a great mountain*" (Kandell 2006).

? I have also read something rather unbelievable with respect to remembering. It was a brief note about butterflies remembering their life as a caterpillar. Must be a myth, probably?

No, as exciting as this sounds, it is far from being a myth! The retention of learned information across metamorphosis was originally demonstrated in amphibians. Hershkowitz and Samuel (1973) conditioned the larvae of crested newts to snap food from a surface marked with either a triangular or circular form. Metamorphosis in newts changes bodily appearance, choice of environment (larvae live in the water, adults are terrestrial), and the kind of nutrition—there is an impressively comprehensive renewal of morphology and behavior. Yet the adult animals kept the conditioned preference of food choice they were trained for as larvae (triangle or circle). The authors did not provide a neuropsychological explanation of their findings. However, follow-up research in insects confirmed memory retention across metamorphosis (Blackiston et al. 2008). The neural substrate appears to be the brain's mushroom bodies, whose connectome allows the larvae to form amnesia-resistant memories (Thum and Gerber 2019).

Where Can We Read More?

Budson, A.E. and Solomon, P.R. Memory Loss, Alzheimer's Disease, and Dementia. A Practical Guide for Clinicians. Elsevier: New York, 2021.

Genova, L. Remember: The Science of Memory and the Art of Forgetting. Harmony: Chatsworth, CA, 2021.

Kandell, E.R. In Search of Memory. The Emergence of a New Science of Mind. WW Norton & Co.: New York, 2006.

Luria, A.R. The Mind of a Mnemonist. A Little Book About a Vast Memory. Harvard University Press, 1987 (originally published in 1965).

Schnider, A. The Confabulating Mind (2nd edition). Oxford Academic Press: Oxford, 2017.

Schacter, D.L. The Seven Sins of Memory: How the Mind Forgets and Remembers. Updated edition (1st edition 2002). Mariner Books, HarperCollins: Boston, MA, 2021.

CHAPTER 7

What Are Agnosias?

The term "agnosia" refers to the inability to process sensory information at a higher level of analysis. It derives from the Greek *gnosis* (knowledge), emphasizing that agnosias do not arise from dysfunction of the sensory organs themselves, nor from disturbances in the primary stages of sensory processing. Instead, they reflect a breakdown in perceptual cognition: a stage at which a normal percept is stripped of its meaning (Teuber 1968, cited in Mesulam 2000).

Agnosias are modality-specific. Common types include:

* visual agnosias
* auditory agnosias
* tactile agnosias
* olfactory agnosia
* gustatory agnosia
* vestibular agnosia

? **Who coined the term "agnosia"?**

It was Sigmund Freud. In his pre-symbolic phase, his focus was on problems of language, perception, and other neuropsychological functions (Freud 1891). Around the same time, Lissauer (1890) differentiated between "apperceptive" and "associative" levels of object recognition. He labeled high-level disturbances in perception "soul blindness."

In the visual, auditory, and tactile modalities, the plural form "agnosias" remains appropriate. These senses consist of multiple submodalities, and one can selectively be agnosic for specific aspects of higher-order perceptual processing. For instance, an individual may show agnosia for color, motion, faces (prosopagnosia), or whole-scene integration (simultanagnosia) (see p. 150). Sometimes alexia—the acquired inability to read—is interpreted as a visual agnosia specific to script (Barton et al. 2022).

Likewise, among the auditory agnosias, we must distinguish between agnosia for environmental sounds (e.g., thunder, vacuum cleaners, and traffic noises) and auditory-verbal agnosia, the inability to recognize spoken words, historically referred to as "word deafness."

DOI: 10.1201/9781003384991-7

Vestibular agnosia may exist in only one well-documented form. Here, peripheral vestibular stimulation reliably triggers a nystagmus response, but produces no—or only attenuated—perception of self-motion. It is observed in conditions with bilaterally disrupted cortical network function, typically in the context of neurodegenerative disease or traumatic brain injury (Hadi et al. 2022).

A clinically informed specialist in olfactory or gustatory perception might object to the singular form used above and argue that these domains, too, merit subdivision. For example, olfactory processing includes not only odor identification but also the affective valuation of smells (via sensory-limbic pathways), their role in social cognition, and their strong links to memory processing (Eichenbaum 1998). A similar case could be made for gustatory subdomains.

? **Is there such a thing as "interoceptive agnosia"?**

In rats, "interoceptive agnosia" has been reported after hippocampal lesions (Clifton et al. 1998). In humans, interoception remains an underrepresented topic in neuropsychological textbooks—including the present one. Unlike the exteroceptive senses of vision, hearing, or smell, interoceptive processing concerns stimuli from within the body: signals from cardiovascular, gastrointestinal, and respiratory systems (Tallon-Baudry 2023). Interoception maintains homeostatic balance and depends on multiple neuroanatomical pathways. It is deeply interwoven with exteroceptive processes and implicated in several neuropsychiatric conditions (Khalsa et al. 2018).

High-level disorders of interoception of the agnosic type are to be expected, yet they remain underreported and are rarely addressed in standard neuropsychological evaluations. Such evaluations should include careful observation and questioning regarding behavior related to eating, drinking, and sexuality. Psychometric approaches may assess heartbeat discrimination, judgments of bladder fullness, or other internal monitoring tasks. A multidimensional self-report tool is also available (Mehling et al. 2012).

"Agnosia" may also refer to the loss of knowledge in a non-perceptual domain, or at least concern the disruption of supra-modal perception or recognition. Examples include:

* Topographagnosia (also called environmental agnosia): The inability to recognize familiar places, often accompanied by a loss of spatial orientation (see p. 162).
* Anosognosia (formerly noso-agnosia): The inability to recognize one's own illness (see p. 163).
* Autotopagnosia: The inability to localize body parts on one's own body (see Chapter 5, p. 162).
* Finger/toe agnosia: The inability to distinguish individual fingers or toes (see p. 162).
* Emotion agnosia (alexithymia): The inability to recognize affective tone (see below, p. 163).

Can one be agnosic for pain?

Yes. The appropriate term for pain agnosia is pain asymbolia. First described by Schilder and Stengel (1931), this condition spares the bottom-up sensory processing of pain—mediated by the insula and anterior cingulate cortex—but impairs top-down, affective evaluation. In other words, patients report sensing the pain stimulus but do not experience it as unpleasant or threatening.

This condition should not be confused with the congenital insensitivity to pain (Manfredi et al. 1981), which reflects a primary failure of nociceptive transmission. See Gerrans (2020) for a discussion of pain asymbolia that moves beyond agnosia-based nosology.

The neuropsychology of agnosias encompasses such a wide terrain that it is useful, for practical reasons, to focus on selected aspects in detail. Monographs recommended at the end of this chapter may guide the reader toward other domains not covered here, but essential for any clinically informed neuropsychologist.

This chapter concentrates primarily on visual agnosias. This focus reflects both the dominant role of vision in human cognition and the historical fact that visual agnosias have been more extensively investigated than agnosias in any other modality. In addition, some visual disturbances that are not strictly agnosic in nature are discussed here to aid in their recognition during clinical examination. Certain "productive" visual phenomena—those involving false or superadded percepts—will also be addressed, though only tangentially.

To appreciate the clinical relevance of visual agnosias and to understand why a broad repertoire of testing is necessary, it is indispensable to begin with a brief overview of core principles of visual perception and cognition.

What Should Neuropsychologists Know About Visual Perception and Cognition?

Visual perception begins in the retina, as soon as light enters the eye and passes through the eyeball. The sensitivity of retinal cells is extraordinary: just a few photons can elicit an electrical response that informs the brain of incoming visual information (Nelson 2016). Strictly speaking, even when only a couple of photons hit the retina, we might already speak of "seeing." Yet, the mere reception of light is not sufficient for "perception." Even millions of photons do not constitute perception unless they come from a structured environment that conveys information about shape, color, or motion.

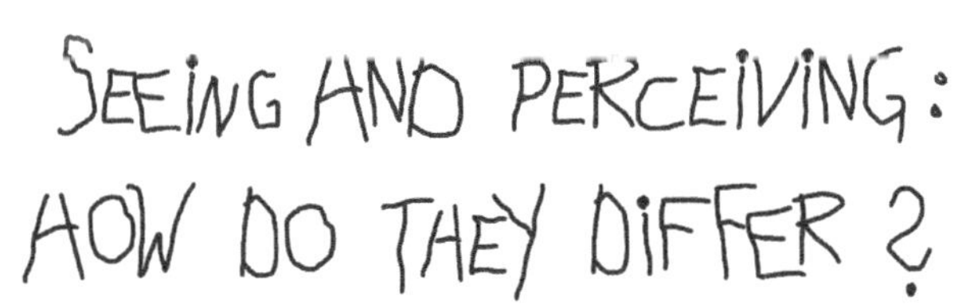

Figure 1: The retinal act of seeing does not distinguish between the starting letters of the words AND and HOW. Your brain automatically disambiguated them by perceiving the first as a "A" and the second as an "H."

The boundary between seeing and perceiving is blurred. But the essence of the distinction is captured in Figure 1.

Inspection of Figure 1 reveals that letters are composed of individual line segments. Each segment activates the retina by creating contrast with neighboring regions on the page. Retinal cells then inform thalamic and cortical neurons about this activation. These neurons are already highly selective: some respond only to vertical lines, others only to horizontal or diagonal orientations (Hubel and Wiesel 1959; see Wurtz 2009, for a historical overview of this Nobel Prize-winning work).

Take the letters "H" and "A" in the words "HOW" and "AND," respectively. Some neurons respond to the vertical bars, others to the crossbars or slants. What observers of Figure 1 often fail to notice is that they automatically interpreted the ambiguous line configuration as two distinct letters. This "top-down" interpretation of ambiguous visual input is what we call perception. It is guided by prior knowledge of word forms, letter shapes, and expectations about context.

So, to answer the figure's question: "Seeing" refers to the bottom-up components of visual percept formation, while "perceiving" also includes the top-down components.

? What are the bottom-up and top-down processes of perception, and why is it important for neuropsychologists to know the difference?

The physical characteristics of a stimulus entirely drive bottom-up processes; top-down processes are shaped by prior knowledge, expectations, and context. The distinction is essential in neuropsychology because agnosias are disorders of higher-order perceptual processes—they arise despite intact sensory organs (eyes, ears, skin, nose, and inner ear) and intact primary sensory projection areas.

Perception is not a unidirectional, serial process moving from lower to higher levels of analysis. Instead, bottom-up and top-down processes are tightly interwoven and mediated by reciprocal feedback across multiple levels of the visual hierarchy. For instance, perceptual feedback from the visual association cortex to primary visual cortex (V1)—and even to precortical stations like the lateral geniculate nucleus—is essential for coherent percept formation (Zeki 1993; Sillito et al. 1994).

Figures 2a and 2b illustrate how these processes jointly shape our experience of reality in the well-known checker shadow illusion (Adelson 1995/2005). For a comparably striking illusion involving chess pieces, see Anderson and Winawer (2005).

? I heard that visual perception also involves a "radiant component"—that is, a beam of light leaves the eyes before re-entering them to allow the brain to perceive. What should I think about this idea?

"Extramission theories" of visual perception propose that pneuma, a kind of soul-energy, is emitted from the eye, touches objects in extrapersonal space, and returns to the eye carrying information about the world. Ancient Greek

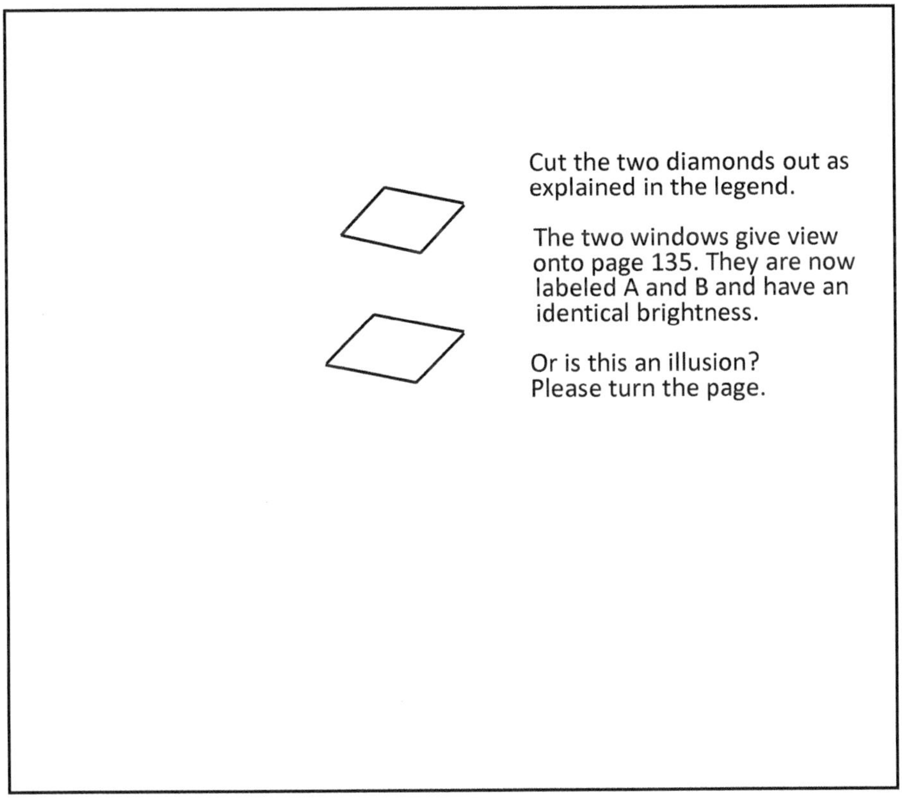

Figure 2a: A brightness illusion. Readers of the print version of this book: Please take a Japanese knife, put a thick cardboard underneath this page, and cut the two diamonds cautiously out. Remove the cardboard and view two patches, now labeled A and B. Their brightness is the same.
This is not the illusion yet. The illusion is that after turning the page, Figure 2b shows that A and B have different brightness. Read on there.

philosophers had long noted the distorted percepts (phosphenes) produced by pressure on the eyeball (cf. p. 156). This "fire in the eye" laid the groundwork for the physiology of extramission as proposed by Plato, Aristotle, and later early Arab scholars (Grüsser and Hagner 1990).

The idea of extramission is specific to vision because the eye—unlike the ear or the nose—is also an active organ (think of the oculomotor system), and because eye gaze conveys emotional and social meaning. From a modern scientific standpoint, extramission has been thoroughly discredited, but it remains of historical and philosophical interest.

Implicit notions of extramission persist today. Children's and adults' drawings of visual perception often include arrows or rays emanating from the eyes (Winer et al. 2002), and belief in the "evil eye" remains widespread (Galt 1982). Experiments on adults' intuitive physics have shown that many people still assume that gaze involves a physical force (Guterstam et al. 2019). Some fringe-science authors (e.g., Sheldrake and Smart 2023) even claim empirical support for scopaesthesia, the alleged sense of being stared at from behind. Such claims are best explained by methodological shortcomings and a general will to believe—hallmarks of pseudoneuroscience (Brugger 2024; Carpenter 2005).

Is the Visual Cortex "Functionally Specialized" in a Neuropsychologically Relevant Way?

Absolutely, yes. Color perception, for example, is processed in a distinct cortical region—area V4—located in the lingual and fusiform gyri of the prestriate cortex. In contrast, visual motion is processed in a different region near the temporo-parieto-occipital junction, area V5 (also known as MT). Even in the posterior thalamus, color and motion information are separated: color is relayed via smaller neurons (parvocellular pathway), and motion via larger neurons (magnocellular pathway). The resulting disorders—color agnosia and motion agnosia—therefore have anatomically distinct correlates. Zeki (1990) offers a detailed account of these two types of visual agnosia, including their divergent research histories.

Other submodalities of vision also show further specialization. For example, some regions within the motion-processing network are tuned specifically to biological motion—i.e., movements of body parts (Stevens et al. 2000). Human body parts, even when static, are processed in a specialized subregion of lateral occipitotemporal cortex known as the extrastriate body area (EBA; Downing et al. 2001). This area is also involved in monitoring one's own limb actions and may contribute to the distinction between self and other. The EBA is not responsive to faces; facial stimuli are processed in adjacent regions, most notably the fusiform face area (FFA; Kanwisher et al. 1997) and parts of the inferotemporal cortex.

Face perception would merit a chapter of its own. Beyond the clinical richness of facial perceptual disorders (facial agnosia, prosopagnosia, face-specific distortions and hallucinations, or emotion-specific impairments), the human face plays a central role in social behavior. One phenomenon from face perception research—though not clinically essential—is particularly instructive for understanding general principles of visual processing: face adaptation. It has been called "the psychologist's microelectrode" (Frisby 1979) because it allows selective manipulation of neural cell populations in specific cortical regions using purely behavioral means. See Figure 3.

Many standard neuropsychological textbooks devote only a few pages to disorders of visual perception and cognition. For example, Schoenberg and Scott (2011), in their otherwise comprehensive 900+ page *Little Black Book of Neuropsychology*, allocate fewer than five pages to agnosias. The reason for this brevity lies in visual perceptual deficits: assessing these deficits demands clinical expertise in behavioral testing, and for many of these functions, normative data are simply not meaningful.

Figure 2b: The different brightness of the patches A and B is a contrast illusion. A appears darker than B because its surroundings are brighter. This kind of "simultaneous contrast" reflects a bottom-up inhibitory effect already taking place in the retina. The shadow cast by the cylinder on B, but not on A, enhances the apparent brightness difference. This is a top-down effect from what we know about light and shadows. Some visual illusions are so compelling that a bit of work is needed for disillusionment – thank you, readers of the print version of this book, for taking the pain to do the little cutting! Thank you, readers of the eBook version, for recognizing that the digital world imposes principal limits to experiencing the world we live in. The illusion is from Edward H. Adelson,©1995, and reproduced here with permission (see https://persci.mit.edu/gallery/checkershadow).

To document a patient's recovery from cortical blindness, you do not need a standardized test with population-based norms. What you need are structured procedures, based on clinically informed and precise questions, that can reveal the extent and nature of functional impairments. Such qualitative assessments allow clinicians to anticipate the real-world consequences of a disorder and to make meaningful rehabilitation decisions.

For those who lack (or mistrust) clinical judgment, a few psychometric aids are available, though each has limitations. These tools can help quantify specific deficits and monitor recovery over time, but they cannot replace the kinds of diagnostic tasks recommended in this book (see Figure 5 and also Chapter 12). Two online tools may serve as examples: the Leuven Perceptual Organization Screening Test (Torfs et al. 2014) and a more narrowly focused assessment of visuo-perceptual deficits in Parkinson's disease (Weil et al. 2018).

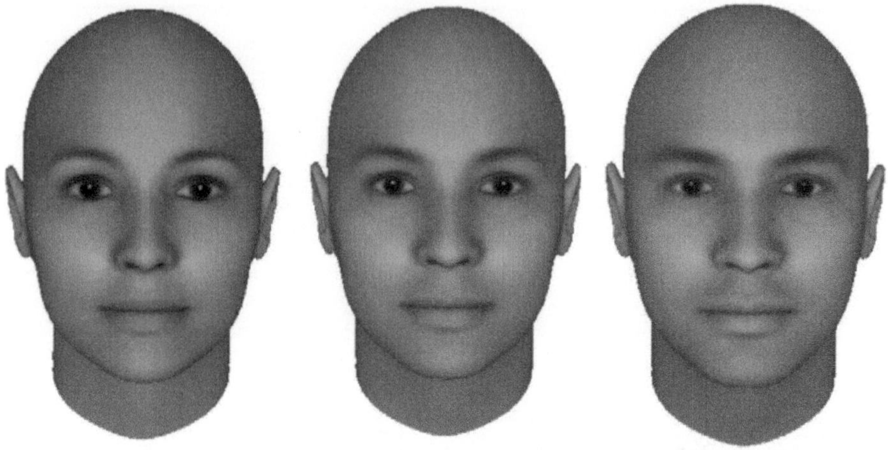

Figure 3: A female face (left), a male face (right), and a gender-neutral face (center). Fixate on either the female or the male face for about 30 seconds, covering the other two faces with a cardboard. Then quickly shift your gaze to the center face. It will appear more male or female, respectively. Face adaptation, as demonstrated here with gender, can also be shown with features such as age, race, or emotional expression. Adapted from Javadi and Wee (2012, Figure 2).

How are the Two Visual Fields Represented in the Brain?

At higher stages of processing, visual information is handled quite differently depending on its origin in the visual world. These functional hemispheric asymmetries (FUHEMAS; see Chapter 11) become evident primarily when a person's gaze is fixed at the center of the visual field. Divided visual field (DVF) studies using lateralized tachistoscopic stimulation are among the most powerful tools for identifying FUHEMAS (Beaumont 1982; see Bourne 2006, for methodological caveats).

To understand DVF studies, one must first grasp the anatomical route visual information takes—from retina to optic chiasm, through the thalamus, and into the cortex (see Figure 2 in Chapter 11). DVF methodology is clinically relevant, especially when there is a need to determine language dominance in left-handed individuals. This task falls naturally within the neuropsychologist's domain—who else is better trained in behavioral laterality tasks and their advantages over high-tech (e.g., fMRI) or invasive (e.g., WADA) procedures?

In What Way are the Ventral and Dorsal Visual Streams Functionally Distinct?

Visual recognition is only one aspect of seeing. Just as essential is the localization of objects in space and the guidance of action based on visual input. As visual information ascends the cortical hierarchy, it is processed along two distinct pathways, specialized for different functions:

* The ventral pathway, projecting from the visual cortex to the temporal lobes, is primarily concerned with object recognition. It is often called the "WHAT"

pathway (Ungerleider and Haxby 1994). Neurons along this route become progressively tuned to complex physical features, from early occipital to inferotemporal regions.

* The dorsal pathway, projecting to the lateral parietal lobes, is primarily involved in spatial localization and visuomotor control. It is known as the "WHERE" pathway.

While the two-stream model has proven clinically useful and conceptually influential, it is now being critically re-examined (de Haan and Cowey 2011), and extended models are under discussion (Wurm and Caramazza 2022).

> That perception and action can dissociate was famously demonstrated in the case of patient DF, who suffered profound form agnosia following carbon monoxide poisoning (Goodale and Milner 1992). DF had extensive damage to the ventral visual stream, but her dorsal pathway remained intact. She was entirely unable to recognize the size, shape, or orientation of visual objects. Even when asked to estimate the width of an object using her thumb and index finger under conscious control, she failed to approximate it.
>
> However, when she was asked to reach for the same object, her hand shaped itself appropriately—the grip aperture adjusted with remarkable accuracy. Similarly, DF could not indicate the orientation of a visually observed slot using her hand alone. But when asked to act, for instance, by sliding a card into the slot, she performed as well as healthy individuals.

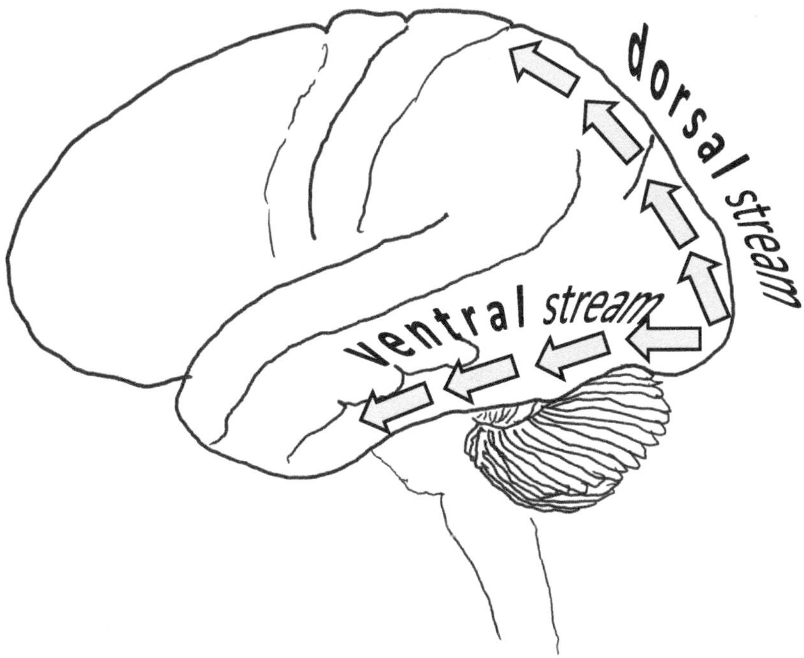

Figure 4: Two pathways of visual information diverging from the secondary visual cortex (V2) and projecting to the superior parietal and inferotemporal cortex, respectively.

? The functional segregation of WHAT and WHERE is central to neuropsychology—does it only hold for visual information processing?

Historically, the distinction between two processing streams—ventral for recognition ("what") and dorsal for localization ("where")—was first described in studies of the visual system, initially in monkeys and later in humans. However, a comparable segregation has also been demonstrated in the auditory system (Romanski et al. 1999). Interestingly, this dual-stream architecture may originate at the level of the brainstem rather than in the primary auditory cortex.

In the domain of speech perception, it has been proposed that the dorsal auditory stream supports the analysis of verbal content, while the ventral stream aids in speaker identification. Particularly intriguing is the question of how this dorsal–ventral distinction interacts with auditory FUHEMAS (functional hemispheric asymmetries; see Chapter 11).

A similar distinction exists in the tactile modality: here, researchers differentiate between touch for perception and touch for action. An influential article in Behavioral and Brain Sciences[1] suggested that the posterior parietal cortex is involved in both perceptual and motor aspects of touch, while the insular cortex is more specifically involved in perceptual recognition, including the hedonic (affective) dimensions of tactile experience (Dijkerman and de Haan 2007; see also de Haan and Dijkerman (2020) for a more recent refinement).

For a brief overview of dual-stream segregations across sensory modalities, see Sedda and Scarpina (2012). On a supramodal level, the two-stream framework is also widely discussed in the context of language (see Hickok 2022 for an overview).

? I heard that the ventral-dorsal stream findings led to the claim that a bottom-brain vs. top-brain dichotomy would be more useful for neuropsychology than the classical left-brain/right-brain dichotomy. Does this make sense?

You heard right—but the claim is unsubstantiated. In their popular book *Top Brain—Bottom Brain* (Kosslyn and Miller 2013), the authors criticize several crude dichotomies derived from low-quality laterality research. For instance, they rightly debunk the popular myth that the left hemisphere is the "rational"side of the brain, while the right hemisphere is the "creative" one. In doing so, however, they are targeting tabloid neuroscience, not the more nuanced and empirically grounded research relevant to clinical neuropsychology (see Chapter 11).

[1] *Behavioral and Brain Sciences* (BBS) is a highly recommended resource for clinical and experimental neuropsychologists. Each issue features a target article that typically presents a novel perspective on a topic of broad relevance. This is followed by a series of short commentaries from experts in the field, to which the author(s) of the target article respond. The result is a dynamic exchange of viewpoints within a single journal issue.

Ironically, Kosslyn and Miller then proceed to replace the left–right simplification they mock with an equally reductive typology based on ventral vs. dorsal stream thinking. They suggest that individuals who operate in a "perceiver mode" rely more on the bottom brain, while those in a "stimulator mode" prefer the top brain. Despite its marketing appeal, this cognitive-style framework lacks serious empirical support. It remains in the realm of pop psychology and does not offer a scientifically grounded alternative to established models.

What Other Senses Show Strong Links to Visual Perception?

There is a "spectacular symbiosis" between the visual, vestibular, and optomotor systems (Schwarz 2002). Particularly intriguing are the intimate connections between the receptive and the expressive functions of vision: both seeing and looking—i.e., visual reception and eye movement control—are performed by the same organ, and often at the same time. This dual role of the eyes links perceptual and motor functions in a way that engages nearly the entire brain.

This deep integration may help explain the enduring popularity of pseudoscientific beliefs about visual perception (e.g., Sheldrake and Smart 2023; see Brugger 2024 for a critique). The social and emotional significance of gaze, combined with its perceptual-motor basis, creates fertile ground for such misinterpretations.

Sometimes labeled "vestibular cognition," the interplay between vestibular, visual, and multisensory perception involves widely distributed cerebral networks (Ferrè and Haggard 2020). Stimulating the vestibular organ in the inner ear can produce significant effects on visual perception and cognition—effects that are relevant both diagnostically and therapeutically (see also Chapter 9).

The neuropsychiatric relevance of visuo-vestibular interactions was already noted in the early work of Levy et al. (1983). And as I once suggested (Brugger 1999), "listening to one's inner ear" may offer access to previously unheard messages from the growing field of oto-neuropsychiatry.

What Major Types of Visual Agnosias Are There?

For clinical practice, familiarity with at least some of the major types of visual agnosia is indispensable. Before listing them, however, it is useful to introduce a distinction that is central both in historical context and for clinical reasoning: the difference between apperceptive agnosia and associative agnosia.

? What is the distinction between "apperceptive" and "associative" forms of visual agnosia?

The distinction dates back to Lissauer (1890), who proposed that apperceptive processes involve the initial stages of object recognition—such as detecting contrast, contours, or basic shapes. See Serino et al. (2014) for a modern case study illustrating this type.

In contrast, associative agnosia reflects impairments at later, semantic stages of visual recognition. These patients can accurately copy or match drawings of objects—tasks that apperceptive agnosics struggle with—but they have difficulty identifying the object's function, typical context, or name. In other words, the percept is intact, but its meaning is inaccessible.

The boundary between apperceptive and associative agnosia is often blurred, and mixed forms occur. As a rule of thumb:

* Right posterior lesions are more likely to produce apperceptive agnosia.
* Left-sided or bilateral damage involving the ventral visual stream is more commonly associated with associative agnosia.

Ultimately, the contribution of perceptual and semantic deficits must be carefully assessed in each individual case.

Figure 5 (p. 141–143) presents a set of visual stimuli essential for classifying a patient's visual complaints (often described simply as problems with "eyesight"). It is not intended that all stimuli be administered to every patient. Instead, the examiner should tailor the selection to the nature of the patient's complaints, using their clinical expertise to determine which apperceptive components require evaluation.

In brief, the templates of Figure 5 test:

* Basic line orientation and contour perception (Figure 5a–d)
* Object recognition under noncanonical conditions, e.g., unusual perspectives, degraded backgrounds, distortions, or removal of use of extreme contrast levels (Figure 5e–n)
* Letter and word recognition (Figure 5h, k, l)
* Face recognition (Figure 5m–p)
* Recognition of clock faces and body parts (Figure 5q, r)

For an extended legend to Figure 5, see below.

Extended Legend to Figure 5

Figure 5a and 5b show stimuli for testing low-level visual functions, particularly useful when a patient is recovering from cortical blindness. Simple line orientation can be assessed using Figure 5a by presenting it at different angles of rotation. To avoid requiring a verbal response, you can present two cards—one with a vertical, the other with a horizontal line—and ask the patient to "point to the horizontal/vertical line."

Figure 5b tests form discrimination. Questions must be adapted to the patient's level of understanding (e.g., "Do you see a round shape?" or "Can you point to the circle?" or "How many shapes are there?"). Brightness discrimination can also be evaluated. Many of these stimuli also serve as informal tools to assess language comprehension. For example, "which is brighter, the circle or the square?" or "touch the circle, then touch the darker of the two shapes."

Figure 5c tests illusory contour perception. You may first show a simple Kanizsa triangle (Kanizsa 1979), then the more complex version presented here, which also

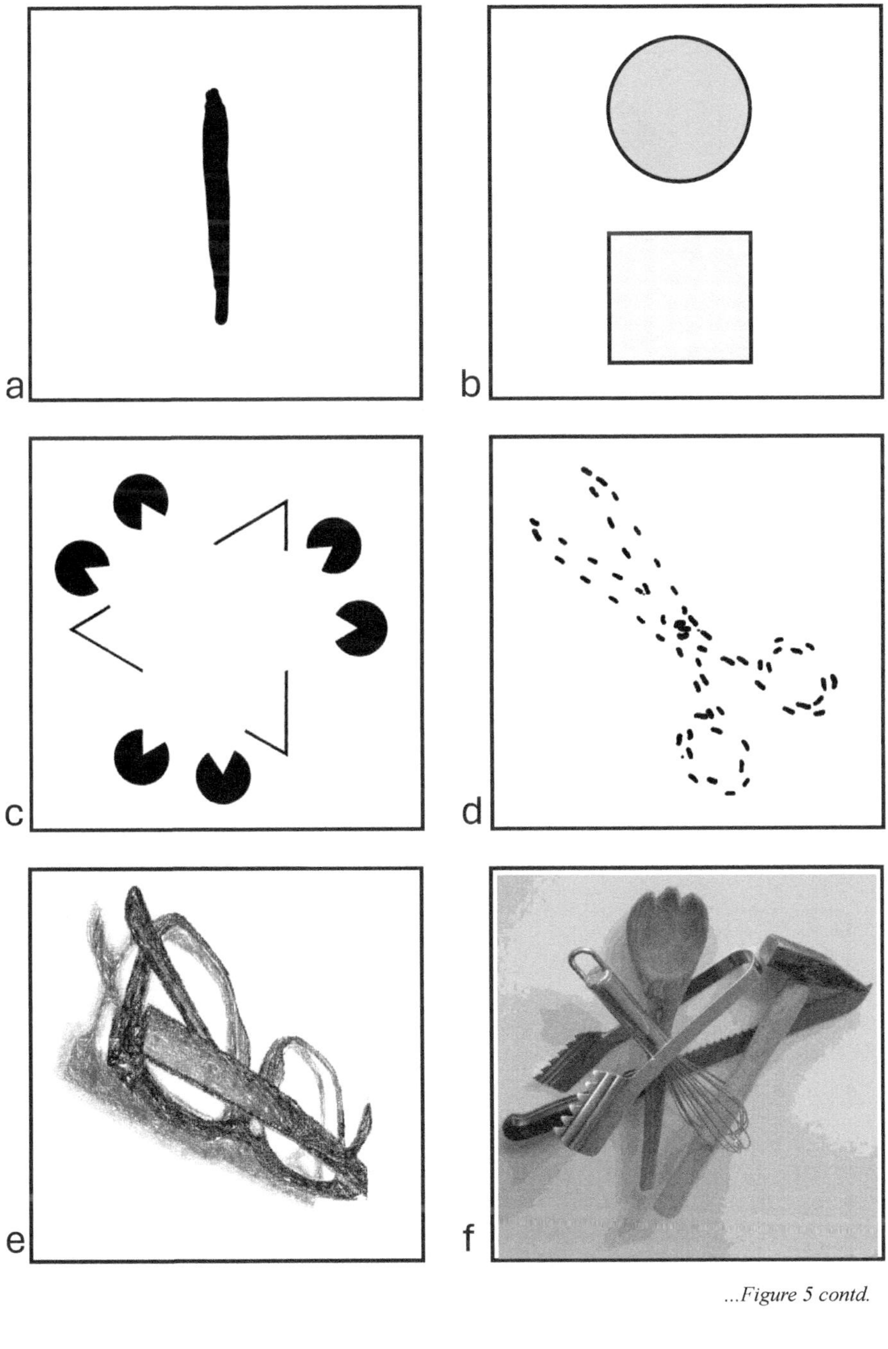

...Figure 5 contd.

...Figure 5 contd.

Figure 5 contd. ...

...Figure 5 contd.

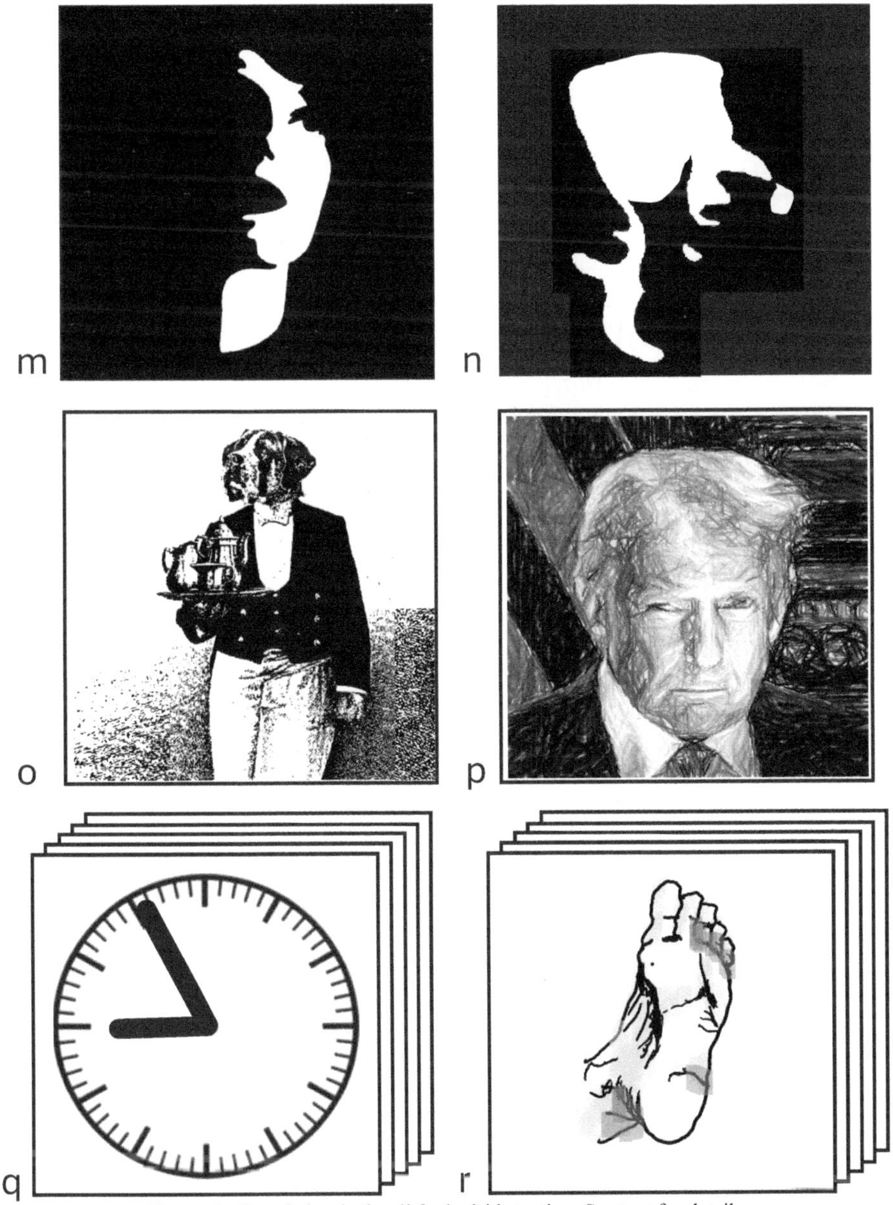

Figure 5: Set of visual stimuli for bedside testing. See text for details.

probes sensitivity to depth cues ("Which triangle appears to be on top?"). The patient is asked to trace the triangle(s) they see (after Scrivener 1983).

Figure 5d shows an object composed of dotted outlines, which is often difficult for patients with Balint's syndrome to recognize (see p. 150). The original Gollin (1960) test features a validated series of such objects, each presented in 10 levels of increasing completion (i.e., decreasing gap size). It is especially useful for tracking changes in recognition ability over time.

Figure 5e displays an everyday object in a noncanonical view—a classic test for object agnosia, even in subtle cases. As an alternative, you may show the patient a real object, like a pair of glasses, held in an unusual orientation directly in front of their face.

Figure 5f is a photograph of multiple household items that need to be recognized. This is slightly more difficult than high-contrast line drawings with overlapping contours (Poppelreuter 1917; see Chapter 12). Instead of overlapping objects, one may also test structural perception using a scrambled object drawing (not shown here; see Davidoff and Landis 1990).

Figure 5g requires Gestalt perception to be recognized as a sailing boat. A full series of similar degraded images—featuring human figures, animals, and objects—was published by Street (1931).

Figure 5h, 5k, and 5l test the recognition of letters and words under degraded conditions. Stimuli include shadow letters (Kanizsa 1979), broken-line letters (Landis et al. 1982), and cross-hatching (Goldstein and Gelb 1918). These variations help uncover subtle cases of alexia, especially when applied to handwritten material.

Figure 5i and 5j are especially difficult for patients with impairments in global-local integration. These images require the perceptual segregation of local details from the overall structure.

Figure 5m to 5p relates to face perception and recognition. Patients with face agnosia (p. 148) may be unable to decide whether 5m or 5n depicts a face. They also tend to fail in recognizing animals with human-like features, such as those shown in Figure 5o. For famous face recognition (Figure 5p, depicting a popular poetaster's face; see Sears 2017), clinicians must evaluate whether the patient had prior familiarity with the pictured individuals. A more informative test involves asking a relative to bring photographs of family members and close friends for recognition testing.

Figure 5q and 5r present two specific categories of visual stimuli: clock faces and body parts. A sequence of such images is needed for reliably diagnosing clock face agnosia (p. 147) or impairments in limb laterality judgments ("Is this a left or a right limb?").

Standardized test sets containing some of the perceptual subtests included in Figure 5 are available in the VOSP (Visual Object and Space Perception Battery; Warrington and James 1991) and the CORVIST (https://www.corvist.org, last accessed March 1st, 2025). Both offer validated procedures for testing object shape recognition, fragmented pictures, noncanonical views, words, and face recognition.

Not included in Figure 5 are items targeting color perception (e.g., Ishihara 1932; Farnsworth 1943; Ghose et al. 2014) or dynamic visual processing, which typically require computerized presentation. However, especially in low-level recognition impairments, any non-recognized object—or its 2D representation—should be shown again in motion. Paradoxically, recognition may improve when a stimulus is moved across the visual field.

Form Agnosia

(Visual) form agnosia is the most severe variant within the spectrum of agnosic disorders. It is characterized by a predominance of apperceptive deficits, such that even basic perceptual operations like contrast detection and line orientation become difficult. In contrast, brightness, color, and motion are typically preserved.

Object shapes are complicated to recognize when:

* Contours are fragmented
* Objects overlap
* They are presented from noncanonical (unfamiliar) perspectives
* Only local features are available and must be integrated into a global contour

In short, nearly all sample steps of perceptual integration illustrated in Figure 5 are impaired in this condition.

Form agnosia is commonly associated with lesions of the primary visual cortex, often due to hypoxia, and may represent an early stage of recovery from cortical blindness (see below). When visual field defects are present, they may obscure the true nature of the agnosia. However, a diagnosis is possible when stimuli are presented within an intact region of the visual field.

Interestingly, even patients with multiple and severe impairments in object recognition—documented through various items in Figure 5—may still be able to navigate complex, familiar environments. This apparent paradox underscores the multifaceted nature of visual processing and the partial independence of object recognition and spatial orientation.

> We observed a 66-year-old right-handed farmer who had become cortically blind following hypoxic brain damage. Although his basic vision eventually returned, moderate to severe visual form agnosia was still evident on clinical testing. For many years, the farmer had offered carriage rides through the countryside—a hobby he resumed without apparent difficulty or complaint, personally guiding clients along a multi-kilometer route.
>
> Somewhat reluctantly, I once accepted his invitation to take such a ride. I was struck by the stark contrast between his highly practiced, overlearned visuomotor routine and the very poor visual performance he had demonstrated just days earlier during formal neuropsychological assessment.

What should be assessed are potential cueing effects by color. A patient may not recognize drawings of four-legged animals except by tracing the outlines ("a dog? a horse?") but immediately identify a lion once the yellow color is added (Landis et al. 1982).

? Can visual agnosias be restricted to only a part of the visual field?

Yes. In hemiagnosias (Pflugshaupt et al. 2016; Rennig et al. 2018), patients with unilateral ventral pathway lesions are agnosic only within the contralesional visual half-field.

Two case reports illustrate the neuropsychological complexity of such hemiagnosias. Meichtry et al. (2018) described a patient with a rare combination of right visual field object agnosia and right-sided optic ataxia. Structural MRI revealed white matter damage affecting both dorsal and ventral streams in the left hemisphere.

Rennig et al. (2018) reported a stroke patient with a unilateral lesion of the right fusiform and parahippocampal gyri. This patient was unable to recognize familiar everyday objects in his left field unless they belonged to a domain in which he had acquired expertise. The study elegantly demonstrated that top-down mechanisms, such as visual expertise, influence object recognition even at early stages.

Object Orientation Agnosia

First described by Best (1917), this form of agnosia does not impair object recognition or identification, but instead affects the ability to perceive an object's orientation in space. The ventral visual stream is spared, and lesions typically involve the parietal lobes, most often on the right or bilaterally. In some cases, Balint's syndrome (p. 150) is also present.

Best's original case could correctly identify faces, whether upright or inverted, yet was unable to determine their orientation. He confused letters like *u* and *n*, even though his general letter recognition was intact. A bullet wound had damaged both inferior parietal lobes.

Modern case reports are provided by Turnbull et al. (1997) and Fujinaga et al. (2005). Impairment in recognizing mirrored object orientations following a right temporoparietal stroke was described by Priftis et al. (2003).

Agnosia for Heading Direction

This rare condition is only partly visual—the other component relates to a supramodal sense of direction. A person with heading direction agnosia experiences a conflict between what is seen and what is felt. For instance, they may see that they are approaching a lake—the intended destination of a hike—but simultaneously feel that they are moving away from it.

The condition was described in French and German neurology as early as the early 20th century (Janet 1908; Pick 1909), and even earlier in English (Binet 1894). Though originally considered a pathological phenomenon, later reports noted that healthy individuals may also occasionally experience it (Wollenberg 1925).

The top-down component may reflect a disruption in vestibulo-visual integration, while the bottom-up perceptual basis could involve dysfunction of head direction cells in the bilateral (para)hippocampal regions (Hartley et al. 2014). This might also explain the severe memory difficulties often reported: information learned while facing one direction (e.g., north) may become difficult to retrieve when facing the opposite direction.

At present, it remains unclear whether individuals with agnosia for heading direction have an impaired or perhaps hypersensitive magnetoreception system (Nordmann et al. 2017), or whether such a perceptual system even reliably exists in humans (Chae et al. 2022).

> A 39-year-old right-handed college student (CB; formerly a secretary) was referred by her general practitioner due to "serious problems with spatial fixation." In standard classroom settings, CB had no difficulties taking exams and earned good marks in all subjects. However, on one occasion when a test had to be taken in a different classroom with a changed seating arrangement, she became visibly distressed. She requested permission to rearrange the desk and chair to match the spatial orientation she had during learning.
>
> CB reported that she could, out of curiosity, induce brief reversals of front and back (e.g., by rapidly turning around). This ability appeared deliberate and exploratory.
>
> Her medical history was unremarkable for neurological or psychiatric illness, except for a longstanding history of migraine. She had never experienced seizures. CB also reported occasional out-of-body experiences. For many years, she had experienced brief episodes—lasting from seconds to minutes—of left/right switching, occurring with a stable frequency of several times per day or hour. During these episodes, the world appeared to her as if mirror-reversed. These mirror phenomena were not under voluntary control, and CB perceived no relation between them and the reversible front/back heading she could induce intentionally.
>
> Neurological examination, including EEG, was normal. On formal neuropsychological assessment, CB performed within normal limits across all tested domains: language, praxis, learning and memory, attention and executive function, and visual-spatial perception. Borderline performance was observed only in a mental rotation task.

Unfortunately, interactions between body orientation (heading direction) and learning/memory performance were not assessed—a missed opportunity given the suspected orientation-dependent learning. For further discussion of this patient, see Chapter 6, p. 125.

Clockface Agnosia

Evolution has certainly not equipped us with a dedicated module for reading analog clock displays. Yet, clockface agnosia—the inability to read time from an analog clock ("Uhrzeitagnosie" in German)—may be experienced by patients as the only noticeable sign of brain damage, particularly after a stroke.

Further testing often reveals additional deficits, typically pointing to right parieto-occipital lesions (e.g., constructional apraxia, topographagnosia). Though less common, left posterior damage has also been associated with this condition (see Gloning et al. 1958).

> A 61-year-old right-handed woman (secretary) had suffered a right posterior stroke. She was entirely unable to tell the time from the analog watch worn on her left wrist. Neuropsychological testing revealed mild deficits in verbal and nonverbal learning

and memory, along with slightly reduced interference control (Stroop task). All other functions—including the perception of objects and faces, spatial relations (no constructional apraxia, no topographagnosia, no neglect)—were normal.

Following our advice, her son gave her a digital watch, worn on the right wrist, allowing her to read the time flawlessly. Reading analog clock faces remained impossible at discharge. This case illustrates a double dissociation between agnosia for analog and digital clock reading. After left posterior strokes, many patients show the opposite pattern: they can no longer read digits on a digital watch but retain the ability to read analog clockfaces. Such hemisphere-specific deficits might tempt one to conclude that the left hemisphere specializes in digital and the right in analog modes of processing. This generalization, however, is not justified—see Chapter 11, p. 268.

Face Agnosia

This is not the same as prosopagnosia (see below). It is surprisingly difficult to find a proper definition online, as the term "visual agnosia for faces" is often automatically replaced by prosopagnosia. However, face agnosia refers to the inability to recognize a stimulus as a face at all. Familiarity plays no role.

Degraded or abstract representations of faces are particularly difficult to identify, whereas symbolic face-like depictions (e.g., emojis) may be recognized, including their emotional expression. Mooney faces are well-suited to test face agnosia (Figure 5m; Schwiedrzik et al. 2018). Drawings of human bodies with animal faces—often found in advertisements—are also useful. When asked what profession the depicted person might have, patients infer answers based on clothing or accessories, without recognizing that the face is not human.

The neural substrate of face agnosia involves face-selective cells in the inferotemporal cortex (Logothetis and Pauls 1995). Some of these neurons are also sensitive to gaze direction and emotional expression (Perrett et al. 1985).

Prosopagnosia

Prosopagnosia is the inability to recognize familiar faces. Unlike face agnosia, a face is perceived as a face—yet its identity is not accessible. The condition has been known for centuries, but the term was coined by German neurologist Joachim Bodamer in 1947.

Prosopagnosia results from damage to the right or bilateral occipitotemporal cortex, including the fusiform and lingual gyri. It is often associated with topographagnosia and sometimes with achromatopsia or left-sided hemiachromatopsia. It may also occur in the context of seizure disorders, epilepsy, or as a persistent symptom after COVID-19 infection (Kieseler and Duchaine 2023).

Autoprosopagnosia refers specifically to the failure to recognize one's own face.

Covert Familiarity Recognition in Prosopagnosia

While the ventral stream—crucial for conscious face identification—is damaged in prosopagnosia, the dorsal stream often remains intact. This allows for covert recognition: patients may show autonomic responses (e.g., skin conductance changes) to familiar faces they cannot overtly recognize (Bauer and Verfaellie 1988).

Congenital Prosopagnosia

Some healthy individuals suffer from congenital (or developmental) prosopagnosia. This condition may be as severe as acquired forms, but individuals often learn to compensate by relying on accessories (glasses, facial hair, and piercings) or voice and movement cues. It tends to run in families, suggesting a heritable component.

At the other end of the spectrum are super-recognizers—individuals with exceptionally accurate face recognition (Ramon et al. 2019). A 20-item screening test for congenital prosopagnosia is available (Shah et al. 2015). Online tests (e.g., UNSW Face Test, University of Sydney) also assess face memory and can help identify super-recognizers. These tests are available only on a PC or a laptop.

Color Agnosia (Achromatognosia or Achromatopsia)

Color agnosia arises from dysfunction in the parvocellular pathway (p. 134), especially the color-processing areas of the ventral visual stream (Zeki 1990).

Terminology in this domain is often confusing. Some authors refer to "color anomia" when describing the inability to name colors. However, this is better understood as an optic aphasia selective for color, not an agnosia: patients can match hues correctly even if they cannot name them.

Others use the term "color amnesia" (Varney 1982) to describe the loss of knowledge about typical object colors—e.g., not knowing the color of a lemon. True color agnosia (or central achromatopsia) is frequently associated with prosopagnosia.

Assessment must include perceptual, associative, and linguistic aspects of color processing. Tools include Ishihara's pseudoisochromatic plates (1932), the Farnsworth-Munsell Hue Test (Farnsworth 1943), and more recent computerized tools (e.g., Ghose et al. 2014).

Motion Agnosia (Akinetopsia; Cerebral Motion Blindness)

In akinetopsia, the perception of visual motion—speed and direction—is impaired. Form and color are usually unaffected. The prototypical lesion site is the bilateral occipito-temporo-parietal junction (Zihl et al. 1983). Some cases appear selective for biological motion, pointing to specific involvement of parietal or temporal regions.

Simultanagnosia and Balint Syndrome

Patients with simultanagnosia cannot form a coherent visual overview of a scene. They may identify individual elements, but fail to integrate them into a whole. Two forms are distinguished:

Ventral Simultanagnosia. Most often caused by left inferior occipital lesions, patients perceive that multiple objects are present but can only identify one at a time. They fail to integrate the parts into a meaningful whole. This contrasts with associative agnosia, in which patients see the whole scene but fail to recognize individual objects.

Dorsal Simultanagnosia. Caused by bilateral occipitoparietal lesions, these patients exhibit severely narrowed visual attention. They see the world piecemeal, describing isolated elements rather than integrated scenes. Ocular apraxia (inability to fix and shift gaze effectively) and optic ataxia (difficulty grasping visual targets) are often present. Remarkably, patients may perform better with eyes closed or when pointing to auditory rather than visual targets.

Balint Syndrome

Balint syndrome refers to the triad of:

* Dorsal simultanagnosia
* Ocular apraxia
* Optic ataxia

The condition was first described by Rudolf Balint (1909) and later studied in more depth by Gordon Holmes (1918), leading to the alternative term Balint-Holmes syndrome.

A striking feature: handwriting is often illegible when the eyes are open, but becomes well-formed when written with the eyes closed—an inversion of what one might expect (see Chapter 12, pp. 286).

What Other Types of (Non-Agnosic) Visual Disturbances Should Neuropsychologists Know?

Not all visual disturbances are agnosic in nature. For example, if you suffer an accident and lose both eyes, you are blind—but not agnosic for any aspect of visual perception. And if you see tiny fairies dancing around, waving and smiling, you are certainly seeing things—but what you are experiencing are hallucinations, not agnosias.

These types of productive visual phenomena—including distortions, illusions, and hallucinations—do not reflect a failure to recognize what is seen. Rather, they involve false or altered perceptual experiences and must be clearly distinguished from agnosias, which are defined by a loss of meaning in the face of otherwise normal sensory input.

Cortical Blindness

Cortical blindness is characterized by the ability to detect only gross differences in brightness. It results from a bilateral lesion of the occipital cortex, typically involving near-total destruction of primary visual areas.

If only one hemisphere is affected, the condition is restricted to the contralateral hemifield—a relatively common presentation known as hemianopia. More circumscribed damage can lead to a scotoma, and if a full quadrant of the visual field is affected, the result is called quadrantanopia.

Importantly, patients are not always aware of these blind spots. This is why systematic assessment by perimetric testing—typically carried out by neurologists or optometrists—is essential.

Nonetheless, neuropsychologists would benefit from learning basic finger perimetry, which can help detect the presence of a hemianopia or quadrantanopia during clinical evaluation.

? Are neuropsychologists allowed to assess a patient's visual field with finger perimetry?

Why should they not? In many cases, a patient's chart lacks clear or current information about visual field integrity. Sometimes the available data are weeks old, which justifies a re-test. For the diagnosis of certain agnosias—and especially of alexia—it is crucial to know whether the visual field is intact or impaired.

The procedure, often referred to as confrontation visual field testing, is simple and quick. Have the patient sit directly opposite you and ask them to close one eye. Then test monocular field perception by showing one, two, or three fingers in each of the four quadrants, using a pseudorandom order. Occasionally, show no fingers at all, to control for guessing.

In a dynamic variant, bring your hand in from the periphery toward the center in each quadrant and ask the patient to report as soon as they see your hand. Wiggling your fingers may help with detection.

This is a crude screening method, but it often suffices to detect a left or right hemianopia, or an upper/lower quadrantanopia. If findings are ambiguous or if formal documentation is needed, you can refer the patient to a neuro-ophthalmologist or optometrist for professional perimetric testing.

Blindsight

The term "blindsight" was coined in a seminal paper by Weiskrantz et al. (1974) to describe the paradoxical observation that some patients with cortical blindness can correctly guess properties of visual stimuli they insist they do not see. When asked to guess the color, direction of motion, or shape of an unseen stimulus, these patients often perform well above chance.

This dissociation between vision with and without awareness had been known earlier under the term Riddoch's phenomenon—named after George Riddoch, who observed that cortically blind patients could detect motion of a stimulus they could not consciously perceive (Riddoch 1917).

The functional neuroanatomy of blindsight demonstrates that multiple visual pathways project from the retina to the brain. While conscious vision typically relies on signals traveling from the retina → lateral geniculate nucleus (LGN) → primary visual cortex (V1), alternative projections exist. These include pathways from the retina to the superior colliculus, a midbrain structure especially tuned to motion and capable of driving behavior outside of awareness.

Blindsight has been studied extensively in both humans and nonhuman primates and has inspired wide-ranging debates about visual awareness and, more broadly, the neural correlates of consciousness (Danckert et al. 2021). Related findings come from hemianopic patients who, despite being unaware of stimuli in their blind field, show affective responses—such as to emotional faces or threatening objects—suggesting a non-conscious visual-limbic pathway from the retina via the superior colliculus to the amygdala (Morris et al. 2001). Experimental simulations of blindsight have also been conducted in healthy volunteers—both through behavioral paradigms (Schärli et al. 2003) and transcranial magnetic stimulation (TMS) applied to V1 (Railo and Hurme 2021).

"Anton's Syndrome", i.e., Denial of Blindness

Some patients with cortical blindness deny their blindness as adamantly as hemiplegic patients deny their paralysis. This anosognosia is referred to as Anton's syndrome, after the Austrian neurologist Gabriel Anton, who described it in 1889. (Interestingly, earlier cases of denial of cortical blindness had already been documented; see Adeyemo and Nesathurai 2013. For noneponymous alternatives, see the table in the Appendix to Chapter 4.) Denial is most commonly observed in the early stages of blindness. Patients often construct elaborate rationalizations for bumping into objects and may synaesthetically integrate auditory information into their imagined visual experience. For instance, when presented with a pen, a patient might report seeing "keys" if the examiner simultaneously jingles a keychain.

Optic Ataxia

Optic ataxia is a visuomotor deficit characterized by a marked inability to reach and grasp visual targets under visual guidance. A typical everyday example: the patient fails to pick up pieces of food from a plate directly in front of them. After repeated failures, they might attempt the same task with eyes closed—and, to their surprise, succeed, as proprioceptive guidance remains intact.

The disorder reflects damage to dorsal stream fibers, particularly in the superior parietal lobule, the intraparietal sulcus, or the occipito-parietal sulcus. Clinical patterns vary:

* Right hemisphere lesions → misreaching into the left visual field by both hands ("field effect")
* Left hemisphere lesions → misreaching into either field, especially by the right hand ("hand effect")
* Bilateral lesions → misreaching into both hemifields with both hands

Optic ataxia is one of the three core symptoms of Balint syndrome (see above), alongside simultanagnosia and ocular apraxia.

Inverted Vision and Mirror Vision

Inverted Vision

Seeing the world upside down—a phenomenon also called the "room-tilt illusion"—has been reported in patients with lesions to the vestibulocerebellar system, brainstem, or parietal lobes. A classic review is provided by Solms et al. (1988), who also describe a frontal lobe case: the patient copied the Rey-Osterrieth figure upright but reproduced it inverted from memory. Most inversions are 180°, but cases of 90° rotations and oblique distortions have also been documented. The experience is rarely chronic, often fluctuating, and may represent an early symptom of visual disturbance. Interestingly, when the initial direction of torsion is reported, it is almost always clockwise.

In healthy individuals, inverted vision has been observed during parabolic flight (Graybiel et al. 1967) or experienced spontaneously during spaceflight. It was famously investigated by American psychologist George M. Stratton (1896), who wore inverting goggles for extended periods. Remarkably, an earlier version of this experiment was conducted by Italian philosopher Roberto Ardigò, who anticipated Stratton's results by several decades (Giora and Büttemeyer 2020).

The topic of inversion continues to intrigue psychologists:

* Contrary to popular belief, retinal images are not "inverted" in any meaningful perceptual sense (Baden and Nilsson 2022). The brain receives neural firing patterns, not "pictures" to rotate.
* Partial scene inversion—such as face or body inversion—is more disorienting than global inversion. This is well studied in face recognition research (Gerlach et al. 2023).
* The Thatcher illusion, in which facial features are inverted within an upright face, remains one of the most striking visual phenomena bridging science and art (Lobmaier and Mast 2007).

Mirror Vision

What dental students must learn through extensive training (Neumann 1988) may, in rare cases, occur spontaneously in patients with visual system damage: they see the world mirror-reversed.

Mirror vision is extremely rare—even compared to vertical inversion. It typically co-occurs with mirror writing and mirror reading (Heilman et al. 1980). This can be assessed easily: have the patient read a short passage once in normal print and once in mirror font. In mirror vision, the reversed text is read fluently, while the normal print appears distorted and difficult to read.

A literary description of postictal mirror vision is discussed by Andermatt (2007). A notable clinical case—mirror vision following cerebral hypoxia—is described by Pflugshaupt et al. (2007). This study is exemplary not only for its clinical insight but also for its use of mental chronometry and eye movement tracking to quantify symptoms—a paradigm paper for neuropsychological assessment methodology.

Figure 6: Place a mirror perpendicular to this page and trace the star, starting from the point marked *S* and proceeding clockwise back to *S*. Your hand should be hidden so that you can see its movement only in the mirror. The patient described by Pflugshaupt et al. (2007) would likely have found this task easier than you do (see text).

? The perceptual aberrations just mentioned—mirror vision and inverted vision—remind me of the old question: Why do mirrors reverse left and right, but not up and down? Has this ever been discussed in the neuropsychology literature?

Your association is not far-fetched! The question "Why do mirrors reverse left and right but not up and down?" has long been debated in philosophy. When it appeared—printed in mirror script—as the title of a 1974 article in the Journal of Philosophy (Block 1974), the author did not frame the issue in neuropsychological terms. His discussion focused on the linguistic meaning of "to reverse" and the fact that left and right are arbitrarily assigned in a bilaterally symmetric organism, whereas head/foot and front/back are not.

Over the following decades, many cognitive scientists have contributed to this puzzling topic. Among the most neuropsychologically relevant contributions is the one by Michael Corballis (2000), who clarified several key points:

Mirrors reverse along the axis perpendicular to their surface—not along a left–right or top–bottom axis.

To mentally simulate a "mirror reversal," you would have to imagine your front side pushed backwards, so that your back oozes forward (Corballis 2000). Only then would left and right remain in place.

A bilateral observer compares herself to her mirror reflection by aligning top-down, front-back, and left-right axes. But the left-right axis is the least anatomically constrained, determined only after top–down and front–back are anchored—hence its apparent "reversal."

Corballis's account is neuropsychologically valuable because it touches on how the brain encodes egocentric spatial axes. It helps explain:

Why mental rotation tasks involving left-right distinctions are particularly difficult, why left-sided neglect often applies primarily to front space (Viaud-Delmon et al. 2007; see Chapter 9, p. 197), and perhaps even why inverted vision is more frequently reported than mirror vision. The top-bottom and front-back axes are neurologically privileged—so when disturbances of spatial orientation occur, they more often manifest as vertical or depth inversions than as true left–right reversals.

Illusions, Hallucinations, and Other "Productive Visual Symptoms"

The variety of visual disturbances that may arise following focal brain damage, neurodegenerative disease, seizure disorders, migraine, or even in times of perfect health is too extensive to cover fully here.

As a rule of thumb, patients will spontaneously report visual anomalies as long as they believe the cause lies in the eye. They become much more hesitant to describe complex or multimodal visual experiences—especially if they fear such reports might lead to a referral to a psychiatrist.

In other words, symptoms like "seeing stars" are often mentioned without prompting, while "seeing things" requires active inquiry. A highly useful reference work is Blom's *Dictionary of Hallucinations* (2023).

Photopsia

Photopsia refers to the experience of phosphenes—simple flashes of light, either colorless, monochromatic, or multicolored. These may indicate:

* Retinal pathology, such as detachment or macular degeneration
* A blow to the head (the origin of the classic "stars" in comic strips)
* Optic neuritis, often an early sign of multiple sclerosis
* Seizure or migraine activity

An important clinical question: Are the phosphenes seen in one eye or both?

? **Can I induce photopsia in myself?**

Absolutely. Nothing could be easier. Just rub your eyes and observe the emergence of lights and colors. Pressure on the eyeballs mechanically stimulates the photoreceptors—rods and cones—in the retina, causing them to fire. This "fire in the eye" is a fascinating phenomenon to experience (watch for the drift of pineapple-like slices after rubbing).

On the downside, it may have misled ancient philosophers into fundamentally flawed theories of visual perception—see the discussion on extramission (p. 132).

Fortification Patterns

Many people suffering from migraine have made drawings of the patterns they "see" during the aura of an attack. Typically, these are zigzags reminiscent of ancient fortified town walls. Their individual form, formation, and time course of appearance allow insights into the type of migraine a person suffers.

Kaleidoscopic Vision

This is a transient distortion that makes you see the world as broken up in pieces, mostly highly colored, sometimes moving around, as if you peered through a kaleidoscope. It can be monocular or binocular (which always has to be inquired about with this and other distortions). It is almost always a migrainous phenomenon (with or without accompanying headache) but can also occur as a symptom of a transient ischaemic attack (TIA).

Visual Allaesthesia

Probably the first description is from the German neurologist Ernst Beyer (1895), who suffered from migraine and experienced a (visual) replacement of objects in extrapersonal space. For instance, what is located on the right side of space is transposed to the left visual field and vice versa. Tactile allaesthesia (also "allochiria") is observed in cases of neglect and means a transposition of touch from the affected left side to the spared right side (Halligan et al. 1992). Meador et al. (1991) provide a valuable overview of these phenomena appreciated in past centuries. They are rarely reported in modern times, probably because patients are rarely inquired about.

Palinopsia

This is a form of visual perseveration. An object looked at in one moment reappears later in a newly observed scenery. Sometimes it is multiplied ("poliopsy," or more specifically "poliopic palinopsia"). The reappearing object may smoothly match the underlying real object, as in the ictal palinopsia of a knee joint radiograph "melting" with a table leg (Smith et al. 2003). Nine months after a stroke in the territory of the right posterior cerebral artery, a 62-year-old lady took a bus ride. She briefly looked at a poodle sitting on the lap of a fellow passenger (Figure 7a). When she looked away, she noticed that all passengers had now poodle faces (Figure 7b). After getting out of the bus, all the people had poodle faces. The experience lasted for about half an hour, and the poodle's face was exclusively seen on other people's faces (not on other body parts or non-living objects).

Micropsia and Macropsia

These distortions involve altered size perception (not to be confused with changes in distance perception). They often occur as part of Alice in Wonderland syndrome (Blom 2023) or during seizures. In one seizure patient, time was judged to run faster

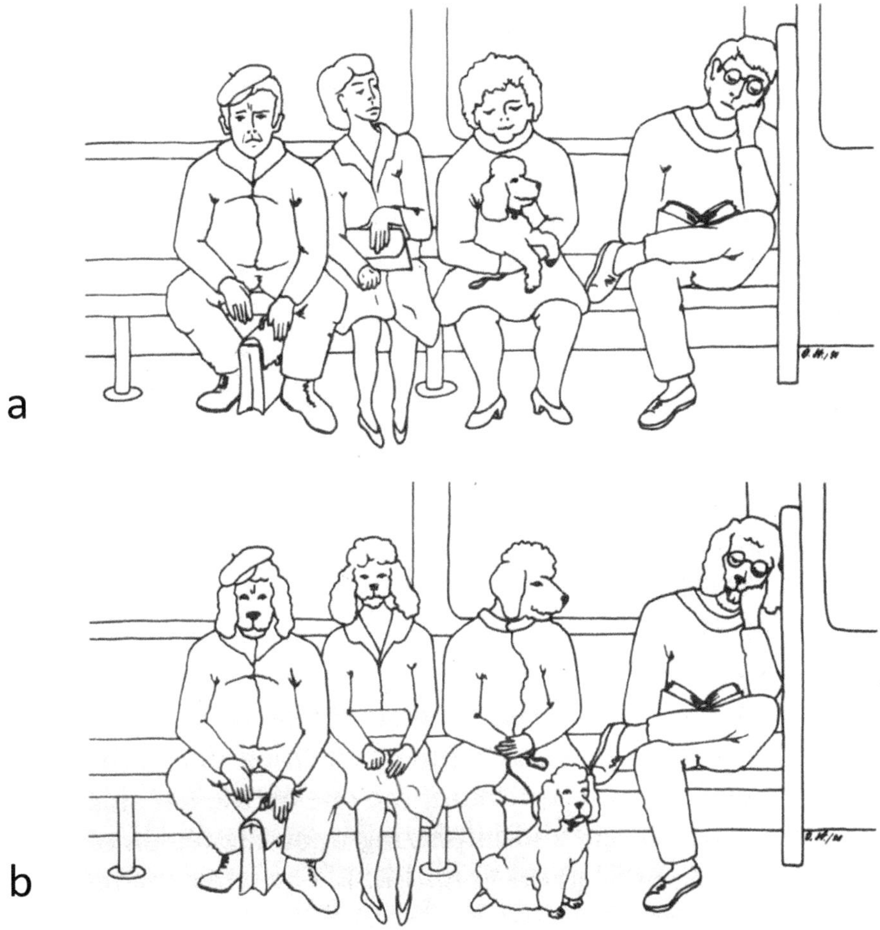

Figure 7: Palinopsia in a female patient with prosopagnosia. See text for explanation. Drawings by D. Starke. Reprinted, with permission, from Grüsser and Landis 1991, Figure 14.6.

during micropsia (Loetscher and Brugger unpublished); similar time dilation is hypothesized for macropsia. These links highlight the intimate relationship between spatial and temporal perception (Walsh 2003; Zäch and Brugger 2008).

Teleopsia and Pelopsia

Here, objects appear abnormally far away (teleopsia) or unusually close (pelopsia), reflecting impaired distance perception. Such symptoms may arise from parieto-temporal lesions (left or right), seizure activity, or migraine.

Metamorphopsia

Metamorphopsia is the visual distortion of objects or faces. It may be due to retinal damage (e.g., macular degeneration—Simunovic 2015) or to central causes (Critchley 1949). If limited to faces, the term is prosopometamorphopsia (Blom et al. 2021).

Paraprosopia

This refers to face-specific distortions in which normal faces abruptly appear monstrous or threatening (e.g., demonic or vampire-like). It is most commonly observed in schizophrenia, especially in children and adolescents (Dewdney 1973). It differs from prosopometamorphopsia in its intense emotional valence and its sudden onset. One patient with paraprosopia had fully normal face perception on neuropsychological testing (Kemp et al. 2003).

Pablo Picasso Syndrome

A migraine-associated form of prosopometamorphopsia in which facial features appear misaligned—ears on top of heads, mouths on foreheads, and cubist displacements. The term was coined after a patient's description (Suri and Ali 2020). There is no evidence that Picasso himself suffered from migraine (Lisotto et al. 2015; Haan and Ferrari 2011).

Hyperfamiliarity for Faces

This syndrome involves the false conviction of recognizing unfamiliar faces. Patients may insist on prior meetings with strangers—"We sat together last weekend in the Beevers, don't you remember?"—even though no such encounter occurred. Unlike delusional misidentification, patients may recognize their error and feel embarrassed. Performance on face perception tasks is normal (Vuilleumier et al. 2003), and the phenomenon is usually associated with left temporal lobe lesions and seizure disorders. There are also auditory variants (Mole et al. 2021), but the same individual rarely experiences both.

Delusional Misidentification Syndromes (DMS)

DMS involves persistent, monothematic delusions about identity. Two principal forms:

* Capgras delusion: a familiar person has been replaced by an impostor
* Fregoli delusion: multiple people are believed to be the same person in disguise

These syndromes often follow parietal or frontal lobe damage. Other variants exist (Christodoulou et al. 2009). DMS are rooted in psychiatry but are mirrored in neurology through phenomena such as reduplicative paramnesia or *déjà vu*.

? Is *déjà vu* a perceptual phenomenon?

No. It is an illusion of memory, occasionally also labelled reduplicative paramnesia. It involves the conviction of having experienced a situation before, down to the most minute detail. This conviction arises from an overwhelming feeling of familiarity with what is being experienced at the moment. The term *déjà vu* (French for "already seen") was popularized by the English neurologist John Hughlings Jackson (1835–1911). Its wording may be responsible for the common misassumption that it is a perceptual phenomenon. A more appropriate label might be *déjà vécu* ("already lived through"), which is often used synonymously, although some authors reserve it for more prolonged or pathological variants.

Neuroanatomically, *déjà vu* and *déjà vécu* are associated with activity in networks involving the medial and lateral aspects of the temporal lobes (Bancaud et al. 1994). Incidence is increased in people with temporal lobe epilepsy (TLE), who may also report prescience—a distinct experience involving the feeling of being able to predict what will happen in the immediate future (Sadler and Rahey 2004).

The opposite of *déjà vu* is *jamais vu*—a lack of familiarity in an objectively familiar situation. *Jamais vu* is rare in the general population. It occurs in TLE as well, but also in conditions involving depersonalization and derealization. For an experimental attempt to elicit jamais vu in healthy individuals, see Moulin et al. (2021).

Pareidolias

Pareidolia is the illusion of perceiving structured meaning in random stimuli (clouds, stains, and foliage). Though not face-specific, most clinical research focuses on face pareidolia due to the biological salience of faces (Abad and Zwaan 2025). Even 10-month-old infants and nonhuman primates exhibit it. Normed pareidolia tests exist (Mamiya et al. 2016). Gosselin and Schyns (2003) developed a method to reveal the mental templates driving pareidolic perception. See Chapter 12 (Figure 8) for a pictorial case report.

Eidetic Imagery, Aphantasia, and Synesthesia

These illustrate the close ties between imagery and perception. Eidetic imagery is the vivid and detailed visual recall of past perception. In a classic experiment, Stromeyer and Psotka (1970) demonstrated photographic recall using random-dot stereograms presented days apart. Eidetic imagery borders on hyperphantasia. This is the opposite of aphantasia (Keogh et al. 2021)—the inability to voluntarily generate visual imagery (Galton 1880; Zeman et al. 2025). Eidetic imagery and aphantasia are rarely encountered in clinical neuropsychology, but their existence confirms the wide range of visual mental imagery.

Synesthesia involves cross-modal perception (e.g., colored audition). While traditionally regarded as rare, it is now recognized in a substantial portion of the neurotypical population. The most common type is grapheme-color synesthesia, in which digits or weekdays evoke specific colors. Though the experience appears unidirectional at the phenomenal level (e.g., 7 is associated with blue but blue does not evoke the concept of "sevenness"), neural bidirectionality can be demonstrated experimentally (Knoch et al. 2005).

The value of simple behavioral or observational testing over high-tech approaches is elegantly demonstrated in a method for verifying claims of synesthesia—particularly relevant given that some individuals falsely claim to have it because it sounds intriguing. Consider a person who reports seeing the digit 2 in one color and the digit 5 in another. This person is shown the pattern below, depicted on the left.

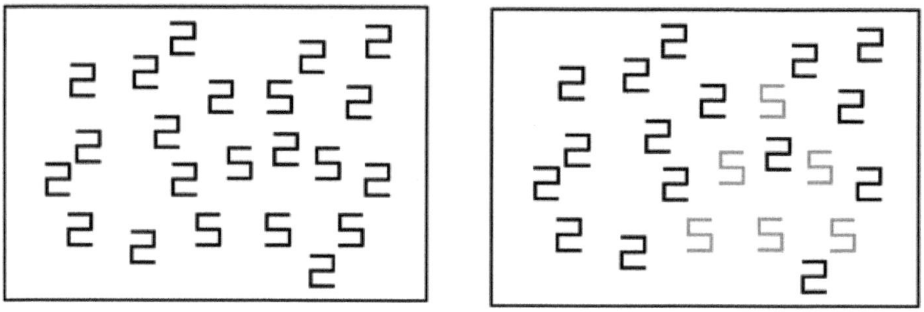

If the synesthesia is not genuine, identifying the triangular arrangement formed by the 5s will be difficult. If the synesthesia is real, however, the shape will automatically "pop out," just as it does for non-synesthetes in the right panel (where the distinction is marked in visible contrast). The figure is adapted from Ramachandran and Hubbard (2001).

Visual Illusions and Hallucinations

While optical illusions involve stimulus-driven misperceptions, visual illusions more broadly reveal how the brain interprets ambiguous input. An illusion is a misinterpretation of a real stimulus, whereas a hallucination occurs without any external input.

Visual hallucinations may accompany posterior strokes, Parkinson's disease, Lewy body dementia, epilepsy, and more. Notably, the occipital cortex is not always involved; subcortical lesions (e.g., in peduncular hallucinosis) can produce rich visual experiences (van Bogaert 1927; Taylor et al. 2005).

Extracampine hallucinations (Bleuler 1903) involve perceptions beyond the visual field—for example, "seeing" behind one's head. Unfortunately, this term has been misapplied in modern literature to describe the sense of an invisible presence (see Chapter 10, p. 228).

What Types of Non-Visual Agnosia Are There?

We briefly introduce auditory and tactile forms of agnosia. As in the visual modality, it is useful to distinguish apperceptive from associative variants. In apperceptive agnosia, primarily low-level sensory analysis is impaired, whereas associative agnosia affects higher-level integrative or semantic processes.

Auditory Agnosia

Auditory agnosia is the inability to make sense of *sound information* despite intact hearing. The auditory modality comprises different categories of sound, which must be differentiated diagnostically:

* Agnosia for environmental sounds refers to the failure to recognize common nonspeech noises such as thunder, a cricket's chirping, keys jingling, or mechanical sounds (e.g., cars, motorcycles, vacuum cleaners). Diagnosis requires confrontation with recorded environmental stimuli. In this form, speech perception is intact.
* Amusia is the inability to process musical elements such as melody and intonation, which renders music unrecognizable or unpleasant. As with aphasia, we may distinguish receptive, expressive, and amnestic variants. There are also forms analogous to alexia and agraphia, affecting musical notation (Kawamura and Miller 2019). Lesions may be unilateral or bilateral, and lesion laterality may differ between professional musicians and non-musicians. Congenital amusia is also described (Williamson and Stewart 2013).
* Auditory verbal agnosia—formerly termed word deafness—refers to the inability to comprehend spoken language, despite preserved reading, writing, and recognition of nonverbal sounds. A striking dissociation may be observed between oral and written command comprehension. The condition may show familial clustering (Kim et al. 2023b).
* Phonagnosia is the inability to recognize familiar voices. It is the auditory analogue of prosopagnosia. Notably, voice recognition is often a compensatory strength in patients with face recognition deficits, making this dissociation particularly interesting.

Tactile Agnosia

Tactile agnosia is the failure to recognize objects by touch, despite normal somatosensory function, intact vision, and preserved auditory object recognition. To qualify as agnosia, the deficit must occur despite standard elementary tactile processing—including 2-point discrimination, proprioception, and hylognosia (i.e., the ability to discern material properties such as texture, softness, or weight).

* An apperceptive tactile agnosia is often referred to as astereognosia—a failure to integrate tactile features into a coherent 3D percept. The patient cannot identify objects by touch, even when exploring them thoroughly.
* An associative tactile agnosia occurs when the patient can integrate the sensory features of the object and even distinguish between different items through touch or draw them adequately, but cannot name or verbally identify the object (Endo et al. 1992; Reed et al. 1996).

Lesions causing tactile agnosia typically involve the parietal somatosensory association cortex, but may also affect the parietal operculum or posterior insula. Roughly speaking, right hemisphere lesions more often result in apperceptive forms, while left hemisphere lesions are associated with associative tactile agnosia.

In What Other Types of Recognition Disorders Does the Term "Agnosia" Pop Up?

As mentioned above (p. 130), the term agnosia is also used in the context of multisensory or amodal recognition disorders. We briefly touch upon some of these forms, fully aware that each of them would merit a monograph of its own.

Topographagnosia

This loss of "whereness" is essentially a disorder of familiarity (Landis et al. 1986). Also referred to as environmental agnosia, it denotes the inability to recognize familiar places or to find one's way to and from them. While place recognition typically relies on visual input, patients with topographagnosia do not suffer from primary visual agnosias. However, there is a statistical association with prosopagnosia and with agnosia for noncanonical object perspectives. Most cases occur after occipitotemporal infarctions in the territory of the right posterior cerebral artery (Landis et al. 1988). Tseng and Tzeng (2013) proposed the term dromosagnosia to describe the specific loss of directional sense while driving.

Autotopagnosia

Autotopagnosia is the inability to localize body parts on one's own body. Patients may be able to name the function of a knee, describe its appearance, and even place it in a situational or cultural context (e.g., "kneeling for prayer"), but are unable to point to it. In pure cases, the patient can still point correctly to the examiner's knee(s) (Tamura et al. 2016). When accompanied by heterotopagnosia (errors in pointing to another person's body parts), detailed testing shows that autotopagnosia is more than a disturbance of the body schema (Sirigu et al. 1991). In neurodegenerative contexts, the symptom may evolve within a broader progressive apraxia. Lesions typically involve the parietal lobe of the hemisphere dominant for expressive language.

Finger Agnosia

The term finger agnosia (Gerstmann 1940) is misleading: the disorder is not an agnosia in the strict sense. Patients recognize a finger as a finger, both on their own and on another person's hand. However, when tested systematically (Rusconi et al. 2014), they are unable to:

* Differentiate between the fingers of one or both hands (typically bilaterally after left parietal lesions; more rarely unilaterally after right parietal lesions)
* Name a finger or wiggle the correct finger when one is named
* Wiggle the corresponding finger on the other hand when one is touched

Finger agnosia may also affect the toes (Tucha et al. 1997). It is one component of the classical Gerstmann syndrome, which also includes agraphia, acalculia, and left–right disorientation (Gerstmann 1930, 1940).

? **Why should neuropsychologists know "Gerstmann's syndrome"?**

In 1930, the Austrian neurologist Josef Gerstmann described a combination of four neuropsychological deficits following left parietal damage (English version in Gerstmann 1940): finger agnosia, acalculia, left–right disorientation, and agraphia. He proposed that this tetrad had syndromic character—clinically relevant for lesion localization and theoretically important for understanding the functional organization of the brain.

A sometimes fierce debate followed and has continued for nearly a century. Rusconi et al. (2010) provide both a comprehensive historical overview and a modern reinterpretation, framing Gerstmann's syndrome as a possible disconnection syndrome, supported by combined neuroimaging and behavioral data from neurological patients and healthy individuals (Rusconi et al. 2009). Especially valuable for clinicians is this author team's rigorous scheme for testing finger gnosis (Rusconi et al. 2014).

Emotion Agnosia; Affective Agnosia

Lane et al. (2015) introduced the term affective agnosia to describe a general blunting of emotional awareness and an inability to recognize emotionally salient material—a concept closely related to but broader than the older term alexithymia. Lesions to the prefrontal lobes or insular cortex may produce such deficits (Hogeveen and Grafman 2021).

Anosognosia

Previously referred to as noso-agnosia by some authors, the term anosognosia was coined by Babinski (1914) to describe the nonrecognition of one's own illness or neurocognitive deficit. Distinct lines of research have developed around different manifestations, including:

* Anosognosia for hemiplegia
* Anosognosia for cortical blindness (Anton's syndrome, see p. 152)
* Anosognosia for memory disorder (especially in the context of dementia)
* Anosognosia for mental illness (Streiff 2023; Lehrer and Lorenz 2014)

The last two forms are particularly interesting because they cannot be easily explained by the same mechanisms proposed for anosognosia for hemiplegia (Berti et al. 2005).

By linking corporeal awareness with the recognition of one's own self, anosognosia serves as a bridge between neurological and psychiatric models of self-representation. This form of non-visual agnosia is discussed in more detail in Chapter 10 (p. 226).

Where Can We Read More?

Blom, J.D. A Dictionary of Hallucinations (2nd ed.). Springer: New York, 2023.
Grüsser, O.-J. and Landis, T. Visual agnosia and other disturbances of visual perception and cognition. Vol. 12 of Cronly-Dillon, J. (ed.). Vision and Visual Dysfunction. Macmillan: London, 1991.
Hoffman, D.D. Visual Intelligence. How We Create What We See. WW Norton & Co.: New York, 1998.
Introduction to Sensation and Perception Copyright © 2022 by Students of PSY 3031 and Edited by Dr. Cheryl Olman is licensed under a Creative Commons Attribution 4.0 International License, except where otherwise noted.

CHAPTER 8

What Do The Frontal Lobes Do?[1]

1000

In the primate brain, the frontal lobes are the most evolutionarily recent addition. In humans, they are home to the highest intellectual functions: abstract thought and reasoning, creativity, and insight. On the affective and emotional side, they allow us to express and regulate emotions, interact socially with competence, and develop a personal identity—shaped by varying degrees of empathy and moral awareness. In close cooperation with the temporal and parietal lobes, the frontal lobes also support the development of religious feelings, with all their benefits for social bonding—and all their risks of self-deception and irrational belief.

The primary motor cortex and premotor cortex are distinct, bilaterally symmetrical regions of the frontal lobes, located just anterior to the central sulcus. Broca's area is another well-circumscribed region of frontal cortex; anatomically, it corresponds to the inferior frontal gyrus—or, in Brodmann's terms (see Chapter 2, p. 15), to area 44. This chapter does not include a discussion of speech and language (see Chapter 4). Nor does it address the specific functions of the motor strip and premotor areas,[2] even though the borders between "motor" and "cognitive" are blurred—and the frontier between bodily motion and abstract thought is, to a large extent, located in the frontal lobes. The field of embodied cognition deals with these borderlands and may well deserve its own chapter in a neuropsychological textbook.

In what follows, we focus on the part of the frontal cortex that is exceptionally large in humans relative to other primates—representing roughly one third of the entire cortical surface. This is the region Korbinian Brodmann, in 1912, referred to as the regio frontalis, now known as the prefrontal cortex.

The prefrontal cortex is typically divided into three parts: a ventromedial portion (which includes the anterior cingulate cortex), a dorsolateral portion, and an orbitofrontal portion. In a nutshell: the ventromedial/cingulate region is involved in regulating general arousal and motivation; the dorsolateral region supports cognitive

1 Try to screen out (a classic frontal lobe function...) the framed numbers from here to page 170. An explanation how to deal with them is found on page 177.

2 But see Chapter 12, p. 276, for the assessment of hand motor sequencing as a screening tool for frontal lobe functioning.

DOI: 10.1201/9781003384991-8

and behavioral flexibility and is most closely associated with the notion of "executive functions;" the orbitofrontal cortex contributes to inhibition, emotional control, and the maintenance of a coherent personality. In psychiatric contexts, dysfunction of the orbitofrontal cortex is often held responsible for various "pseudopsychopathic" states—including personality disorders and antisocial behavior (David 1992).

40

The behavioral spectrum observed in patients with prefrontal cortex damage (including damage to its subcortical connections) provides a preliminary answer to this chapter's central question: the frontal lobes are involved in virtually everything, and this is reflected in how patients behave during a neuropsychological examination. Examiners should be aware that patients necessarily recruit their prefrontal cortex:

* to remain wakeful, attentive, and motivated to cooperate
* to show insight into their own deficits and understand the need for examination
* to sustain attention over an extended period
* to shift flexibly from one task or problem to the next 1000
* to ignore irrelevant information
* to suppress perseverative actions and thoughts
* to avoid overinclusive thinking and confabulation
* to adhere to basic norms of social interaction

Thus, regardless of the domain being tested—language, memory, praxis, perception, numerical cognition, or corporeal awareness—the frontal lobes are always being co-examined, at least implicitly.

Because the frontal lobes are exceptionally well-connected to other brain regions, a clinical impression of "frontal" behavior does not necessarily imply damage to the prefrontal cortex itself. The vague label "frontal behavior" is often used by nurses and physicians to express an intuitive sense that something is wrong with the patient's frontal lobes. But there is no single, unified frontal lobe syndrome. So-called "frontal" symptoms can take on different forms.

Medial Prefrontal/Cingulate Syndrome (Keyword MOTIVATION)

The medial prefrontal cortex is densely interconnected with the anterior part of the cingulum (ACC in the inset of Figure 1). The behavioral syndrome associated with lesions in this region—including its projection fibers to and from the thalamus—is sometimes referred to as the anterior cingulate syndrome. Its hallmark feature is a marked reduction in arousal and a general lack of motivation. Self-initiated movements are rare, though actions can often still be executed when prompted. Bilateral lesions may result in akinetic mutism.

Patients often show limited insight into their situation. For instance, the urinary incontinence frequently observed in this syndrome may not bother the patient at all and has to be pointed out by others. The ventromedial prefrontal cortex also has direct

«MOTIVATION»	«COGNITION»	«IMPULSIVITY»

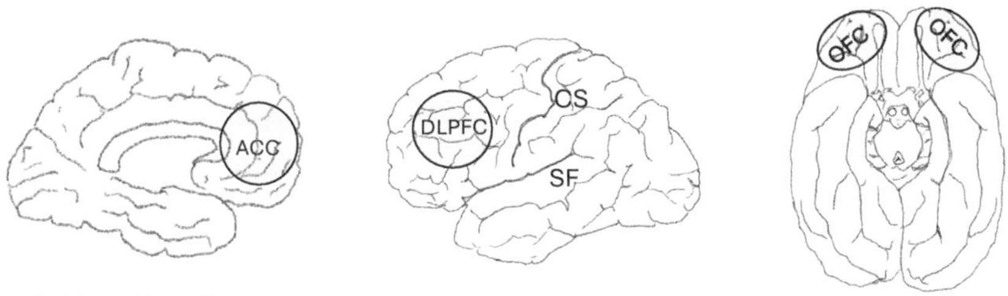

Medial frontal/anterior cingulate cortex	**Dorsolateral prefrontal cortex**	**Orbitofrontal cortex**
Ventral striatum (n. accumbens)	Dorsal striatum: caudate body	Dorsal striatum: caudate body
Dorsal pallidum (rostrolateral)	Dorsal pallidum (lateral)	Dorsal pallidum (medial)
Substantia nigra (rostrodorsal)	Substantia nigra (rostral)	Substantia nigra (rostromedial)
Thalamus (n. medialis dorsalis)	Thalamus (n. ventralis anterior, n. medialis dorsalis)	Thalamus (n. ventralis anterior, n. medialis dorsalis)

Impact of Loop Disruption (Clinical Picture)

Apathy (akinetic mutism when bilateral lesions)	Impaired problem solving, reduced set-shifting	Hyperactivity, disinhibition, poor impulse control
Indifference, lack of concern	Psychomotor slowing	Failure to screen out irrelevant stimuli, emotional lability, lack of empathy, inadaequate social behavior
No speech unless prompted, emotional blunting	Poor concept finding, perseverations	

Utility of Tests to Capture the Clinical Picture

Fair	Good	Poor

Figure 1: Three fronto-subcortical loops, associated clinical pictures and utility of standard neuropsychological tests to capture the symptomatology. Note the slight differences in projection sites and the vastly different clinical symptoms. ACC: anterior cingulate cortex; CS: central sulcus; DLPFC: dorsolateral prefrontal cortex; OFC: orbitofrontal cortex; SF: Sylvian fissure.

connections with medial temporal lobe structures, including the hippocampus—explaining the often pronounced memory deficits. These typically involve both anterograde and retrograde components and are marked by a notably flat learning curve. Qualitatively, the amnesia often includes spontaneous confabulations, resulting from a failure to correctly remember the temporal sequence of events (Schnider 1996; see Chapter 6). 30

Dorsolateral Syndrome (Keyword COGNITION)

The symptoms of dorsolateral prefrontal cortex damage are commonly summarized under the term dysexecutive syndrome. Patients struggle with problem-solving and abstract thinking. Working memory is often impaired, and there is increased susceptibility to interference. Performance on fluency tasks is quantitatively reduced and qualitatively marred by repetition and rule breaks (e.g., a patient may repeatedly offer words that do not begin with the required letter—such as producing knife in an S-word fluency task—even while knowing the rule). Perseverations in both verbal and nonverbal responses may occur. Flexibly switching between different task demands is often difficult. Motivational deficits frequently accompany the cognitive impairments.

Does It Matter Whether Damage is to the Left or the Right Hemisphere?

The above-mentioned fluency impairments may be asymmetrically pronounced, depending on lesion laterality. Verbal fluency is typically more impaired after left hemisphere damage, whereas design fluency tends to be more affected after right hemisphere damage. A similar asymmetry can appear in working memory: digit span backwards may be more impaired after left-sided lesions, while block tapping backwards may be more affected after right-sided ones. Performance on certain analytical reasoning tasks also appears more severely compromised after left-hemisphere damage (Goel et al. 2004).

1000

? **Talking about fluency tasks: is a patient's fluency performance not extremely dependent on vocabulary and years of schooling?**

Let us clarify this for readers unfamiliar with the standard verbal fluency task. It tests speeded word retrieval to a phonemic cue (i.e., a letter of the alphabet). Typical instructions go something like this: "Tell me as many words as you can think of that begin with a certain letter of the alphabet. You can say any kind of word—nouns, verbs, adjectives—but no proper nouns of people or places ('Nathan,' 'Norway,' or 'New York' would not count). Also avoid words with the same stem. For example, if the letter is R, 'rain' would count, but adding 'rainfall' and 'raincoat' would not give extra counts. Ready? Now tell me as many words as you can think of that begin with the letter..."

20

Your assumption is correct: verbal fluency performance depends significantly on educational background. That is exactly why normative values are stratified by education level.

Personally, I am a fan of verbal fluency tasks. But I often think—what a shame that their evaluation is usually limited to just three variables: total number of correct responses, repetitions, and "rule breaks" (such as naming a city, saying "snow" during a B-letter task, or saying "lioness" after already having said lion as an L-word).

There is much more to explore in verbal fluency—see Chapter 12, p. 299 f. If you are concerned about educational bias, consider choosing fluency tasks less dependent on schooling than letter fluency. Category fluency requires patients to name as many items as possible within a given semantic category. These can include animals, fruits, vegetables, supermarket items, actions, first names—you name it.[3]

Even less dependent on education is nonverbal fluency. There are several variants of design fluency tasks, one of which is introduced in Figure 2.

1000

The most conspicuous feature of Figure 2 is how the patient proceeded across the test sheet. In Western cultures, the typical strategy is to work line-by-line from

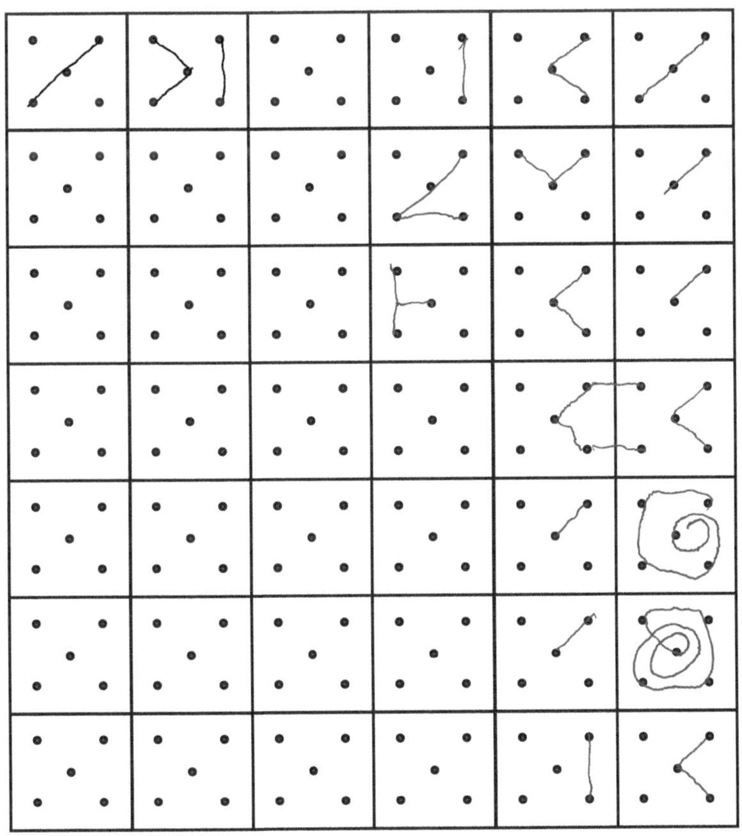

Figure 2: Test sheet from a design fluency task (the 5-Point Test by Regard et al. 1982), showing the productions of a fictitious patient. Instructions require the participant to draw as many different designs as possible by connecting two or more dots within each cell using only straight lines. Designs must not be repeated, lines must remain within the borders of each cell, and only straight connections are allowed. The two sample designs in the upper left corner are standard and presented during the instruction phase. Now make a guess: where is the patient's brain lesion most likely located? See text.

[3] As neuropsychologists, you might wish to name as many categories used in category fluency tasks as you are aware of—normative data would not so much have to consider schooling, but expertise in your field.

left to right. This patient, however, began in the upper right corner and proceeded vertically, column by column. At the end of the three-minute test (note: productions per minute, not shown here, are usually recorded), the left half of the sheet remained blank. This is a clear sign of left-sided neglect and points to dysfunction in the right hemisphere.

The high number of repetitions—seven in total, i.e., one more than the number of correct designs—suggests poor error monitoring and supports the assumption of a frontal lobe lesion. Note: the first repetition is pointed out to the patient; later repetitions are not (same rule applies to rule breaks).

10

Within the right frontal lobe, the orbitofrontal cortex is likely to be involved (Milner 1964). This is suggested by the frequent rule violations—four in total. Although the sheet in Figure 2 represents a fictitious case, the two spiral-like figures are real: they were produced by a patient with a right orbitofrontal lesion, who made more rule breaks than correct designs on the 5-Point Test. While drawing the first spiral, she remarked, "I know, you said 'straight lines,' but it is too late now, I'm afraid... ." The second spiral followed immediately, accompanied by the comment: "I know I should not do that"—yet the action went ahead regardless.

A detailed description of the 5-Point Test, including normative data, can be found in Lezak et al. (2004), Goebel et al. (2009), and Matias-Guiu et al. (2022).

?

Orbitofrontal Syndrome (Keyword IMPULSIVITY)

The prefix *orbito*-refers to the anatomical location of this part of the prefrontal cortex—just above the orbits that house the eyes. The orbitofrontal cortex (OFC) is interconnected with other prefrontal regions, particularly the ventromedial portion. Its functional specialization arises from its distinct pattern of connectivity (not shown in Figure 1), especially via the ventral striatum and its rich connections to the amygdala and hypothalamus (Du et al. 2020). These circuits support the OFC's role in fear reactions and in modulating aggressive behavior. More broadly, the OFC mediates hedonic experience and reward-related behavior.

The pure orbitofrontal syndrome is best characterized as a state of excess and emotional disinhibition, with reduced empathy but relatively preserved cognition. The spectrum of psychiatric and psychopathic disorders linked to orbitofrontal dysfunction in neuroimaging studies is so broad that it resists narrow classification—but it clearly illustrates that the OFC lies at the crossroads of neurology and psychiatry.

Does It Matter Whether Damage Is to the Left or the Right Hemisphere?

The OFC's connectivity to other prefrontal regions and to the anterior cingulate cortex is largely bilateral. However, the lateral OFC connects with the inferior frontal gyri, introducing a potentially significant asymmetry: the left OFC may play an as-yet unspecified role in language, while the right OFC appears crucial for impulse control (Du et al. 2020). It is this latter function—right orbitofrontal modulation of impulsivity—that becomes visible in the frequent rule-breaking behavior observed in the 5-Point Test (Figure 2).

One of the most famous cases in neuropsychology—the case of Phineas Gage (Macmillan 2000)—illustrates the dramatic and isolated personality changes that can follow traumatic damage to the orbitofrontal cortex (among other regions and disconnections), while leaving general cognitive abilities relatively intact. A tractographic reconstruction of Gage's lesion, published by Thiebaut de Schotten et al. (2015), is highly recommended reading.

Are "Executive Functions" a Hot Topic or Just a Cool Term?

Use of the term executive functions has exploded over the past two decades. A comprehensive overview of the term, its conceptual foundations, and the associated theoretical debates is provided in Bilder (2012). Another focus—entirely neglected in this chapter—is explored by Logue and Gould (2014), who examine the genetic and neuropharmacological bases of various meanings of executive functions. Ardila (2008) goes even further and speculates about the evolution of executive functions across primate and hominid lineages.

There is general agreement that executive functions refer to a set of brain-based mechanisms that allow individuals to regulate their own behavior, particularly when adapting to situational or environmental demands. Whether these functions are labeled hot (when self-regulation is emotionally charged) or cold (when applied to emotionally neutral tasks), executive functions are indeed both a hot topic in neuropsychology—and a cool term beyond it. In fact, it has become something of a hype term, used by neuropsychologists, psychotherapists, and kindergarten teachers alike.

The heterogeneous set of self-regulatory abilities typically associated with executive functioning is metaphorically illustrated in Figure 3 and includes:

* goal-setting and planning
* decision making
* sustaining and deploying attention
* using and updating working memory
* spotting hidden regularities and deriving rules
* ensuring cognitive flexibility
* switching between multiple task demands
* inhibiting maladaptive behavior; monitoring for errors
* concept finding and concept formation
* coordinating cognition and emotion
* maintaining the coherence of personality
* sending and interpreting social signals

Figure 3: A Swiss army knife metaphor to illustrate the diversity of executive functions. On top is a variant which is multipurpose to an unhelpful degree: it is bulky in offering too many specific functions. On the bottom is a more modest variant which, however, presents only few general-purpose instruments that lack specificity of application. See the text for elaboration.

One influential attempt to distill the multitude of so-called executive functions into a smaller set of relatively independent metafunctions comes from Miyake et al. (2000). Using a latent variable approach, the authors proposed that three key components underlie executive functioning:

* Shifting (between mental sets)
* Updating (and monitoring working memory contents)
* Inhibition (of prepotent responses)

These three functions demonstrate both the unity and the diversity of the executive construct: all three contributed to the test performance of their study population, yet the three latent factors—shifting, updating, and inhibition—were only loosely intercorrelated.

From the perspective of experimental neuropsychology, clarifying what exactly unites these executive components—and what accounts for their divergence—remains a significant challenge. One can only hope that the theoretical frameworks still to come will remain close to the real-world practice of clinical neuropsychology.

Until we arrive at a unifying theory of executive functioning, we run the risk of racking our patients' brains with ever more elaborate batteries of tests.

? Why not resort to test batteries when dealing with a domain as heterogenous such as executive functions?

Ouch! "Test battery"—what an ugly term! It evokes images of artillery units with mortar systems and rocket launchers aimed at an unseen enemy hiding behind a hill. The idea behind test batteries is, of course, understandable: no single test can capture a complex dysfunction in all its nuances. Batteries promise breadth where precision fails. They offer comfort to those who, unable to firmly grasp a patient's true difficulties, settle instead for psychometric success.

In military terms, a battery may hit the target—but only at the cost of firing off most of its ammunition into space.

My teacher, Marianne Regard—to whom neuropsychology owes several highly original clinical and research insights—abhorred the very concept of test batteries. To mock it, she kept a model car battery on her desk with the label: "my only test battery."

Does One Need Frontal Damage to Show Signs of "Frontal Behavior"?

Not at all. Even we "neurologically intact" individuals sometimes behave in ways that are strikingly frontal. And what's more, this "frontal behavior" may appear either as a trait or a state variable. Think of certain acquaintances who are so chatty that their verbal output seems almost unstoppable. This comes close to logorrhoea seen in patients with right frontal lesions and disinhibited language production. By contrast, others among your healthy friends rarely speak unless directly asked—and even then, it may take effort to draw words from them. This recalls mutism after damage to frontal-subcortical circuitry (Thompson and Thompson 2023) or cerebellocerebral interactions (Catsman-Berrevoets and Patay 2018).

In such examples, we are dealing with personality traits—stable positions along dimensions like talkativeness. But there are also situational states in which even the most disciplined person may exhibit frontal-like behavior: we become distractible, perseverative, short-sighted in reasoning, or unable to think "outside the box." In the following sections, we briefly consider three such features.

Distractibility

A core task of the prefrontal cortex is to keep attention focused on a relevant target—even when distractors are present. Frontal damage does not cause distractibility from scratch; rather, it amplifies the baseline distractibility that healthy individuals already show. In everyday life, we are often unaware of how distractible we are—because many distractors are so well camouflaged.

Interestingly, hypnosis can help reduce distractibility in healthy volunteers (Terhune and Brugger 2011). Even if you're not in a hypnotic state right now, you may experience the power of suggestion by briefly looking at the two images below (Figure 4), taken from an advertisement for dental floss. Just follow the instruction in the figure legend.

? Do you want to suggest that hypnosis is more than just humbug?

Absolutely! I am not endorsing the esoteric mumbo-jumbo that still surrounds hypnosis in some circles—but yes, I believe hypnosis has real value in both basic cognitive science and neuropsychological rehabilitation. A better

Figure 4: Two photographs taken from an advertisement for dental floss. Inspect them for a few seconds and report what is most conspicuous about each picture.

understanding of hypnotic suggestibility may help us understand functional neurological disorders. Hypnosis may even enhance the effectiveness of established therapies for attention deficits (Eide et al. 2024). For the link between hypnosis, executive functions, and hypnotic responsiveness, see Parris (2017) and the work by McGeown et al. (2015).

Many readers may have been distracted by the seemingly casual note that the images come from a dental floss advertisement. Your attention was likely drawn to the spot on the gentleman's teeth. That may have prevented you from noticing the real surprises: the third left arm of the woman in the top image—and the six fingers on her hand in the bottom one. Everyday distractibility can have surprising consequences, and clever advertisers know exactly how to exploit this for commercial purposes.

Figure 5a: Select one of the six cards. Focus your attention on that card; concentrate on the colors on it, the patterns, and the symmetry. Do not let yourself be distracted by the magician's eyes that observe you. Do this for approximately 20 seconds and then turn the page to look at Figure 5b.

Figure 5b: I have removed your card! See text for explanation.

Working Memory

Working memory has a limited capacity. We can usually hold just a few items in our mental workspace—and only for the short time we need to "work" with them. A driver might try to rehearse the street names just called out by a passerby; a chess player holds a few planned moves in mind; or someone mentally revisits the pattern on a dress they think they have seen before.

Some individuals with neurodevelopmental conditions but exceptional domain-specific talent (e.g., in savant syndrome) appear to retain far more information than the average person. But for most of us, the limit is close to what Miller (1956) famously called the magical number seven—give or take two items.

Speaking of magic: have you ever come across the card trick that proves a magician can read your thoughts—even while you are studying a neuropsychology textbook? Figure 5a presents six cards. Look at them. Let the magician's eyes meet yours. Then follow the instructions in the figure legend and move on to Figure 5b.

Go back to Figure 5a and you will discover the real trick: the magician did not just remove your chosen card—she removed all five of the others as well. She simply replaced them with similar cards. The king of clubs had been the king of spades; the queen of hearts was actually the queen of diamonds; and so on, with analogous exchanges between black and red suits.

While your attention was focused on the one card you had selected—and you were accordingly surprised to find it gone—you may also have tried to keep the other cards in memory: their colors, their suits, their approximate layout. But the exact pairings of suits and colors exceeded your working memory capacity. That is why you recognized the remaining cards as unchanged, even though none were the same.

Of course, this trick could just as well have appeared under the heading of distractibility. Frontal executive functions are diverse—but they overlap.

Perseveration

Perseveration refers to the inappropriate continuation of a behavior even when the situation clearly calls for a change. It comes in many forms. In the context of frontal executive dysfunction, these have been systematically described by Sandson and Albert (1984). Cohen and Dehaene (1998) proposed a clever method to track the temporal course of perseverative naming behavior.

Perseveration is not confined to motor or verbal output. Palinopsia and palinacusis refer to the lingering of visual and auditory impressions, respectively. Flashbulb memories can be interpreted as persistent mnestic intrusions—forms of memory perseveration—not unlike the earworms (intrusive musical fragments) many of us experience (see Sierra and Berrios 1999 for review).

In everyday conversation, too, people differ along a perseverance dimension. Some flip from one topic to the next with dizzying speed. Others linger on a subject long past the point of relevance. But perseveration is not just anecdotal—it can be experimentally triggered. One example is the following serial addition task, which invites all readers to perseverate—at least if you are among the 70% of the "normal" population who do (Brugger and Gardner 1994).

Simply add the numbers printed in the rectangles on pages 165 to 170, one after another. When you reach the question mark, say the result quickly—without rechecking or hesitating.

Do it now—before reading on.
(An explanation follows the next question.)

? Why can a behavior as devastating as perseveration remain an element of normal human behavior? Why is it not progressively eliminated over generations by Darwinian selection?

A very good question—and one that touches on a broader evolutionary puzzle. Like the type of perseveration demonstrated here, other seemingly non-adaptive behaviors persist as population-level traits. If we assume a genetic predisposition to perseverative tendencies, we can resolve this Darwinian paradox by positing that the maladaptive trait is linked to a closely related adaptive one.

In this case, the adaptive counterpart to perseveration could be perseverance—the capacity to pursue a goal or task even under adverse or highly demanding conditions. This trait, undeniably advantageous in many contexts, could offset the costs of carrying a gene set that also increases the risk of maladaptive perseveration.

In neuroscience and evolutionary psychology, the "Darwinian paradox" is often discussed in relation to various neuropsychiatric disorders—and also in the context of homosexuality, another trait with unclear direct reproductive advantage but potential indirect adaptive value.

Starting from one thousand, the task required three consecutive jumps in the thousands range: two thousand (and something), three thousand (and something), four thousand (and something). The "something" varied each time but always involved only tenths. No hundreds were involved at all.

This repeated focus on changing thousands primes the mind—and, as a result, the final step often triggers a perseverative error. The brain continues the previous pattern, even though the final calculation no longer justifies it.

Thinking In and Out of the Box

We tend to think in boxes—that is, context, conventions, and expectations shape our thoughts. These boxes provide structure: they make our thinking square, straight, and well-packed for collective shipment. Most of the time, that is useful.

But stepping outside of these boxes requires flexibility and creativity. Creativity is an umbrella term for forms of thinking that are novel, unexpected, or unconventional—but nonetheless appreciable by others who are used to more standard paths.

In neuropsychology, creative thought has been approached from multiple perspectives. In most of them, the frontal lobes—and particularly the prefrontal cortex—play a central role.

? So, you are saying that the frontal lobes house creativity?

Not quite. Creativity is not "housed" in any one part of the brain. Nor is it just one among many executive or integrative functions of the prefrontal cortex. But there is no doubt that the prefrontal cortex plays an important role in the flexible selection of problem-solving strategies—and that includes knowing when to think "in the box" and when to step out of it.

For a deeper dive into theories of hemispheric contributions to creativity, see Chapter 11 (p. 268). The popular notion of the "creative right hemisphere" may be oversimplified, but it is not entirely baseless. On the flip side, many efforts to dismiss right hemisphere involvement are themselves unscientific—they reject hypotheses too sweepingly and do not even attempt to falsify them.

This reluctance to falsify—and the tendency toward confirmation bias—does more than stall scientific progress. It also brings us to one final, and very human, example of everyday "frontal" behavior…

Reasoning Biases

Humans show a striking tendency to prefer confirmation over falsification of their hypotheses about the world. This confirmation bias is widespread and deeply ingrained. It affects scientific reasoning as much as it does everyday decisions.

Behavioral, lesion, and neuroimaging studies suggest that successful falsification in inductive reasoning tasks requires recruitment of the left prefrontal cortex—specifically when the task demands go beyond generic verbal reasoning (see Goel et al. 2004, for review).

One classic tool to reveal your personal bias toward confirmation—possibly even now, after reading this paragraph—is Wason's card selection task (Wason 1966).

Figure 6: Wason's 4-card problem. Shown are four cards taken from a deck in which each card has a letter on one side and a digit on the other. The rule to be tested is: **"If a card has a vowel on one side, then it has an even number on the other side."** Which card(s) must you turn over to determine whether this rule is true or false?

Read the following lines only after you have tried to solve the problem stated in the legend of Figure 6!

Statistically, most people correctly choose the card showing the *A*—since it allows a falsification of the conditional statement. If the reverse of the A is an odd number, the rule is violated.

However, many also choose the card with the 4, even though its reverse cannot provide evidence to disconfirm it. A consonant would merely be compatible with the rule, and a vowel would only serve to confirm it—for that card. Yet the task is not to verify examples, but to test the rule.

Only a minority selects the card showing the 7. But this card is crucial: if it has a vowel on the other side, the rule is violated. The 7-card thus offers another chance at falsification—and its neglect exemplifies our natural bias for confirmation.

Puzzles, conundrums, and magic tricks offer deep insights into the workings of the frontal lobes. Many neuropsychological tasks—tests of fluency, flexibility, planning, sequencing, and concept formation—are essentially structured puzzles. The difficulties even healthy individuals show when confronted with such challenges are educational (Macknik et al. 2008): they reveal both the diversity and unity of frontal lobe functions—and their susceptibility to failure.

Where Can We Read More?

Barkley, R.A. Executive Functions. What They Are, How They Work, and Why They Evolved. Taylor and Francis: New York, 2020.

Goldberg, E. The Executive Brain: Frontal Lobes and the Civilized Mind. Oxford University Press: Oxford, 2001.

Levin, B. and Craik, F. (eds.). Mind and the Frontal Lobes. Cognition, Behavior, and Brain Imaging. Oxford University Press: Oxford, 2011.

MacPherson, S.E. and Della Sala, S. (eds.). Handbook of Frontal Lobe Assessment. Oxford University Press: Oxford, 2015.

CHAPTER 9

What Is (Hemi)Spatial Neglect?

To begin with, let us clarify what hemispatial neglect[1] is not. It is not a motor deficit. It is not a sensory deficit. And it is not a disorder of visual perception.

A useful and widely cited definition was provided by the American neurologist Kenneth Heilman:

"Neglect is the failure to report, respond, or orient to stimuli presented to the side opposite a brain lesion, when this failure cannot be attributed to either sensory or motor deficits." (Heilman 1979, 2004)

Hemispatial neglect is thus best understood as a disorder of attention—specifically, a deficit in directing attention toward the contralesional side of space. The inattention may affect visual, somatosensory, or auditory stimuli—either separately or in combination. It may also extend to imagined space, such as when a patient fails to report stimuli located on one side of a remembered scene. Neglect can even concern stimuli that are only associated with a particular side, or that depend on the mere notion of "left" or "right."

Patients with clear signs of neglect tend to orient spontaneously toward the ipsilesional side of space (e.g., to the right, after a right hemisphere lesion) and fail to explore the contralesional hemispace. They may apply makeup or shave only one half of the face, eat only from one side of the plate, write exclusively along one margin of the page, or reproduce only one side of an object when copying or drawing from memory. Not all signs will be present in every patient. Instead, it is the pattern of signs that provides diagnostic insight into the type of neglect and the likely site of the lesion.

Describing that pattern is one of the most rewarding challenges for a neuropsychologist. In fact, the syndrome of neglect illustrates more clearly than most conditions the need for a multiplicity of assessment tools. I deliberately use the word "task," not test—see the question below and Chapter 12. Because neglect symptoms fluctuate and may spontaneously recover, the clinical picture must be updated repeatedly. The portrait you sketch today may need to be redrawn tomorrow.

[1] The terms "(hemi)spatial neglect," "hemineglect" or simply "neglect" are used here interchangeably and synonymous with the term inattention to one side of sapce.

DOI: 10.1201/9781003384991-9

? How come that many textbooks on clinical neuropsychology neglect the syndrome of hemispatial neglect?

I think a major reason is the lack of proper "neglect tests" discussed in this chapter. One popular neuropsychology textbook—almost 1,000 pages long and organized by syndrome—does not mention neglect at all in its index. Also absent are associated terms like synchiria and alloaesthesia (Bonato and Cutini 2016), which should be routinely asked about in any standard neuropsychological exam. That omission reflects a strong bias toward disorders that can be quantified using standardized instruments. This quantification bias is of limited value to students of neuropsychology.

First, signs of neglect are seen in about half of all patients with right hemisphere damage (Stone et al. 1992). Second, neglect is a paradigmatic disorder for understanding the interplay between hierarchical and parallel processing in the brain. Ignoring it reduces neuropsychology to little more than a branch of psychometric assessment.

Strictly speaking, everything discussed so far applies to cases of pure neglect—that is, attentional deficits without simultaneous visual field loss. In real-world clinical settings, however, neglect is frequently accompanied by hemianopia. Distinguishing between these two disorders is therefore essential.

By hemianopia, we always mean its homonymous form: a visual field defect that affects the same half of space in both eyes. For details on variants, see Goodwin (2014).

? What are the differences between hemianopia and hemineglect?

Hemianopia is a sensory deficit involving loss of vision in the half-field opposite a unilateral brain lesion located posterior to the optic chiasm (see Figure 2 in Chapter 11). Patients are blind in that field, and visual evoked potentials (VEPs) are abnormal. Hemianopia is a retinotopic disorder—everything on one side of the fixation point is not perceived, regardless of head or body position. It does not affect non-visual stimuli or imagined space. Patients are usually aware of the deficit and often develop compensatory strategies (e.g., turning the head).

Hemineglect, by contrast, is an attentional disorder with normal VEPs in both visual fields. Its severity does depend on head and trunk position—turning toward the lesioned side often reduces neglect. It is therefore a non-retinotopic, hemispace-defined disorder that can affect non-visual and even imagined stimuli. Patients with neglect typically show reduced awareness of their impairment.

For additional differences and strategies to distinguish between neglect and hemianopia, see the discussion below.

What Is Meant by the "Relativity of 'Left' and 'Right'"?

In hemianopia, the distinction between left and right is clear-cut: it is defined retinotopically—that is, strictly relative to eye fixation. The left visual field is what falls to the left of the gaze; the right field, to the right. Simple and stable.

In neglect, however, the definition of left and right becomes relative. It depends not just on where the eyes are pointing, but also on where attention is focused, on trunk orientation, and on the salience or competition of other stimuli. What counts as "left" may shift with context.

This becomes especially evident in perimetric comparisons between hemianopia and neglect patients (see Figure 1). In a study where all stimuli were presented within the intact right visual field, stimulus location within that field still mattered, depending on the syndrome.

Intriguingly, then, signs of neglect can also manifest within the non-affected, ipsilesional visual field or hemispace. In addition to the findings illustrated in Figure 1, another compelling demonstration of the relativity of left and right comes from visual field assessments involving moving stimuli.

While stimulus direction is irrelevant for measuring hemianopic deficits, it plays a significant role in the assessment of neglect. Specifically, patients with neglect show less inattention to stimuli that move toward the neglected side, compared to those that move away from it. In other words, when attention is engaged, a stimulus

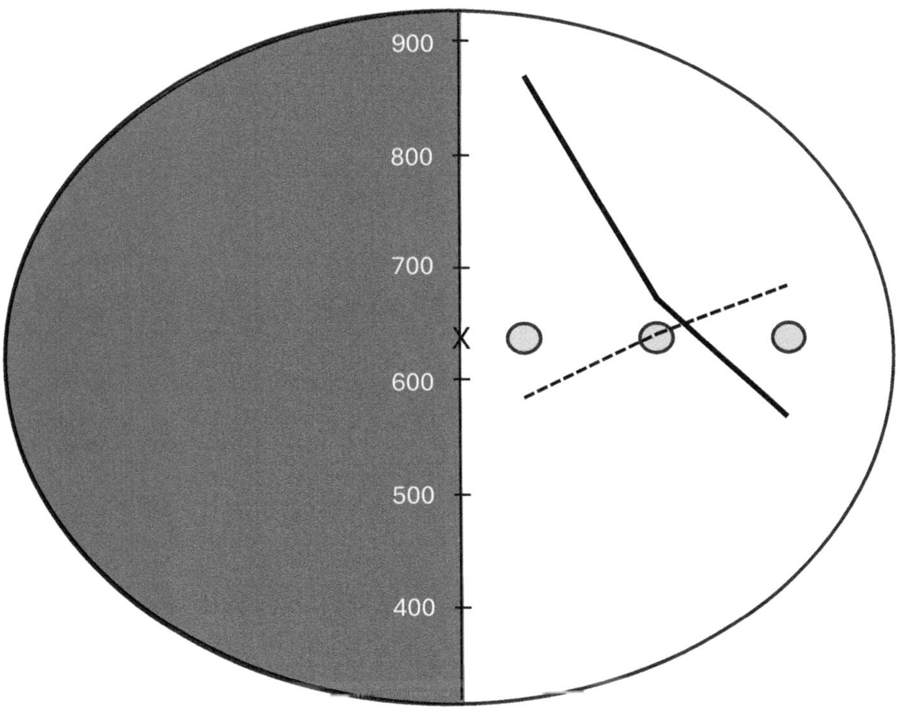

Figure 1: Reaction times to visual detection [ms] in patients with right hemisphere damage and either left-sided hemianopia (dashed line) or left-sided hemineglect (solid line). All stimuli were presented within the intact right visual field (patients fixated at "X"), at three different eccentricities (grey circles). Hemianopic patients were fastest to detect stimuli closest to the center of gaze, consistent with a pure sensory deficit. In contrast, neglect patients were slowest for the same near-central stimuli—because they were relatively leftward from all others.
(Visualization of data from Kerkhoff and Schindler 1997)

light can be followed across the vertical midline into the neglected side of space. This is not possible in hemianopia.

This phenomenon highlights an important dissociation: cueing can temporarily alleviate neglect—but it cannot overcome a sensory loss such as hemianopia.

There are many more instances where neglect targets stimuli that are *relatively* left- or right-sided. We will return to this issue repeatedly throughout this chapter.

? Left- and right-sided hemianopia occur with comparable frequency. Why is spatial neglect much more frequent on the left side of space?

Indeed, left hemisphere damage (whether in temporal, parietal, frontal, or subcortical regions) *can* produce inattention to the right side of space. But analogous lesions in the right hemisphere result in more conspicuous, more severe, and more persistent neglect—typically directed toward the *left* hemispace.

One common explanation appeals to hemispheric specialization: the right hemisphere is specialized for spatial attention more broadly, just as the left hemisphere tends to dominate in language tasks.

A more nuanced account, originally proposed by Mesulam (1981), emphasizes asymmetry in spatial attention deployment. The right hemisphere can direct attention to both sides of space—primarily contralateral (left), but also to a lesser extent, ipsilateral (right). In contrast, the left hemisphere directs attention only to the contralateral (right) hemispace.

As a result, patients with left hemisphere damage can still benefit from the right hemisphere's residual spatial functions. Patients with right hemisphere damage, however, are left without any mechanism to compensate for left-sided inattention.

What Is "Extinction"?

One neglect-related phenomenon in unilateral spatial attention is extinction. According to Benton (1956), the German physiologist Jacques Loeb had already described it in the late 19th century: extinction is the failure to detect a contralesional stimulus when a simultaneous stimulus is presented on the ipsilesional side. Crucially, the contralesional stimulus is perceived when presented alone—but it is "extinguished" when paired with its competitor.

Neglect and extinction are not the same. While some authors mistakenly describe extinction as a "milder" form of neglect, they are distinct phenomena—though they frequently co-occur, likely due to overlapping neural circuits.

Testing for extinction should be routine in every patient with a unilateral hemispheric lesion. Lesions that most often accompany extinction are located in the intraparietal sulcus and at the temporo-parietal junction (de Haan et al. 2012).

To test visual extinction, the examiner faces the patient and wiggles fingers on the left and/or right side of the patient's visual field, ensuring central fixation (usually by asking the patient to look at the examiner's nose).

To test auditory extinction, the examiner stands behind the patient and snaps fingers—sometimes left, sometimes right, sometimes both simultaneously.

Tactile extinction is assessed with the patient's hands resting palms-down on a table and eyes closed. The examiner touches the left, the right, or both hands simultaneously in pseudorandom order.

Extinction can also be crossmodal. For example, a tactile stimulus to the face can be extinguished by the simultaneous presentation of a visual stimulus (e.g., a light) near the same region (Làdavas et al. 1998). This suggests that *vision and touch are mapped to a shared peripersonal space*—especially around the face.

What Frames of Reference Determine Left and Right in the Somatosensory System?

Ways to differentiate hemianopia from visual neglect have been outlined above. A comparable differentiation within the somatosensory modality, however, is far more challenging.

? **How can we test, in the somatosensory modality, whether a patient with right parietal damage has a left-sided sensibility deficit or a left-sided neglect?**

Here is a pictorial answer:

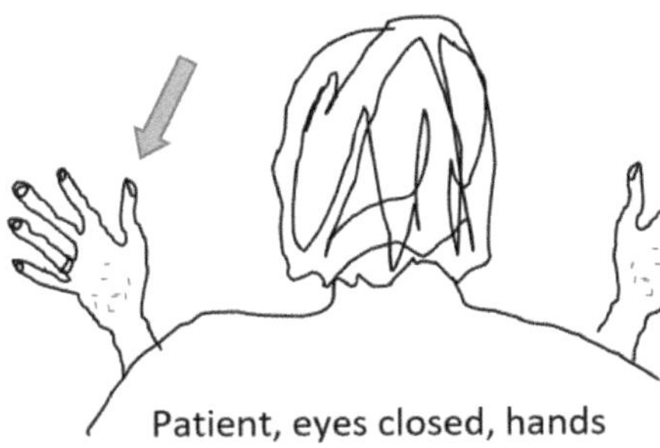

Patient, eyes closed, hands in corresponding hemispace

In the standard test situation, the patient (viewed from behind) is touched on either the dorsum of the left or right hand in pseudorandom order. At this stage, somatotopic and spatial frames of reference are confounded. A failure to report touch on the left hand (indicated by the grey arrow) could result from either a primary sensory deficit or from somatosensory neglect.

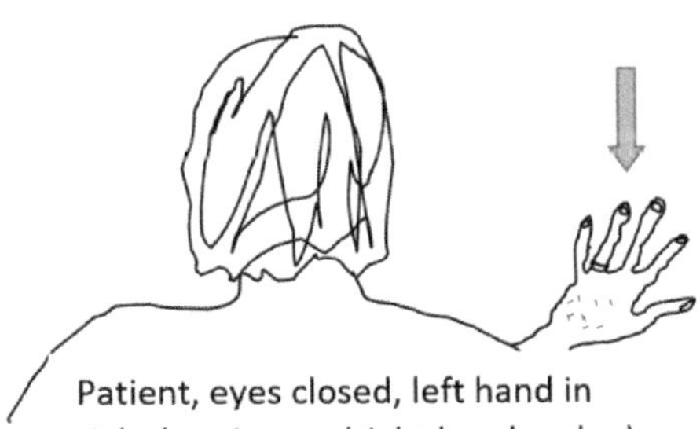

Patient, eyes closed, left hand in right hemispace (right hand on lap)

Now the left hand is placed into the right hemispace. If the patient still fails to detect touch on this hand (grey arrow), the deficit is somatotopic—that is, sensory in origin. But if the patient now detects the stimulus, or the error rate decreases significantly, then the failure in the standard condition was due to neglect, and the deficit is spatially defined.

A jokingly intended rule of thumb sometimes used to teach children how to distinguish left from right is: "Left is where the thumb is right."

But as experiments in somatosensory extinction show, the distinction between left and right is not so easily pinned down—even in adults with brain damage. In a study by Moscovitch and Behrmann (1994), extinction was tested by simultaneously stimulating the 2nd (index) and 4th (ring) fingers on the same hand. With the palms facing downward, stimulation of both fingers led to the extinction of the 4th finger on the left hand, and the 2nd finger on the right hand. In both cases, it was the relatively left-sided finger that went undetected.

When the same test was repeated with the palms facing upward (supination), the pattern reversed. These findings clearly show that extinction does not operate at the level of primary somatosensory representation (i.e., the hand's internal map), but rather reflects a higher-order, spatially defined body schema.

They also confirm: "Left is where the thumb is right" may be a cute mnemonic for kids—but it's a terrible rule of (any) thumb in neuropsychology.

Some authors distinguish "sensory neglect" from "motor neglect." But we just read here that neglect is defined as *not* being due to either sensory or motor deficits. Do you understand my confusion?

Yes, your confusion is understandable—and you are not alone. The *terms "sensory neglect" and "motor neglect"* can indeed be misleading. They do not imply that neglect is a sensory or motor deficit per se, but rather describe the predominant mode of expression of neglect symptoms in a given patient.

Let us use the line bisection test to illustrate. When asked to mark the midpoint of a horizontal line ("make the two segments equal in length"), a patient with left-sided neglect typically bisects the line *too far to the right* (i.e., ipsilesionally):

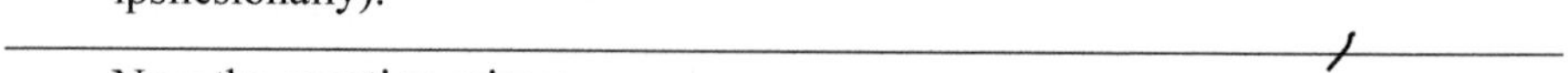

Now the question arises:

Is this rightward displacement due to a visual-judgment error (the left half must be perceptually larger to appear equal to the right), or is it due to a motor-exploratory failure (an inability to move the hand far enough left—what is called directional hypokinesia)?

One elegant way to disentangle these possibilities was proposed by Bisiach et al. (1990). Using a pulley system, the patient moves a handle (dark rectangle) to the right in order to shift an arrow (light triangle) leftward. If a patient still displaces the arrow to the right of center in this setup, directional hypokinesia can be ruled out. The error must instead reflect a deficit in sensory comparison.

Patients with predominantly sensory components of neglect tend to have posterior parietal lesions, while those with stronger motor components usually show more anterior, often pre-rolandic (mostly frontal) lesions (Bisiach et al. 1990).

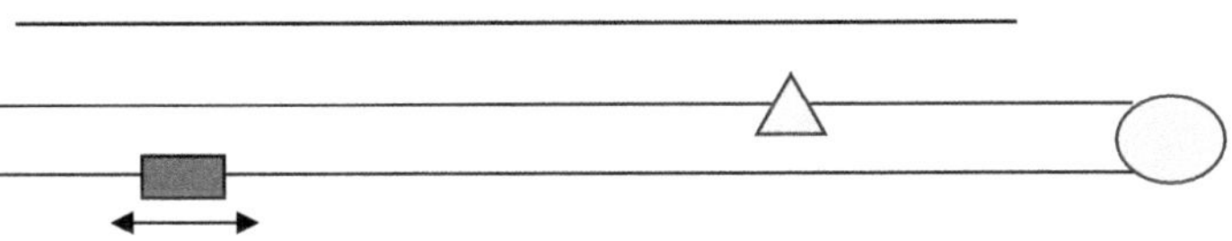

Other methods to disentangle motor and sensory components of neglect use a laterally placed mirror.

Neglect of the Left Side of Space or the Left Side of an Object in Space?

One of the most intriguing aspects of hemispatial neglect once again highlights the relativity of "left" and "right." When asked to copy a complex scene—such as the one shown in Figure 2 (top; one of many variations of the classic "Ogden scene," Ogden 1985)—patients may display two distinctly different types of neglect. One patient may reproduce the right side of the sheet in fine detail, omitting the left side entirely. This pattern reflects space-based neglect: it concerns the left half of external space, relative to the viewer.

Another patient may sketch all the objects on the template sheet, but consistently omit the left half of each object. This indicates object-based (or object-centered) neglect: the inattention concerns the left side of each individual object, regardless of where it appears in space.

Space-based neglect is most commonly associated with lesions to the right parietal lobe. Object-based neglect is often linked to temporal lobe regions involved in visual object analysis (Medina et al. 2009). However, Pedrazzini et al. (2017) reported a somewhat different association—connecting space-based neglect to inferior parietal lesions, and object-centered neglect to superior parietal areas.

In practice, these two types of neglect are rarely observed in pure form, but one form may be clearly dominant in a given patient.

A clever test to quantify object-based and space-based components of an individual patient's neglect is discussed in the following paragraph.

How to Quantify Neglect?

Neglect is such a multifaceted disorder that no single test can capture it adequately. You will always need a set of tasks to identify and quantify its various components.

Among the most common bedside assessments is the line bisection task. But beware of so-called "standardized" versions (e.g., Schenkenberg et al. 1980). Attempts at standardization are doomed to fail, because too many factors influence

Figure 2: Two variants of neglect. Top: the template to be copied by the patient. Bottom left: copy by a fictitious patient with "space-based" neglect. Bottom right: copy by a fictitious patient with "object-centered" neglect. The two variants of neglect are here depicted in their pure form. In reality, mixed forms are the rule rather than the exception.

a patient's bisection behavior: the hand used, the patient's age (and possibly sex), cultural background (reading/writing direction), room illumination, the line's color, length, orientation, distance from the eyes, and the presence of flankers at one or both ends—to name only a few.

In fact, line bisection may not be a good predictor of everyday neglect behavior. It appears to reflect damage specifically to the temporo-occipital junction rather than neglect in general (Rorden et al. 2006).

The influence of line length can be substantial. A well-known paradox is the cross-over effect: for short lines, some patients paradoxically err leftward. If only short lines are used (under 10 cm), a rightward shift might be absent, and neglect could be falsely ruled out. The origins of this effect remain debated. Interestingly, Chatterjee et al. (2000) found a similar cross-over effect for non-spatial stimuli: for example, patients with right hemisphere damage judged medium and heavy weights as heavier in the right hand, but very light weights as heavier in the left.

? Apart from its application as a bedside test for neglect, what should neuropsychologists know about line bisection?

The line bisection task was introduced in the late 19th century by Theodor Axenfeld as a test for hemianopia (see Kerkhoff and Bucher 2008 for translation and historical commentary). Importantly, hemianopic patients tend to displace the subjective midpoint toward the contralesional side—opposite to what is typically seen in neglect. This matters when evaluating patients with both homonymous field defects and signs of neglect.

Several variants of bisection have been proposed. These include trisection and quadrisection (Zivotofsky 2004); lines with flankers such as dots, numbers, or arrows, which may trigger length illusions even in healthy subjects (Rinaldi et al. 2018); ocular bisection, where the patient uses gaze rather than the hand; and somatosensory bisection, which evaluates attention to various body surfaces—of particular interest is the contrast between front and back of the head (Lenggenhager et al. 2016).

Clinically, line bisection can be followed by a judgment task: patients evaluate whether their own bisections were accurate. Each line is presented twice—once as drawn initially, once rotated 180°. This approach helps disentangle sensory from motor neglect components (see Belli and Fischer 2025). They also report an intriguing modulation of performance by respiratory phase—inhalation vs. exhalation.

Healthy individuals show a systematic *leftward* bias in bisection—this is known as pseudoneglect (Jewell and McCourt 2000), and it extends into number space (Loetscher and Brugger 2007a). The degree of this spatial bias correlates with various cognitive, emotional, and neuropsychiatric traits (Taylor et al. 2002).

Other formats include radial bisection, with lines extending from the body outward—healthy individuals tend to err toward far space—and the "endpoint task," where the patient draws a line so that a pre-placed dot marks its midpoint. Researchers have also used planes, circles, or words for bisection, and the number line bisection paradigm is discussed later in the section "Spaces in one's head..." (p. 198).

One underexplored variant asks participants to divide a line not at the midpoint, but according to the Golden ratio (Livio 2002). In an unpublished case study, we tested an architect with extensive right parietal and temporal lobe damage. His standard bisections showed severe rightward displacements—but when asked to produce Golden sections, his markings were remarkably close to the true midpoint (see Chapter 12, p. 290 for more detail).

Finally, neuropsychologists should recognize pseudoneglect as a subtle confound in behavioral research. For example, questionnaires using left-to-right Likert scales may yield more favorable answers when positive options

> are placed on the left side. In a study titled "Satisfaction Guaranteed," Nicholls et al. (2006) showed exactly this effect. Similarly, emotional word processing is influenced by spatial bisection biases (Tamagni et al. 2009), reinforcing the deep links between language, emotion, and spatial attention.
>
> Please excuse this lengthy answer—but line bisection is such an excellent tool for dissecting behavior across perceptual, cognitive, emotional, and even aesthetic domains (see Fischer 2001 for more inspiring examples)!

In a typical cancellation task, patients are asked to cross out all instances of a specific symbol—or to cancel target items among distractors—on a sheet of paper or a screen. The number and distribution of omissions across the sheet are used to assess space-based neglect. More omissions on the side contralateral to the lesion suggest a directional inattention toward that side of space.

To help differentiate between space-based and object-based neglect, Bickerton et al. (2011) developed the so-called "Apples Test." Patients are instructed to cross out all the intact apples scattered across the sheet while ignoring those with a bite mark. The bite marks appear on the left or right side of the apples.

In pure space-based neglect, patients tend to omit intact apples on the neglected side of the sheet—but do not mistakenly cross out apples with bite marks anywhere. In contrast, patients with object-based neglect may successfully cancel all intact apples across the sheet, but systematically fail to identify those with bite marks on the neglected side of each object—regardless of where the apple appears in space.

Naturally, both line bisection and cancellation tasks require manual responses and some degree of visual exploration. They are therefore unsuitable for assessing the sensory/perceptual aspects of neglect in isolation.

By contrast, judgments of chimeric stimuli require neither hand movement nor visual scanning beyond initial fixation. In one version, patients are shown chimeric animals—e.g., a left-facing body with the head of a camel and the rear of a bird—or a composite frontal face with the left half of a wolf and the right half of a sheep. Patients are informed that "unusual animals" may be shown and asked to name both components.

In another task, patients view cards or screens showing two faces, one above the other. One face has a stronger left-sided smile, the other a stronger right-sided smile. Participants are asked to indicate which face looks happier. In half the trials, the smiling face appears on top; in the other half, on the bottom.

In the example shown in Figure 3, healthy individuals tend to choose the bottom face as the happier one—a reflection of pseudoneglect. Patients with left-sided neglect, however, are more likely to choose the top face, failing to process the left-sided emotional expression.

Figures 4 and 5 illustrate additional features of hemispatial neglect, each accompanied by brief notes on the administration of relevant tasks. Figure 4 presents a case of left-sided hemi-inattention, while Figure 5 shows a case of right-sided hemi-inattention.

Figure 3: Chimeric faces task: Which face appears happier, the one on top or the one on the bottom? See text for explanation.

Case LSN (for Left-Sided Neglect), Figure 4

LSN is a 78-year-old right-handed former cheese maker with eight years of education. She lives independently and enjoys traveling by train and boat. She was referred to University Hospital Zurich due to a slowly progressing weakness in her left leg. Despite having experienced gait impairment for months, she only casually mentioned it to her GP—others had already noticed. MRI revealed a large cerebrospinal fluid cyst in the right temporal lobe.

Neuropsychological testing revealed a selective nonverbal memory impairment and executive dysfunction (including low verbal and design fluency, rule-breaking, perseverations, and increased Stroop interference). Core functions—attention, language, praxis, visual perception, and construction—were preserved. Her marked left-sided neglect was accompanied by anosodiaphoria and a striking optimism bias (see p. 201). Referring to her brain scan, she rejected the word "tumor" and proposed instead: "houseplant" or "dripstone cave."

* Figure 4a: Traditional line bisection (n = 36). Strong rightward displacement in all cases.

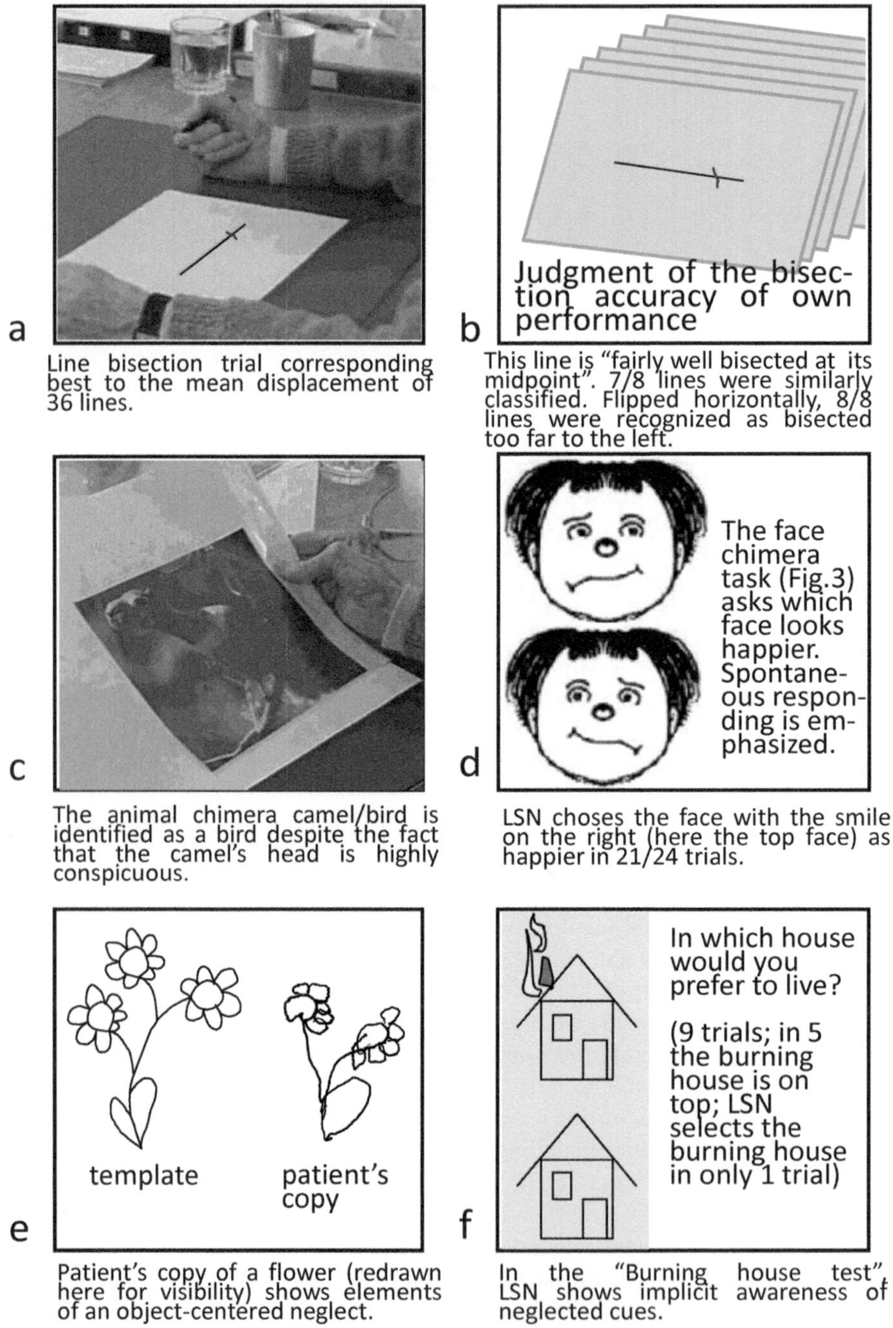

Figure 4: Some features of the left-sided neglect in patient LSN. See text.

* Figure 4b: Judgment of pre-bisected lines. She accepted almost all of her own (rightward) bisections as accurate but rejected the same lines rotated 180° (perceiving the mark as too far left), indicating no clear dominance of motor or sensory neglect.
* Figure 4c/d: Chimeric stimulus identification. LSN consistently based judgments on the right half of stimuli (9/12 animal-object, 21/24 faces).
* Figure 4e: Flower copy task. Object-centered neglect—left-side omissions in petals despite correct verbal counting ("1–5").
* Figure 4f: In a variant of the burning house task (Marshall and Halligan 1988), she typically avoided the burning house (9/10 trials) but could not attribute her choice to the flames. On one occasion (trial 9), she chose the burning house, noticed the flames, and changed her answer—prompting termination of the task.

For unconscious processing in neglect, see Berti (2002) and Driver and Vuilleumier (2001).

Case RSN (for Right-Sided Neglect), Figure 5

RSN is a 61-year-old right-handed handicraft teacher who suffered multiple ischemic strokes affecting the left ACA, ACP, and PICA territories. She presents with a sensorimotor hemi-syndrome affecting the right arm and, to a lesser extent, the right leg. Visual field testing revealed a homonymous right-sided hemianopia.

Neuropsychologically, RSN shows ideomotor apraxia, constructive dyspraxia, and reduced executive functions. Apart from mildly reduced verbal learning, memory is spared. Language is intact, but reading and writing are impaired due to a pronounced right-sided neglect.

The neglect is mainly space-based, as shown in spontaneous drawings and copying tasks, but object-centered components are also present.

* Figure 5a/b: Chimeric stimuli tasks. Of 12 animal/object chimeras, RSN based 9 identifications on left-sided features (2 unidentifiable, 1 misidentified, but also based on the left side). In the chimeric faces task (Figure 3), RSN selected the face with the left-sided smile as happier in all 24 trials, which is consistent with right-sided neglect.
* Apples Test (p. 190): RSN completed only 4 of 16 rows, with a comparable number of errors on the left (3) and right (4).
* Reading (Figure 5c): While she reads full lines, most words are only partially read. Rightmost letters are either omitted or confabulated. Object-centered neglect thus dominates in written word processing.
* Line bisection: Of 24 lines, 14 were bisected within the first third from the left margin.
* Personification of paralyzed arm (Figure 5d).

RSN demonstrates verbalized action monitoring ("Now I first draw this little circle, then…"), and shows a striking personification of her right, densely plegic forearm. She caresses it with her left hand and speaks to it with sympathy and

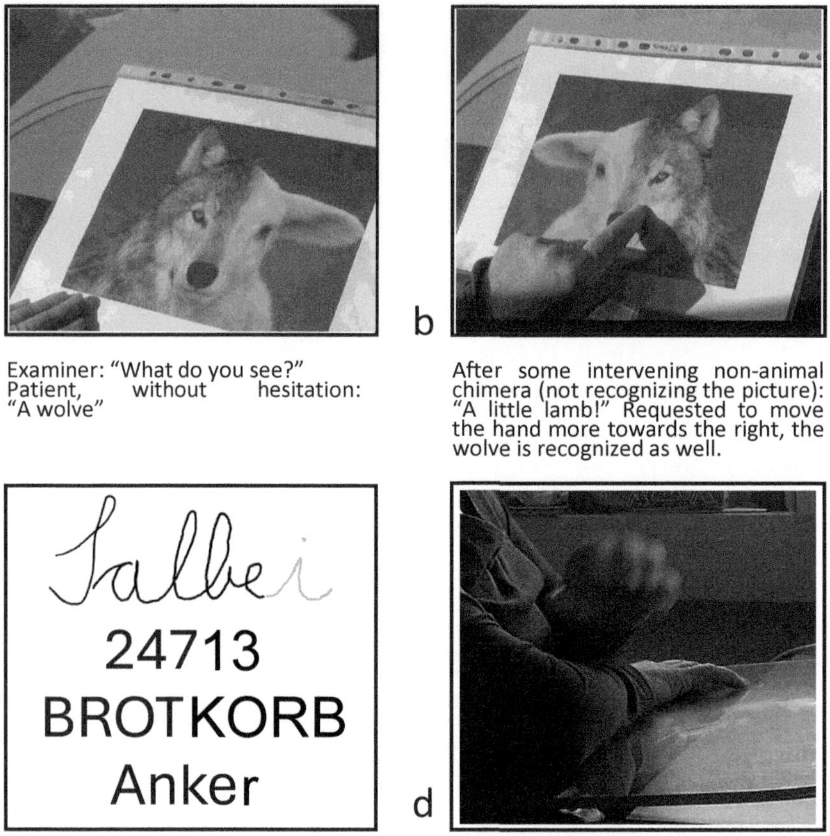

a: Examiner: "What do you see?" Patient, without hesitation: "A wolve"

b: After some intervening non-animal chimera (not recognizing the picture): "A little lamb!" Requested to move the hand more towards the right, the wolve is recognized as well.

c: Reading aloud: the right ending of almost all words and numbers is omitted (grey parts). What is read (black) is always a meaningful word. (Salbe = ointment; Salbei = sage; Brotkorb = a basket (Korb) to serve bread (Brot); Anke = Swiss German for butter; Anker = anchor). The writing style was deliberately diverse (but there was no effect).

d: Dialog with the paralyzed right arm; "You beloved one! You have a hard time these days; How courageous you are, beloved one, I admire you!" As visualized here, the patient often caresses the right forearm with her left hand.

Figure 5: Some features of the right-sided neglect in patient RSN. See text.

reassurance. By contrast, left-sided personifications—when they occur—are treated with hostility, a pattern noted in prior observations (Critchley 1955; Brugger 2007).

Any method aimed at quantifying a qualitatively well-described aspect of neglect should, in the end, be relevant to the patient's daily life. Yet because everyday activities depend on a wide range of spatially lateralized tasks, no single test will suffice. A broad battery of tasks—each tapping different components of spatial attention—is needed to capture the full picture of an individual's neglect.

Some computerized methods have been developed to simulate activities of daily living. Examples include baking cookies or setting a table. But one must ask: how close are these simulations to real life? Consider the example of "virtual table setting."

It typically involves dragging a fork or glass across a screen with a computer mouse or by tapping icons on a touchscreen. The gap between such stylized, interface-dependent tasks and actual table setting in a lived environment is considerable—and raises important questions about ecological validity.

? Is the "Pusher syndrome" a sign of hemineglect?

No. Pusher syndrome occurs in some hemiplegic patients and is characterized by an active pushing of the body toward the plegic (typically contralesional) side (Karnath 2007). In the seated position, the patient's upper body leans markedly and persistently toward the paretic side, and any attempt to upright the patient only leads to stronger counter-pushing. When standing, these patients typically fall—not because of loss of balance per se, but due to a disturbed perception of upright posture.

Pusher syndrome is not a manifestation of neglect, although the two syndromes may co-occur. This co-occurrence can make it difficult to assess the motor components of hemineglect, as both affect posture and spatial orientation in different but interacting ways.

Neglect: A Balance Between Hypoattention and Hyperattention

Unilateral neglect is often understood as a problem of reduced attention to the side opposite a lesion. But it may also reflect an imbalance between two opposing attentional systems. In this view, neglect of the left hemispace can result not just from hypoattention to the left but also from hyperattention to the right—a form of pathological attentional dominance.

This notion is supported by findings of a "magnetic attraction" to stimuli in the right hemispace—particularly pronounced in some patients. If this hyperattention is the driving force of a patient's left-sided neglect, then clinical strategy changes: placing all important stimuli on the right (as is often done) may worsen the problem. Instead, removing stimuli from the right hemispace can reduce attraction and shift attention back to the left.

An elegant experimental demonstration was offered by Mark et al. (1988), who compared two cancellation tasks:

- In version one, patients marked short lines on a transparency with a felt-tip pen.
- In version two, they used an eraser to remove the lines.

In both tasks, patients proceeded from right to left, as neglect patients usually do. But in version one, every mark increased the visual salience of the right side. In version two, each erasure removed such attractors. As a result, patients in the erasure condition often continued much further to the left, revealing the influence of hyperattention.

For an approach to quantifying this imbalance between hypo- and hyperattention in mental space, see Loetscher and Brugger (2007b).

? What is neglect dyslexia?

Neglect dyslexia is a non-aphasic reading disorder characterized by the omission of visual information from the left side—either of a line of text (space-based neglect) or of individual words (object-centered neglect), regardless of where on the page the word appears (Rich and Palmer 2023).

Not all patients with left-sided hemineglect show neglect dyslexia, but roughly one in four do—making it a clinical feature worth testing. Diagnosis involves more than simply having a patient read continuous text. Text lines should begin with variable indents to help distinguish between spatial and object-centered errors. In addition, a reading list should be used—containing both words and digits, and preferably words with multiple internal letter combinations.

For example, the word "FIREPLACE" includes several potential truncations or reinterpretations (e.g., ACE, LACE, PLACE, and REPLACE) when read from right to left. Errors on such items reflect varying degrees of neglect dyslexia severity.

Try this diagnostic list:

FIREPLACE
INDETERMINATION
14577
LIP SERVICE
CARTWHEEL
DECORATE
AUTOCORRELATION
829701
STEPMOTHER
UNDERESTIMATE
CAPRICE
ANTIVIVISECTION
7'384'920

If you allow me a brief elaboration on neglect dyslexia, there are patients who neglect the beginning of a word even when the word is presented vertically (Nichelli et al. 1993). That is, they omit or misread the top of the vertical word, which, from a spatial standpoint, would not correspond to the "left" in any body- or screen-centered reference frame.

This finding underscores that left and right are not fixed anatomical directions but are defined relative to the object—in this case, the word itself. It is another compelling instance of the relativity of LEFT and RIGHT in neglect, demonstrating how deep the brain's spatial processing is embedded in semantic and object-based coordinate systems.

How Many Spaces to Neglect?

Different regions of the space surrounding us are encoded by the brain in distinct ways. Among the most thoroughly investigated distinctions is that between near (peripersonal) and far space.

Near vs. Far Space

These two spatial zones, which exist along a continuous proximodistal gradient, differ in one critical respect: motor action is only possible in near space. Neglect can be restricted either to near space (Halligan and Marshall 1991) or to far space (Vuilleumier et al. 1998). Lesions causing neglect in near space typically involve the dorsal stream, while those affecting far space more often involve the ventral stream (see Chapter 7, Figure 4).

Testing for a near–far dissociation can involve comparing manual line bisection (in near space) with laser pointer bisection projected onto a distant screen. While the beam of a laser is immaterial, a physical stick can transform far into near space—tool use, in effect, extends the body's reach (Berti and Frassinetti 2000). Studies of manual vs. ocular bisection in healthy participants (Rinaldi et al. 2018) suggest that the right hemisphere plays a dominant role in coding space near the body, even though (left hemisphere-dominant) motor control is only effective in near space.

Front vs. Back Space

Neglect research has also shown that the space behind us is represented separately from the space in front. Some patients with severe neglect in the front space show preserved attentional orientation in back space—for instance, when imagining a scene behind them (Viaud-Delmon et al. 2007). Conversely, auditory neglect can selectively target back space (Vallar et al. 1995).

Of the major spatial dimensions—left/right, top/bottom, and front/back—it is left/right that is the most relational. Its definition requires the anchoring of the other two. Neglect in the back space has real-world relevance, as seen in patients who navigate a wheelchair in reverse. In healthy individuals, spatial interactions across near/far and front/back space can be explored with compatibility paradigms that vary hand posture and location (Leuthard et al. 2005).

Physical vs. Imagined Space

Although described clinically long before, scientific interest in neglect of imagined space was reawakened by the classic study of Bisiach and Luzzatti (1978). Neglect patients were asked to close their eyes and imagine themselves in *Piazza del Duomo* in Milan, once with the cathedral in front of them, once with their back against it. They were then asked to describe the square. In both conditions, some patients omitted elements on the left side of the imagined scene, relative to their imagined perspective.

These results revealed that "left" and "right" are not properties of the physical scene, but of the mental representation—and that neglect can affect imagery as much as perception.

Later studies used more abstract stimuli to assess left–right orientation in imagined space across different populations and cultural backgrounds. One of the most fruitful paradigms to emerge involved the use of the mental number line.

Spaces in One's Head: Are There Non-Spatial Representations?

To answer the heading directly: there is hardly any representation—however abstract—that does not include a spatial component (Brugger 2008a). Even binary or conceptual dichotomies are automatically mapped onto a horizontal space, and can thus be subject to hemispatial neglect.

For instance, political categories—"left-wing" and "right-wing"—are preferentially processed by the left and right hand, respectively (van Elk et al. 2010). More systematically studied are left/right associations for numerical magnitude (small vs. large numbers) and temporal sequencing (early vs. late).

Number Space

Numerical magnitude is mentally represented along a "number line"—from left to right in cultures with left-to-right writing systems, and from right to left in those with the opposite script direction (Dehaene et al. 1993). In patients with right hemisphere lesions, Zorzi et al. (2002) observed a consistent rightward bias in number line bisection. For example, when asked which number lies midway between 1 and 9, patients often responded "6" or "7" rather than the correct answer, "5."

Although both spatial line bisection and number line bisection reveal directional biases, these two measures are not correlated within individuals (Doricchi et al. 2005). Interestingly, neglect in number space appears especially frequent in patients with frontal lobe lesions.

Figure 6 provides a suggested protocol sheet for assessing neglect—and pseudoneglect—in number space. Participants are read aloud a pair of numbers and asked to guess the number that lies between them. (The word "guess" is emphasized to discourage calculation.) If a response takes longer than 4–5 seconds, the trial is skipped and repeated at the end.

In left-sided number-space neglect, patients tend to err toward the larger number; in pseudoneglect, the opposite pattern occurs. Right-sided neglect patients also tend to err toward smaller numbers. The average direction and size of errors across the 16 trials differentiate between healthy controls and patients with different types of neglect. The participant's guess is recorded in column 2 of the protocol sheet, and the difference from the actual midpoint is calculated afterward.

The advantage of using three-digit test items over one- or two-digit numbers lies in minimizing the risk of simple calculation strategies. For instance, the common mental shortcut "first digit plus second digit divided by two" no longer applies straightforwardly. This helps ensure that participants are genuinely guessing rather than computing. In earlier studies using 1- and 2-digit numbers, we observed nearly 25% correct responses—suggesting a calculation strategy at play (Brugger et al. 2010).

Each number pair in the stimulus set is presented twice—once with the larger number first, once with the smaller. This design controls for potential asymmetric

The two numbers to be read to the patient		Patient's guess	Deviation from correct number	Correct number
149	735	610	+ 168	442
928	662	750	− 45	795
119	873			496
574	412			493
299	423			361
764	544			654
325	687			506
261	345			303
687	325			506
412	574			493
735	149			442
544	764			654
423	299			361
345	261			303
873	119			496
662	928			795

Figure 6: 16 items to assess neglect in number space. See text for instructions to administer and score.

scanning biases in number space, which have been shown to depend on whether sequences ascend or descend (Loetscher et al. 2008).

Time Flies—From Where to Where?

Time has spatial attributes, and the parietal lobes—among other brain regions—play a key role in temporal processing (Critchley 1953; Chapter 10). Depending on the direction of reading and writing in a given culture, the "arrow of time" is mentally mapped either left to right or right to left (Quellet et al. 2010).

Eye-tracking studies (Rinaldi et al. 2015) have helped to uncover the mechanisms linking space and time, supporting the idea that both working memory and episodic memory are spatially structured. Which leads us directly to the next section:

The Space to Remember

If space, time, and number share a common structure in terms of horizontal spatial representation (Bonato et al. 2012), then serial memory—such as primacy and recency effects in short-term recall (see Chapter 6, p. 124)—should show a spatial component as well. Indeed, primacy appears to map onto a leftward spatial bias, and recency onto a rightward one.

Loetscher et al. (2012) confirmed this by showing that patients with left-sided neglect exhibited a relative preference for recency over primacy in recalling a list of ten words. The degree of this preference correlated with the number of left-sided omissions in a standard cancellation task.

Looking at memory across a broader time frame, Saj et al. (2014) found that left-sided spatial neglect was associated with difficulties recalling remote autobiographical events—as if the left side of memory, like the left side of space, had faded.

In working memory, which spans just a few seconds, Robertson (1990) noted a similar spatial pattern: performance was disproportionately impaired for backward digit span compared to forward—consistent with the need to mentally shift attention "leftward" in number space during backward recall.

In short, there is a "space to remember"—and its foundational structure begins forming very early in development (de Hevia et al. 2014). This structure is further shaped by culture, especially through reading habits and symbolic practices.

This has an important implication for clinical neuropsychologists: cross-cultural comparisons using universal normative data may be misleading. Tasks that seem neutral—such as serial memory, fluency, or digit span—may actually be culturally biased due to differing underlying spatial mappings. The same caution applies to language-based tasks, as the interrelation between language and space is deeper than commonly assumed (Maass and Russo 2003).

? You often point out that we should ask patients about their symptoms during night dreams. Does neglect also affect "dream space"?

Thanks for the reminder. Yes, I sometimes do informally ask patients with hemispatial neglect about their dreams. One curious report came from a patient with neglect dyslexia, who described experiencing the same reading difficulties in dreams as in waking life. However, such observations are rare, and most patients do not spontaneously acknowledge neglect symptoms, whether awake or asleep. This is likely due to the tight link between neglect and anosognosia.

One particularly elegant clinical observation is reported by Doricchi et al. (2007), who studied a 68-year-old right-handed man with right MCA stroke, left hemiplegia, and left-sided neglect (though *not* for imagined scenes). The authors recorded eye movements during REM sleep and found a striking congruence between the dream content and the laterality of rapid eye movements: during dreaming, nystagmoid REMs[2] were markedly suppressed to the left, and dream scenes appeared to be shifted toward the contralesional side.

Even more intriguingly, the patient—who was not anosognosic when awake—showed signs of anosognosia in his dreams, prompting speculation by the authors about the deep link between oculomotor systems, dream generation, and even dream bizarreness.

---

2 "Regular" REMs are thought to be triggered by cholinergic mechanisms of paradoxical sleep. Nystagmoid REMs, in contrast, are physiologically more akin to saccadic eye movements during waking. See the literature cited in Doricchi et al. (2007).

For a broader view of a scientifically grounded "neuropsychology of dreams," see Schwartz and Maquet (2002). For a neuropsychoanalytic perspective that includes aphasia, amnesia, and dream content, see Erdelyi (2023).

Neuropsychiatric Symptoms Associated With Spatial Neglect

Several irrational behaviors and monothematic delusions may accompany hemispatial neglect. For a fuller discussion of these phenomena—especially as they pertain to corporeal awareness—the reader is referred to Chapter 10. We list here only the most frequent clinical presentations:

* Anosognosia: the failure to recognize or outright denial of one's own illness, most often hemiplegia. As a special form of agnosia, anosognosia is also discussed in Chapter 7 (p. 164). Anosodiaphoria is just a milder variant of anosognosia, in which patients acknowledge a deficit but respond with emotional indifference, downplaying or trivializing its significance.

? **Prosopagnosia is the failure to recognize faces; topographagnosia is the failure to recognize familiar places. Both face and place recognition have clear biological relevance—faces are social, places are navigational. But what about anosognosia? What is the function of illness recognition when we are in good health?**

A brilliant question—and one we have also explored. We proposed that the healthy equivalent of anosognosia may be found in a robust psychological tendency called Unrealistic Optimism (UO)—in effect, a kind of prospective anosognosia. UO refers to the tendency to underestimate future threats, risks, or misfortunes (Weinstein 1980). It is a powerful and widespread cognitive bias—diminished in individuals with depression (cf. "depressive realism"), and typically absent in hypochondriacal patients.

Neuroscientist Tali Sharot and colleagues have explored UO in a series of clever neuroimaging studies (see Sharot 2011 for review). In one fMRI experiment (Sharot et al. 2011), participants first estimated their personal likelihood of experiencing various adverse events—ranging from burglary to Alzheimer's disease. They were then shown the real statistical likelihood for someone of their age, sex, and lifestyle.

Some received "good news" (real likelihood lower than expected), others "bad news" (real likelihood higher). When tested again later, participants tended to update their beliefs more effectively after good news than after bad.

Crucially, brain activity during belief updating showed a hemispheric asymmetry:

- Good news led to increased activation in the left inferior frontal gyrus
- Bad news led to activation in the right inferior frontal gyrus

Although Sharot's group did not discuss their findings in relation to anosognosia, the overlap is striking. When someone suffers a sudden, life-altering event—such as a stroke affecting the right hemisphere—this right-sided updating mechanism may be impaired, blocking the incorporation of bad news. Supporting this, Berti et al. (2005) found that in 15 of 17 patients with anosognosia for hemiplegia, the right inferior frontal gyrus was involved.

We approached the same idea differently. In a study by McKay et al. (2013), we applied caloric vestibular stimulation (CVS) to healthy participants after first measuring their baseline level of UO. CVS (in this case, cold water to the left ear) selectively stimulates the right hemisphere's vestibular network. In patients with right-hemisphere lesions, CVS is known to abolish anosognosia—even if only temporarily (Cappa et al. 1987). In our healthy participants, the same stimulation led to a reduction in UO.

These findings suggest that UO and anosognosia may reflect a shared mechanism—a kind of continuum. From that perspective, "nosognosia" (the normal awareness of one's health and vulnerabilities) might be best understood as a form of subclinical anosognosia (McKay et al. 2013).

* Asomatognosia, the sensation that one's body or parts of it are absent
* Somatoparaphrenia, the delusion that a paralyzed limb belongs to somebody else
* Misoplegia, the hatred of hemiplegia

See Chapter 10 (p. 224 and 227) for a discussion.

What Methods Are Useful in Neglect Rehabilitation?

When signs of neglect are subtle, no formal intervention may be needed. However, in cases of severe neglect, the condition can significantly impair activities of daily living. Unfortunately, neglect is notoriously resistant to rehabilitation. This limited responsiveness may partly stem from the fact that patients fail to acknowledge the very asymmetries in their spatial attention, exploration, and internal representation.

Even in the absence of overt anosognosia for motor or sensory deficits, hemispatial neglect invariably includes a degree of "symptom neglect"—a failure to recognize that anything is wrong at all. This intrinsic unawareness complicates both motivation and compliance with therapeutic interventions.

In the special case where neglect dyslexia is the most prominent symptom of a left-sided neglect, patients may suffer enormously, often more from their inability to read a newspaper than from routinely bumping into objects on their left. One simple but powerful trick can help: teach them to "read with their finger." Instruct them to move their finger slowly along each line of text, so far to the right that they can be confident no more letters or words will appear.

Suddenly, the text begins to make sense again.

What is striking is that—despite the general resistance to rehabilitation—this little strategy *is* rarely forgotten. Its effectiveness underscores the strong motivational component that can drive rehabilitation, especially when the strategy offers immediate, meaningful benefit.

In most cases, a combination of interventions is necessary to address the various components of a patient's neglect. A detailed neuropsychological examination is essential to identify the specific manifestations of neglect—whether spatial, object-centered, representational, motor, or affective—in order to guide the optimal combination of therapies. Below is a selection of the most established and emerging approaches:

Visual Exploration Training

This method seeks to enhance voluntary visual exploration by encouraging the initiation of saccades toward the neglected field. It includes eye-movement training and search strategies. Its efficacy is well established in the presence of visual field defects, but remains controversial for neglect. A related method, spatial exploration training, focuses more on strategy learning (Toglia and Chen 2022), but has shown limited long-term success—particularly in patients with anosognosia or reduced self-awareness. A more recent proposal involves augmented reality settings for exploration training (Stammler et al. 2023).

Multisensory Stimulation Training

Simultaneous audio-visual stimulation at identical spatial locations in the neglected hemispace can improve performance on neglect tests. However, the training intensity required is substantial: typically 10 sessions of 4 hours each, over two weeks (Làdavas et al. 2022).

Neck Proprioceptive Stimulation

Vibration of neck muscles and tendons on the affected side can induce an illusory sensation of head rotation toward the intact side. This can recalibrate egocentric spatial coordinates and improve attention to the neglected hemispace. The method shows long-lasting effects, both alone and in combination with visual exploration training (Schindler et al. 2002; Johannsen et al. 2003).

Optokinetic and Vestibular Stimulation

In full-field optokinetic stimulation, a moving background (e.g., drifting dots) induces reflexive eye movements, often accompanied by shifts in spatial attention. Though widely used in experimental settings, evidence for long-term effects remains mixed.

Vestibular Stimulation—either caloric or galvanic—has similar effects on spatial representation. The vestibular system, beyond its role in balance, also contributes to non-spatial functions (Lopez 2016). In neglect patients, vestibular stimulation can alleviate not only spatial inattention, but also anosognosia and somatoparaphrenia (Cappa et al. 1987; Holé et al. 2020). As discussed earlier (p. 202), vestibular stimulation also reduces unrealistic optimism in healthy volunteers (McKay et al. 2013), supporting its role in modulating self-awareness.

Prism Adaptation

This technique involves wearing prismatic lenses that shift the visual field rightward. Patients initially misreach to the right, then adapt to reach targets accurately. After prism removal, they exhibit a leftward aftereffect, which can alleviate neglect. The magnitude and duration of aftereffects depend on prism strength, training duration, and other variables.

Multiple sessions are required for sustained effects. While promising, long-term efficacy remains debated. Chen et al. (2025) caution that meta-analyses are premature due to methodological heterogeneity across studies. Important variables include prism angle, arm visibility, training intensity, task type, and combined treatments. A home-based prism protocol developed by Fortis et al. (2020) showed effects maintained over six months.

rTMS/Theta Burst Stimulation

Low-frequency repetitive transcranial magnetic stimulation (rTMS) applied to the left parietal cortex may reduce pathological hyperactivation of the non-lesioned hemisphere. This is thought to restore interhemispheric balance, disturbed by transcallosal inhibition from the damaged right hemisphere.

TBS, a newer form of magnetic stimulation, offers shorter training times and longer-lasting effects (Nyffeler et al. 2019). Houben et al. (2021) highlight the promise of TBS but stress the need for standardized protocols across labs and clinics.

Additional and Emerging Approaches

* Pharmacological interventions: Dopaminergic therapies, especially in combination with prism adaptation, show promise (Luauté et al. 2018).
* Visual occlusion therapy: Use of glasses that block the ipsilesional visual field has been proposed to shift attention toward the neglected side.
* Virtual reality: This buzzword enters neglect rehab as well—with controversial evidence. See Pedroli et al. (2015) for review.
* Auditory stimulation: Listening to music, particularly with motion cues shifting from right to left, may help orient attention (Geiser et al. 2024).

Where Can We Read More?

Karnath, H.-O., Milner, A.D. and Vallar, G. (eds.). The Cognitive and Neural Bases of Spatial Neglect. Oxford University Press: Oxford, 2002.

Robertson, I.H. and Marshall, J.C. Unilateral Neglect: Clinical and Experimental Studies. Taylor and Francis: New York, 2013.

CHAPTER 10

Is There a "Neuropsychology of Corporeal Awareness"?

Most neuropsychology textbooks touch on somatosensory processing only in passing. They barely address the experience of having a body, and seldom explore the links between body and self. This is puzzling—because distortions of body boundaries and misperceptions of one's own self are among the most frequent clinical manifestations of brain damage.

Moreover, the study of bodily self-consciousness offers an empirical window into how body and mind are interconnected.

Before we turn to the relationship between body and self—and to higher-order disturbances of somatognosia—we begin with a brief consideration of the tactile senses and somatosensory perception.

? What are the differences between touch, somatosensory perception, and somatognosia?

The three terms are closely related, but they form a meaningful hierarchy. Touch, or tactile perception, is just one of several cutaneous senses that inform the brain about what is happening at the boundary between the self and the external world—that is, the skin. Tactile perception can be active: I touch an object to find out haptically whether it is smooth or rough, hot or cold. The haptic sense illustrates how intertwined motor and sensory processes are: I can sense object properties only by moving the receptive part of the skin. This is sensorimotor processing.

Somatosensory perception is the broader term. It encompasses the tactile senses, temperature perception, itch, pain, and the sense of limb position and movement (proprioception and kinesthesia).

Somatognosia is still higher in the hierarchy. It integrates all these distinct bodily signals to generate the experience of bodily self-awareness: it is my body that I feel and my body that I move. Feeling the boundaries of one's own body allows a fundamental distinction between me and not me. Somatognosia is closely related to the older term coenesthesia (or cenesthesia), which refers to the total impression arising from bodily sensations, including feelings of health, vigor, or lethargy.

DOI: 10.1201/9781003384991-10

Tasks of the Somatosensory System

The skin contains different types of receptors: mechanoreceptors for touch and vibration, thermoreceptors for sensing heat or cold, and nociceptors for pain.

Painful Touch

Among the nociceptors, there is further specialization. Mechanical nociceptors respond to intense pressure or pinching. Thermal nociceptors respond to extreme temperatures. Polymodal nociceptors, often bare nerve endings rather than structured receptor cells, respond to multiple noxious stimuli: pressure, temperature, and even chemicals. Nociceptors also mediate itch and tickling. Not all nociceptors are cutaneous—some monitor harmful stimuli from inside the body (see interoception, below).

> Nociception is as important as it is underrepresented in neuropsychology. Negative expectations, or nocebo responses, are often underestimated in their impact on neuropsychological functioning. They play a central role in chronic pain syndromes and pain-related emotional states such as fatigue (Wolters et al. 2019).
>
> This is neuropsychologically relevant: pain and anhedonia impair attention, memory, and executive function. Clinically, it highlights the importance of collaboration between neuropsychological rehabilitation and psychotherapy. Theoretically, the brain's top-down modulation of nociception—through placebo and nocebo effects—is a fascinating area in itself (see Krummenacher et al. 2010a).
>
> After this brief digression to the painful side of touch, let us return to tactile perception in the more conventional sense. We distinguish between passive, active, and affective touch.

Passive Touch

Touch is considered passive when tactile stimulation is applied to a static body part. The stimulus itself may or may not move—what matters is that the person being touched remains still.

Active Touch

In active touch, by contrast, a person uses a cutaneous receptor surface—typically the hands, lips, or feet—to explore an object. Here, touching is an intentional act; it involves interaction with the tactile stimulus. Active touch requires the integration of motor and sensory processing (and to be precise, proprioceptive and kinesthetic input as well). This makes haptic perception (another term for active touch) a highly complex sensorimotor process. Even at the level of the cutaneous receptors, there are specializations for active versus passive touch.

Affective Touch

Affective touch refers to a special form of passive tactile stimulation—namely, slow, gentle, caressing strokes. This kind of touch plays a crucial role in social bonding, from maternal contact to erotic intimacy. It may boost lost feelings of ownership for one's own body parts after right hemisphere lesions (Jenkinson et al. 2020). Remarkably, our skin contains specialized receptors that seem to have evolved specifically for processing such sensual, interpersonal stimulation. The C-tactile system consists of specialized mechanoreceptors located in hairy skin (i.e., not on the palms or soles of the feet), which respond optimally to stroking at slow velocities (around 1–10 cm/s). Unlike A- and B-type fibers, C-tactile fibers are unmyelinated and thus conduct impulses slowly. They project to the insula, modulate pain perception, and contribute to the hedonic dimension of corporeal awareness. Their significance in social neuroscience has led to the idea of the skin as a "social organ" (Morrison et al. 2010).

? Is tickling not a kind of "affective touch"? Why can we not tickle ourselves?

The inability to tickle oneself is a prime example of how the brain predicts the sensory consequences of self-generated movements. When you attempt to tickle yourself, internal models in the cerebellum and related circuits forecast the exact timing and nature of the touch. This predictive attenuation reduces the sensory salience of the tickling stimulus, rendering it ineffective (Blakemore et al. 1998). In contrast, when another person tickles you, the absence of such precise predictions preserves the surprise—and the laughter. This mechanism not only prevents self-generated movements from overwhelming perception but also helps maintain the distinction between self and other.

Psychotic individuals with delusions of alien control do not exhibit the typical sensory attenuation for self-generated touch. They can thus tickle themselves (Blakemore et al. 2000)! And so could you—if your finger movements were triggered by a TMS pulse instead of your own volition. Such "externally caused" self-touch escapes the usual predictive dampening and can lead to a ticklish sensation (Chronicle and Glover 2003).

? If it is not too embarrassing, may I ask why predictive attenuation does not prevent me from reaching orgasm on genital self-stimulation?

A very fair question! After all, the explanation of "why you cannot tickle yourself" rests on the principle that the brain attenuates sensations it predicts. And orgasm, like tickling, is a salient and somatosensory-rich experience. So why does one disappear when self-initiated, and the other does not? This question is not as trivial—or as amusing—as it may first seem. It calls for a closer examination of what kinds of predictions are involved in sensorimotor control, and whether some forms of self-stimulation involve reward anticipation that actually enhances rather than attenuates the signal. As of now, there is no definitive answer. But it is a fascinating topic worth exploring.

Proprioception and Kinesthesia

Beyond touch, temperature, and pain, the somatosensory system also includes special receptors that monitor the position and movement of limbs—both in space and in relation to other body parts. These two neuromuscular senses are known as proprioception and kinesthesia. The relevant receptors are found in muscles, tendons, and joints, and respond to the stretching and contracting of limbs.

Proprioception is sometimes called "joint position sense" or simply "postural sense," while kinesthesia refers to its dynamic counterpart—the perception of movement. Unlike touch, these senses rarely enter conscious awareness in daily life. We tend to notice them only when asked to touch our nose with eyes closed, or when standing on one leg. Viewed differently, the proprioceptive system could be seen as the root of our intuitive grasp of physical laws—a prime example of *embodied cognition* (Smetacek and Mechsner 2004).

As with all sensory modalities, illusions offer a marvelous way to study and teach tactile perception, somatosensation, and corporeal awareness. Figure 1 introduces some illusions you may wish to try yourself—or with a friend. A larger collection can be found in Hayward (2008) and Lee et al. (2015). Barker and Rice (2012; 2019) have compiled a genre of "folk illusions"—bodily games often passed down informally among children. Karmakar et al. (2023) present an impressive range of lesser-known thermal illusions. And Jones (1988) offers a comprehensive summary of motor and proprioceptive illusions.

A Tiny Selection of Bodily Illusions (Extended Legend to Figure 1)

The illusions sketched in Figure 1a to 1f are assembled more intuitively than systematically. What they have in common is that they require no elaborate equipment—and often work best on a curious friend.

Thermal Referral (Figure 1a)

Place your first, second, and ring fingers on three metal plates. Fingers 1 and 3 rest on heated plates (e.g., warmed on a stove); finger 2 rests on a plate at room temperature. You will feel a uniform warmth across all three fingers—including the one on the neutral plate. The same applies if you cool fingers 1 and 3 in a freezer: the middle finger feels cold. This illusion, described by Green (1977), has been interpreted as the result of low-level spatial summation in thermoreceptors. However, Karmakar et al. (2023) present variations that challenge this explanation.

Ellbow Illusion (Figure 1b)

Move a finger slowly along the inner side of a friend's forearm, from wrist toward the crook of the elbow. With eyes closed, the friend says "stop!" when the elbow is reached. On opening their eyes, they are often surprised to discover their estimate was early, frequently by several centimeters.

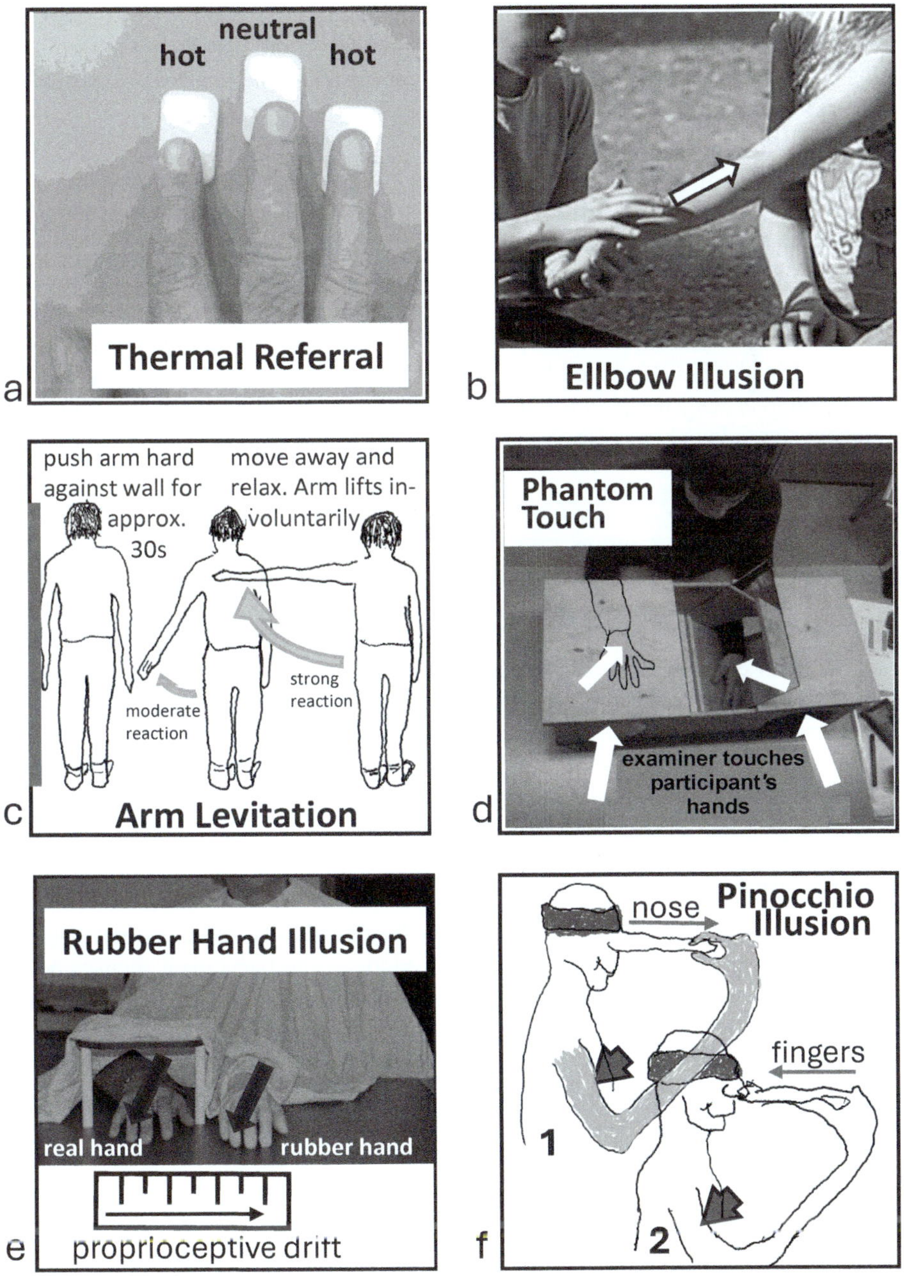

Figure 1: Six illusions of corporeal awareness. See text for explanation.

Brugger and Meier (2015) studied the illusion systematically. The anticipation error was larger in men than in women, and larger on the non-dominant arm. The illusion likely results from a mix of bottom-up sensation and top-down expectation.

Arm Levitation (Figure 1c)

Press the back of one hand firmly against a wall for at least 30 seconds. Then step away. Your arm will lift, seemingly involuntarily, and you may feel as if it is weightless or filled with helium.

This is not an illusion of levitation per se (the arm really moves), but rather an illusion of agency: the movement feels caused by an external force. Known as the Kohnstamm phenomenon (Kohnstamm 1915), it is physiologically complex. De Havas et al. (2017) reviewed over 60 studies, concluding that central adaptation effects interact with peripheral muscle mechanisms. Attention, motor imagery, and visual context also play a role.

Phantom Touch (Figure 1d)

For this illusion, use a simple mirror box setup (as in Ramachandran and Rogers-Ramachandran 1996). The participant places both hands, palms down, inside the box. Only the right hand is visible, reflected in the mirror; the real left hand is hidden. The examiner, sitting opposite, places their own hands and simultaneously presses the middle of each hand. After 15 seconds of steady viewing, the examiner suddenly withdraws their left hand.

Participants often react with surprise—or even a startled scream. They *see* an untouched left hand (the mirror image of their right), but feel a touch on their real left hand. This visual–tactile discrepancy is clinically useful and experimentally revealing (see also Burrack and Brugger 2005).

Rubber Hand Illusion (Figure 1e)

Place a rubber hand (a cotton-stuffed glove will suffice) to the left of the participant's outstretched right hand, which must remain hidden. The participant focuses on the rubber hand while you simultaneously stroke both real and rubber hands in matching locations, using your fingers or soft brushes. Avoid rhythmicity and vary the touch locations.

After 1–2 minutes, many participants will report feeling the touch *on* the rubber hand—and even incorporating it as part of their body.

Botvinick and Cohen (1998) described this as a visuotactile binding illusion, often measured by the proprioceptive drift: a shift in the perceived location of the hidden real hand toward the rubber one. The illusion has inspired a vast body of research, including its use to elicit "out-of-body" experiences in VR (Lenggenhager et al. 2007).

Pinocchio Illusion (Figure 1f)

To induce this illusion, use a physiotherapy vibrator (~ 100–120 Hz) to stimulate the blindfolded participant's biceps. The arm rests, hand near the face. During vibration, the participant gently pinches the tip of their nose. The brain, sensing the hand "extending" away from the body while still feeling contact with the nose, has to resolve the spatial conflict.

Many participants report feeling their nose elongating (version 1 in the figure), especially if told that the illusion is called "Pinocchio." Others report that their fingers feel longer to bridge the gap.

Lackner (1988) introduced the illusion; the underlying factors that determine these individual differences remain unknown. A trivial but elegant fMRI study might compare cortical activation in "nose elongators" vs. "finger elongators." We would expect an enhanced BOLD signal in the face area for the former, and in the hand area for the latter.

? **Does the somatosensory system include interoception?**

British physiologist Charles Sherrington (1857–1952) would likely have said *yes*. He proposed that somatosensation comprises three domains: exteroception (perception of the external world through the body), proprioception (a term he himself coined, for the sense of body position and movement), and interoception—the sense of the internal state of the body.

Contemporary models, however, tend to treat interoception as a distinct sensory system. It encompasses the monitoring of a range of physiological processes, including hunger and thirst (or more precisely, the filling state of stomach and bladder), as well as cardiovascular, respiratory, and visceral regulation—key to maintaining homeostasis.

Regardless of its precise taxonomic status, interoception plays a central role in the development of corporeal awareness and the formation of bodily self-consciousness (Longo et al. 2010; Brugger and Lenggenhager 2014). Intriguingly, even the timing of external stimuli in relation to the heartbeat (e.g., systole vs. diastole) can influence exteroceptive accuracy and the integration of self and body (Aspell et al. 2013).

Interoceptive signals do not primarily project to the somatosensory cortex. Instead, they are processed in the insula, a region buried deep in the lateral sulcus of the temporal lobe, with extensive connections to both limbic and neocortical areas (see Chapter 2, p. 26).

Interoceptive functions are compromised in a wide range of mental and neurocognitive disorders, including anxiety, depression, addiction, as well as developmental and degenerative conditions.

In short, somatosensory and interoceptive pathways are anatomically distinct, but functionally interwoven. Together, they form a dynamic system essential for corporeal awareness and for binding the experience of body and self.

Phantomology: The Neuropsychology of the Bodily Self

"Phantomology" is a term introduced by Polish futurologist-philosopher and science fiction writer Stanisław Lem (1921–2006). It appears as the title of Chapter 6 in his *Summa Technologiae* (Lem 1964/2013), where he lays out a visionary anticipation of virtual reality (VR)—including detailed predictions of cybersex mediated by haptic interfaces. Lem foresaw that VR development would be driven not by science, but primarily by the gaming and entertainment industry.

? **Why should neuropsychologists know the work of Stanislaw Lem?**

Lem might be considered a kind of Jules Verne of the neurosciences—a writer whose speculative visions anticipate scientific and social developments well ahead of their time. His imaginative leaps extend far beyond corporeal awareness and self-consciousness. He explored topics such as cognitive enhancement, reality monitoring, and human–machine interfaces.

Lem also addressed the challenges and promises of artificial intelligence, from both technological and sociological perspectives. In one futuristic short story, he describes a machine called the "Electronic Bard" that writes poetry, with its poetic creativity emerging from having been "fed tons of poetry." This bard is a clear ancestor of systems like ChatGPT: an intelligence amplifier whose output reflects the accumulated work of generations of human minds.

Finally, Lem's philosophical reflections on simulation, evolution, and design offer readers with a scientific mindset robust arguments against intelligent design, the modern face of creationism. His work reminds us that rational thinking is always at risk—and always worth defending.

Lem (1964/2013) used the term "phantomology" in its broadest sense: as a science of virtual bodies and virtual worlds. In this book, we use it more narrowly to refer to empirical investigations of the relationship between body and self.

Living bodies are physical objects—but one's own body is something more. You feel it. You move it. You own it. Your mind is attached to it in a way that produces identification. You seem to *be* your body.

But in reality, you are not your body. Your body is like a perfect avatar in a virtual reality—so familiar, so integrated, that you no longer notice its virtual nature. Your seamless experience of body–self unity feels like reality itself. In the briefest terms, phantomology is the science of the virtual reality of the bodily self.

? **What exactly is the "bodily self"?**

Many selves make you who you are—both for yourself and for others. Your autobiographical self ensures your identity across time. Your social self helps you navigate relationships, roles, and responsibilities (Kelly et al. 2019).

The bodily self is the totality of multisensory signals arising from your own body—signals that make you aware of your bodily existence and enable identification with your body. Its core phenomenological components include:

1. Ownership – the feeling that your body and its parts belong to you.
2. Agency – the experience of being the initiator of your own actions.
3. Self-location – the experience of occupying a specific volume of space.
4. First-person perspective (FPP) – the experience of seeing the world from the location of your own body.

> These aspects are supported by overlapping networks integrating somatosensory, visual, vestibular, and interoceptive input—distributed across both hemispheres. A review of 42 studies on electrical brain stimulation and changes in bodily self-perception (Dary et al. 2023) found that stimulation of the parietal lobe was most often implicated. Hemispheric differences were also noted: stimulation in the right hemisphere more often altered self-location, while left hemisphere stimulation more often altered agency.
>
> Building on the interaction between body, mind, and society, a social neuroscience approach to the bodily self has been proposed (Brugger and Lenggenhager 2014). This broader view may help phantomology to bridge the often artificial divide between neurology and psychiatry.

The virtual reality of the body is perhaps most vividly illustrated in the phenomenon of the phantom limb. For centuries, phantom limbs have inspired philosophical inquiry into the nature of the self and its relationship to the body.

Although phantom limbs are often associated with the keywords amputation and pain, amputation is not required for phantom experiences to occur, and painful phantoms are only a minority.

In what follows, we examine "phantomization" processes in several clinical and experimental conditions:

* Phantom limbs after amputation
* Phantom sensations in congenitally absent limbs
* Phantom legs in paraplegia
* Phantoms of central origin: from alien limbs to "hemiplegic twins" to autoscopic phenomena
* Incarnation without animation: Body Integrity Dysphoria (BID) as a neuropsychiatric condition

The Bodily Self After Amputation: Classic Phantom Limb Phenomena

While the term "phantom limb" is relatively new—it was coined by the American surgeon Silas Weir Mitchell (1829–1914) toward the end of the 19th century—knowledge of the phenomenon is much older. Finger and Hustwit (2003) provide a beautifully detailed account of how phantom limbs have been described over the centuries in both philosophical and medical texts.

> Ambroise Paré (1510–1590) was a French barber-surgeon. This means he did not study medicine but was trained as a barber, becoming skilled with knives and blades. In those romantic times, barbers in peacetime often became battlefield surgeons during war. Paré learned by doing—and quickly gained a reputation for saving more patients after amputation than many formally trained physicians. In his *Ten Books of Surgery*, he addressed the phenomenon of phantom limbs. He also developed sophisticated prostheses for the time: wooden legs with movable joints and artificial hands with gears to grasp objects. Let us not forget—this was in the 16th century!

Early medical theories typically located the cause of phantom limbs in the periphery. It is worth recalling, then, that already René Descartes (1596–1650) recognized that limb sensation—and phantom sensation—originates in the brain, not the limb. Finger and Hustwit (2002) convincingly demonstrate that 17th-century philosophy was in some respects ahead of 19th- and even 20th-century medical understanding.

Phantom sensations can be painful or painless (see Figure 2). Among the painless sensations are postural, kinesthetic, and movement-related experiences. Phantom motion can be involuntary (as in spasms) or voluntary, but not every amputee can move their phantom at will. Interestingly, the ability to voluntarily move a phantom limb may have important implications for pain management, and—if we take a Lem-like perspective—for the future possibility of materializing phantom limb movements into an artificial limb that interacts with the external world (Scaliti et al. 2020).

In the second half of the 20th century, interest in phantom limbs—especially painful ones—rose dramatically. They became a model for studying adult brain plasticity, which had long been considered restricted to early development. Studies in animals (Merzenich et al. 1984) and humans revealed massive, systematic functional reorganization in the somatosensory cortex after limb loss, particularly in relation to phantom limb pain (Ramachandran et al. 1992 for an overview).

There are competing interpretations of the functional reorganization of the brain after amputation. After the loss of a hand, its cortical representation in the somatosensory cortex becomes "silent"—it no longer receives input from the periphery. Early neurophysiological and neuroimaging data suggested that *neighboring* areas in the somatosensory homunculus (or *feminculа*, for our female readership) "invaded" the deafferented hand area. For example, the facial representation might expand into the cortical territory that formerly belonged to the hand (see Figure 3).

Early studies showed that the degree of this intrahomuncular reorganization correlated with the severity of phantom limb pain: the more the face representation intruded into the former hand area, the greater the pain (Flor et al. 1995). Pain, in

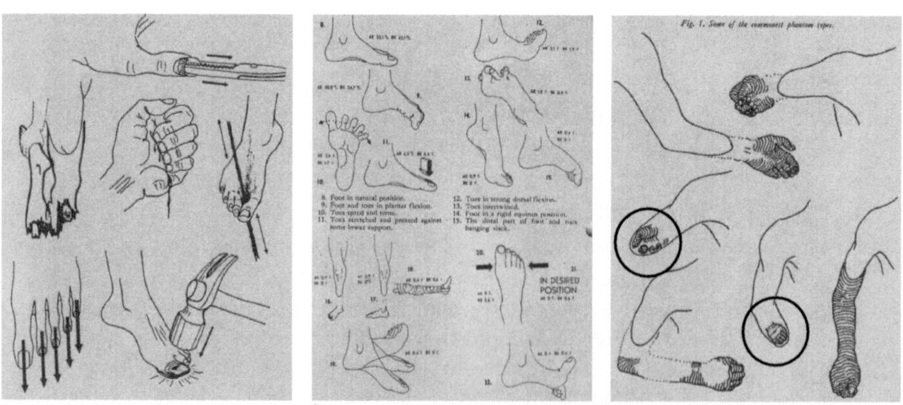

Figure 2: Painful (a) and painless (b) phantom sensations. (c): Long-term changes in the appearance of phantom limbs: telescoping and "within-stump phantom limbs" (circled). Reprinted and adapted, with permission, from Solonen (1962), Figures 1 and 2, Table 7.

this view, was the price mammals pay for possessing a highly plastic central nervous system. This became known as the *maladaptive plasticity* model of phantom pain.

More recent work, however, has challenged this interpretation. An alternative *maintained representation* model has been proposed (Makin et al. 2013, 2015). Functional MRI studies show that the former hand area remains active when amputees move their phantom hand. Rather than being "invaded," the hand representation may persist—and it is this maintenance, rather than maladaptive change, that matters.

What are "homunculus" and "femincula"?

The term "homunculus" refers to the cortical "body map" in the postcentral gyrus (the sensory homunculus) and in the precentral gyrus (the motor homunculus). Homunculus is Latin for "little man." The term reflects both the history and the gender bias of early neuroscience: until quite recently, no equivalent term existed for the female brain. In a welcome shift, femincula—"little woman"—has been proposed as a counterpart (Kinsbourne 1998; see also Di Noto et al. (2013) for the linguistically less satisfying *hermunculus*).

For the sake of inclusivity and linguistic clarity, we propose the gender-neutral term somaculum (plural: somacula) to denote cortical body maps regardless of gender (see Figure 3). A few key points are worth noting:

1. Somaculum topography is not arbitrary, but lawful.In the postcentral gyrus, the hand and face are represented adjacently along the lateral surface, while the lower extremities and genitals are represented medially, also adjacent to one another (Figure 3). These pairings likely reflect fetal body positioning in utero, where hands frequently contact the face, and feet the genital region (Farah 1998). These early contacts may underlie some phenomena of corporeal awareness discussed later, such as referred sensations (p. 217) and the erotic element occasionally associated with Body Integrity Dysphoria (p. 233), specifically after the desire for amputation of a *lower* limb.
2. The original homunculus is based on rigorous, yet limited, data—and imaginative interpolation.
   The famous cortical maps by Penfield and Boldrey (1937) were based on electrocortical stimulation and patient reports. But the "body shapes" we now visualize are interpolated from sparse data sets, with uneven sampling and variable interpretation. There is room—indeed, historical precedent—for imagination (Schott 1993).
3. Beyond binary: consider a "queerunculum."
   If the *homunculus* and *femincula* reflect binary anatomical identities, a *queerunculum* might be proposed to accommodate brain-body mappings that do not conform to the female/male binary. This would reflect the lived experiences and neural realities of gender-diverse individuals.
4. The somaculum is not limited to the cortex and constitutes a low-level body representation.
   While Penfield's maps remain iconic, they are neither the earliest nor the most integrated representations of the body. Subcortical somacula are already present in the thalamus—beautifully illustrated and discussed

by Schott (1993). Conversely, higher-order representations—sometimes referred to as body schema or body image—are too distributed and socially constructed to be mapped onto a specific region. These body constructs incorporate cultural, aesthetic, and emotional dimensions, and show consistent right-hemispheric dominance (e.g., Hilti et al. 2013; McGeoch et al. 2011).

Before neuroimaging ever revealed the adjacency of hand and face representations in the postcentral gyrus, clinical observations had already hinted at this organization. Arm amputees often report that shaving, applying makeup, or being touched on the face—by themselves or others—elicits phantom hand sensations

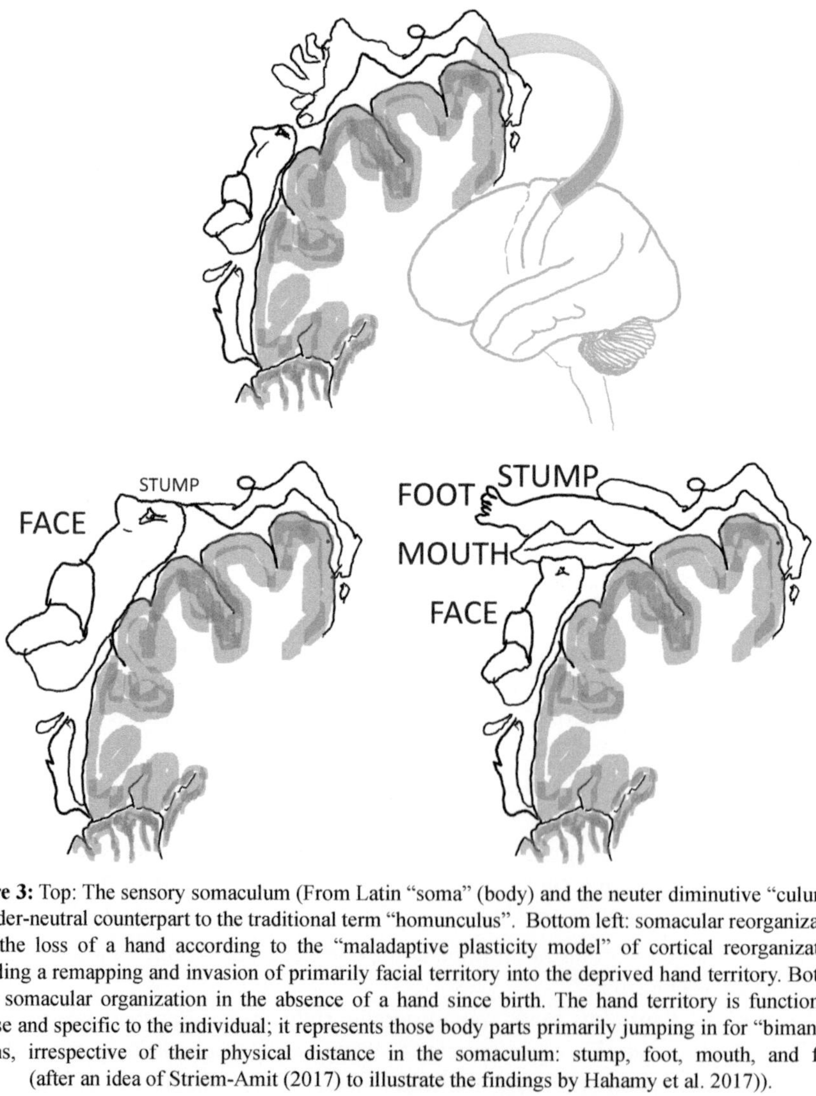

Figure 3: Top: The sensory somaculum (From Latin "soma" (body) and the neuter diminutive "culum"): a gender-neutral counterpart to the traditional term "homunculus". Bottom left: somacular reorganization after the loss of a hand according to the "maladaptive plasticity model" of cortical reorganization, including a remapping and invasion of primarily facial territory into the deprived hand territory. Bottom right: somacular organization in the absence of a hand since birth. The hand territory is functionally diverse and specific to the individual; it represents those body parts primarily jumping in for "bimanual" actions, irrespective of their physical distance in the somaculum: stump, foot, mouth, and face. (after an idea of Striem-Amit (2017) to illustrate the findings by Hahamy et al. 2017)).

simultaneously with the facial contact (Henderson and Smith 1948). Conversely, people with deafferentation of one side of the face sometimes report phantom facial sensations when the hands are touched (Clarke et al. 1996).

More rarely—perhaps due to cultural taboos—lower limb amputees have reported heightened phantom foot sensations during urogenital stimulation (e.g., micturition, defecation, or sexual activity; Henderson and Smith 1948).

These referred sensations—from face to hand and vice versa (Uysal et al. 2021)—are compatible with the maladaptive plasticity model of cortical remapping, but they do not contradict the maintained representation model. In fact, these two models have recently been discussed in a joint paper by their principal proponents, Makin and Flor (2020).

Experimental elicitation of referred sensations in amputees, people born with only one hand, and neurotypical controls (Amoruso et al. 2023) confirmed the existence of the phenomenon, but also raised a note of caution: early studies may have been partially shaped by suggestion effects.

In any case, referred sensations are a beautiful reminder that even seemingly bizarre patient reports can reveal important cartographic details of the brain's body map.

? Every now and then, I experience "referred sensations" on my own body, without any amputation or deafferentation. When I scratch an itchy part of my skin, I feel the touch at the location of the itch—but also on a remote part of my body. How is this possible?

You are describing a fascinating phenomenon known even in English by its German name: *Mitempfindung*—co-sensation. It has been described as a "mysterious form of referred sensation" (Richter 1977), experienced by healthy individuals when scratching or pinching a small region of skin.

Schott (1988) provides an overview of the early literature and reports that *Mitempfindung* is often noted in individuals with sensory impairments at the site of referral. The physiology remains unclear, and the phenomenon itself seems to vary widely: some people experience it frequently, while others never do.

One hypothesis (Schott 1988) is that it reflects hyperexcitability in spinal or thalamic pathways. A striking correlation comes from a study by Burrack et al. (2006), which found that *Mitempfindung* occurs about four times more frequently in people with synesthesia. They suggest that both phenomena reflect atypical—or erratic—neural connectivity.

Anecdotally, healthy individuals (non-synesthetes) who experience strong "phantom touch" illusions (see Figure 1d) are also more likely to report *Mitempfindungen* compared to peers with weaker illusion responses. This observation has not yet been formally published, but it certainly merits systematic investigation.

? Is there any evidence for phantom limbs in animals?

It is difficult to empirically demonstrate phantom limb phenomena in non-human species, since we lack direct access to their perceptual reports. Still, there are indirect indicators—particularly in cats and dogs—that suggest such experiences may occur. Post-amputation, these animals often lick or nibble at their stumps. This could reflect stump pain, but a more specific anecdote comes from a friend of mine: after a forelimb amputation at the elbow, his dog repeatedly licked a spot on the floor—not near the stump, but precisely where the missing paw would have touched the ground.

Searching for "Dog Bites Own Leg" in the channel YouTube will guide you to a video clip showing a dog's "anarchic limb"—a phenomenon discussed below as a potential non-human example of phantom phenomena.

? If phantom limbs are based on perceptual-motor memories of a once-functional limb, then people born without a limb should not experience phantom sensations, right?

Yes, in theory. If phantom limbs are purely the result of remembering previous sensorimotor experience, then people with congenital limb absence should not have them.

And yet—they often do. Which means there must be more to the story than memory. The question is so central to this chapter that it deserves a dedicated section. Just keep reading.

The Bodily Self in Amelia: Phantom Sensations of Congenitally Absent Limbs

Throughout the 19th and much of the 20th century, it was repeatedly asserted—by otherwise independent clinicians—that people born without a particular limb do not experience phantom sensations. The limb, so the thinking went, had never been part of the "body schema," and therefore could not belong to the repertoire of corporeal awareness.

These statements carried authority, but were rarely based on systematic observations. More often, they reflected an a priori conviction: what should not occur was assumed not to occur. Such reasoning is of considerable interest to the philosophy of science—especially to the sociology of knowledge and the plasticity of "scientific facts," as articulated by Ludwik Fleck (1935/1979). Case reports of congenital phantom limbs[1] slowly gained attention across the scientific community. Following Fleck's model, one might say that one early case report (Valentin 1836) remained in a state of "latent discovery:" it existed but failed to shape prevailing theoretical frameworks. Later reports were often force-fitted into existing models—interpreted, for instance, as mirror reflections of the intact contralateral limb, or (in cases of phocomelia with finger buds) as projections of sensory input from the buds into extracorporeal space.

[1] This chapter uses the term "congenital phantom limb" for simplicity. Of course, it is the absence of the limb that is congenital, not the phantom itself. Nonetheless, when asked explicitly, individuals with such experiences often report sensing the missing limb "for as long as I can remember." When I asked AZ, on our first meeting, since when she feels the phantom limbs, she answered with the ironic question "Since when do you see colors?"

A more recent theory views congenital phantoms as the phenomenal expression of inborn motor schemata—e.g., for hand–mouth coordination (Gallagher et al. 1998). These interpretations are imaginative and not without merit. But they overlook key facts: phantom hands have also been reported in individuals without finger buds, and phantom sensations in lower limbs—rendering a hand–mouth schema explanation insufficient.

For a detailed review of the history of congenital phantom limbs—and their long-standing neglect by the medical establishment—see Brugger and Funk (2007).

? Who was the first to demonstrate empirically the presence of phantom sensations in a limb that had been absent since birth?

It is always risky to answer questions about scientific priority. But it seems clear that Gabriel Gustav Valentin (1810–1883), a physiologist in Bern, Switzerland, was among the first to go beyond anecdote.

In one case (Valentin 1836), a woman born without a left hand reported vivid sensations of the missing limb—including all five fingers. Valentin applied pressure to specific nerves in her stump, and the woman described tingling or numbness in the corresponding (phantom) fingers—perfectly aligned with the known nerve distribution.

This elegant experiment, published in 1836, offered empirical support for the claim that congenital phantoms are not just "wishful thinking." Yet many later case reports [reviewed in Melzack et al. (1997) and Price (2006)] lacked Valentin's experimental rigor. It likely delayed broader acceptance of congenital phantoms as a genuine neuropsychological phenomenon.

A particularly striking case is that of subject AZ, a woman born without hands and legs, introduced by Brugger et al. (2000). AZ described detailed and vivid phantom experiences, including body posture preferences that appeared influenced by her phantoms (see Figure 4). A series of experiments aimed to "objectify" her reports—adding behavioral and physiological evidence to her introspective descriptions.

Hand Laterality Decisions

AZ was shown images of hands and feet—either left or right limbs—centered on a computer screen (exposure time: 500 ms). She responded by pressing a key with the left upper-limb stump for "left" and with the right stump for "right." (For an online version of this hand laterality task, see Moreno-Verdú, 2025.)

AZ's responses were significantly faster when fingers or toes pointed upward rather than downward. This pattern is well known from typically limbed observers, who tend to "slip" their own internal limb representation into the visual image. This visuo-postural mapping requires more time when 180° mental rotation is involved. AZ's performance thus indicates an intact postural representation of both hands and feet.

Apparent Motion of Limbs

In a paradigm introduced by Shiffrar and Freyd (1990), Funk et al. (2005) presented AZ with pairs of photographs depicting the beginning and end of simple hand

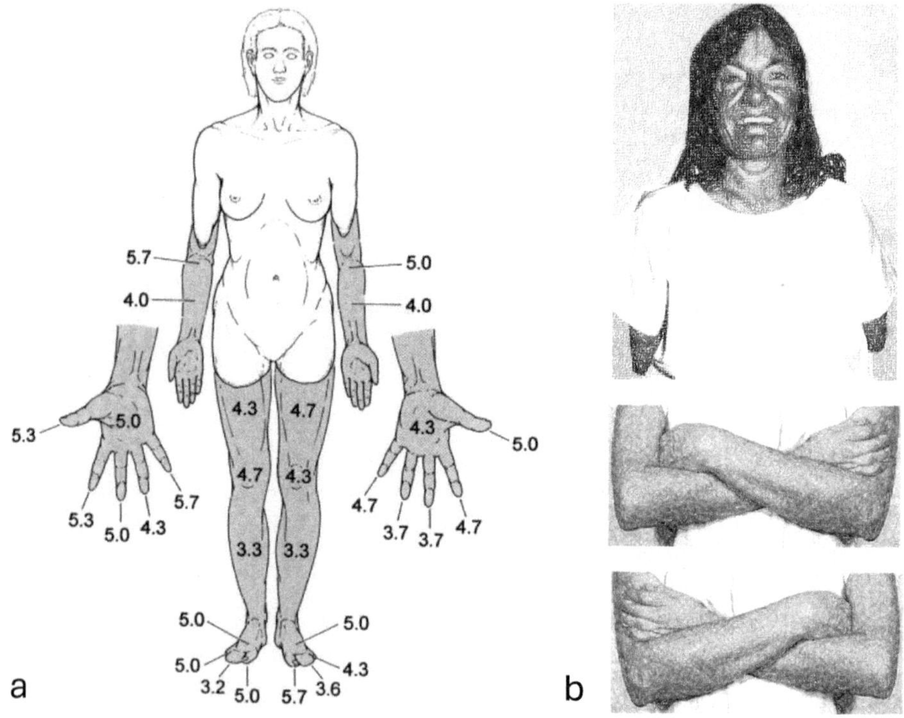

Figure 4: a: AZ was born without forearms and legs. The dark body parts are experienced as phantom limbs. The numbers indicate the vividness of body awareness, from 0 (no awareness) to 7 (most vivid awareness). b: asked to fold her phantom arms and indicate which arm would be on top, she folds them and indicates a left-top posture. When asked to do it the other way around (right arm top) she shakes her body and says she would never do it this way. a: reprinted from Brugger et al. 2000, PNAS Vol. 97, 6167-6172, Figure 1, copyright 2000, National Academy of Sciences.

movements. Each pair differed only in hand position relative to the wrist. When alternated rapidly, the two images induced an apparent motion percept.

In typically limbed observers, very rapid alternation (< 150 ms) yields an illusory short movement path—even when anatomically impossible due to joint constraints. Slower alternation (approximately 1,000 ms) allows perception of a longer, anatomically plausible trajectory.

AZ exhibited the same flash-rate dependency as control participants, suggesting that her brain accesses stored kinesthetic knowledge of hand movement, likely represented in motor and parietal cortices (Stevens et al. 2000; see Saetta et al. 2018 for work with lower limb amputees).

fMRI of Phantom Hand Movements

In an fMRI scanner, AZ was asked to perform phantom finger-to-thumb opposition movements at a pace of approximately 1 Hz. Surface EMG confirmed the absence of residual upper-arm muscle activity.

Phantom finger movements produced consistent bilateral activation in the dorsal premotor cortex and parietal lobes. Additional activation appeared in mesial premotor areas corresponding to the supplementary motor area (SMA). Importantly, there was no activation of the primary motor cortex hand area, nor in the postcentral gyrus.

These findings suggest that AZ's phantom hand is neurally represented in premotor and parietal cortices—regions associated with planning and internal simulation of movement, rather than primary motor execution.

TMS-Induced Phantom Sensations

Transcranial magnetic stimulation (TMS) over the sensorimotor cortex elicited twitch-like sensations in the contralateral stump, with corresponding motor-evoked potentials (MEPs). In contrast, stimulation of premotor and parietal areas produced similar phantom sensations but without measurable MEPs. These were not described as twitches, but as slow movements of phantom fingers, the entire hand, or occasionally the forearm.

Integration and Interpretation

Taken together, the behavioral and neuroimaging findings from AZ suggest that her brain contains stored information about joint constraints and kinesthetic movement, even for body parts she has never physically possessed or moved.

It would be premature, however, to conclude that body representations are fully innate. At least some components—particularly proprioceptive or kinesthetic elements of the body schema—may have been acquired through early observation of others moving their limbs.

The existence of AZ's somatic phantom sensations might reflect activity within a system that matches action observation to action execution (Rizzolatti and Sinigaglia 2007). But the human mirror system cannot be the only source of such phantoms. If it were, then everyone born without limbs would experience them. This is demonstrably not the case.

In a study by Hahamy et al. (2017), none of the 17 participants born with only one hand reported phantom sensations—not even in night dreams. What differentiates individuals like AZ, who show a vivid "animation without incarnation," from others born with incomplete bodies remains an open and intriguing question.

? Everything you wrote here about congenital phantoms rests on observations in a single person. What is, quite generally, the status of single case reports in neuropsychology?

Opinions on this matter diverge widely. At one end of the spectrum are researchers who dismiss single case reports entirely—often because they are preoccupied with designing the next multicenter study. These skeptics may overlook a central principle of scientific reasoning: the critical role of falsification.

At the other end are those so fascinated by a patient's unique symptom that they rush into publication without broader context or replication.

An editorial in *Nature Neuroscience* offered helpful guidance on this question, titled "When once is enough" (Editorial 2004). Three main criteria were proposed:

1. The observation should be prohibitively difficult—not merely inconvenient—to replicate in other cases.
   In the case of congenital phantom limbs, many reports exist, including small group studies, but they are almost entirely descriptive. A notable exception is Valentin's 1836 case (see above, p. 219), which included clear experimental manipulations.
2. The observation should be theoretically significant.
   In AZ's case, it certainly is. If phantom limbs are conceptualized as sensorimotor memories of once-owned limbs, then it matters—a great deal—when someone without such a limb reports a vivid phantom. This directly challenges existing assumptions.
3. Ideally, the case should falsify a widely held theoretical view.
   To disprove the claim that pigs cannot talk, you do not need a population study—you only need one talking pig.

This criterion is especially important in neuropsychology, where single cases have historically advanced core ideas about brain–behavior relationships. In the case of congenital phantoms, the literature consists almost entirely of single case reports (see Price 2006, for a review). This might explain why such phantoms still are not universally accepted as "scientific fact." But theoretical resistance does not negate empirical validity.

Phantom Legs in Paraplegia

In paraplegia due to thoracic spinal cord injury, both legs are paralyzed. This results in deafferentation (loss of incoming sensory input from the legs) and deefferentation (loss of outgoing motor commands to the legs). You cannot feel your legs—and you cannot move them.

And yet, patients may still perceive their legs as separate from their physical bodies—phantom legs without amputation. In an fMRI study of imagined foot movements in completely paralyzed individuals, the vividness of phantom foot movement correlated with peak activations in cortical motor areas (Alkadhi et al. 2005). On phantom sensations in paraplegic patients' night dreams see Brugger (2008b).

But the important lesson from phantom limbs in paraplegia is not just that they occur—it is why they tend to fade over time. Unlike post-amputation phantoms, which often persist for years, paraplegic phantoms diminish. The likely reason is that constant visual access to the motionless legs gradually weakens their phantom vividness.

This effect can even be observed in healthy participants: viewing one's own hand while it is touched reduces somatosensory-evoked potentials in the contralateral primary somatosensory cortex (Longo et al. 2011). Vision calms the somesthetic ghost.

Phantoms of Central Origins: From Alien Limbs to "Hemiplegic Twins"

The phantomology of the bodily self after central nervous system lesions is marked by a bewildering phenomenology. A full treatment of clinical presentations and experimental findings is beyond the scope of this chapter.

What matters here is a qualitative difference between peripheral/spinal phantoms and those arising after brain damage. In the former, phantoms are experienced as "as if" phenomena: patients are aware that they are illusory. In central lesions, this awareness can vanish.

Just as hemispatial neglect (see Chapter 9, p. 184) occurs more frequently after right hemisphere damage, disorders of body and self-awareness also show a right-side predominance. Some authors (e.g., Bartolomeo et al. 2017) have proposed a broad interpretation: that such phenomena reflect confabulations in corporeal and spatial domains, often arising from the left hemisphere's attempts to rationalize distorted or incomplete input from its damaged counterpart.

In this light, we now briefly discuss the following conditions:

* Supernumerary phantom limbs
* Alien and anarchic limbs
* Asomatognosia
* Anosognosia
* Somatoparaphrenia (Critchley's "hemiplegic twin")
* Misoplegia

Supernumerary Phantom Limbs

These are phantom arms or legs experienced in addition to the corresponding physical limb. Often, only partial deafferentation of the duplicated limb is present, as in some cases of multiple sclerosis. A review of stroke patients reported an incidence of > 50%, suggesting that the phenomenon is likely underreported (Antoniello et al. 2010).

While many patients retain insight into the illusory nature of the phantom, some lose that insight and perceive the supernumerary limb as an actual "extra" or "spare" extremity. In rare cases, a supernumerary limb may even be the *sole* experiential content of an epileptic seizure (Millonig et al. 2011).

McGonigle et al. 2002 describe an fMRI study of a patient with a right fronto-mesial lesion who experienced a "ghost arm" that mirrored all movements of her real arm with a delay of approximately one to two minutes. Peak activation during phantom arm movements (compared to rest periods, when the percept was absent) was in the supplementary motor area. This indicates that regions traditionally thought of as purely "motor" contribute significantly to somatognosia.

Alien and Anarchic Limbs

Although not always included under the umbrella of phantom phenomena, alien limbs deserve mention here. The sense of "alienness" can affect either somatognostic or motor experience.

* *Sensory alien limb*
  The alienness concerns belonging: the patient feels the limb is not theirs, though it behaves normally. Brion and Jedinak's (1972) seminal case implicates posterior corpus callosum lesions. Importantly, there is no alienness of action—only of ownership.

* *Motor alien limb*
  Here, the limb *feels* like one's own—but its actions are unintended and unpredictable. This condition is classically associated with anterior corpus callosum lesions and frontal lobe damage.
* *The anarchic limb*
  The anarchic hand is a motor alien limb whose actions are not merely unintended, but resist suppression, acting in direct conflict with the patient's will. Marchetti and Della Sala (2010) use the term "anarchic" to describe this compulsive goal-directedness.

See Brugger (2001) for patient examples—nearly all with right hemisphere lesions—exhibiting varying degrees of hostile behavior, from nipple-pinching to attempted strangulation. Lesions usually mirror those seen in motor alien limb, though rare exceptions include posterior lesion sites typical of sensory alien limb — see this case vignette:

> Patient AL (for Anarchic Limb) is a 57-year-old converted left-hander and bookkeeper. While hiking alone in the Swiss mountains, he suffered a right parieto-occipital stroke.
>
> He vividly recalls the moment it happened: a flock of crows had gathered in a meadow, evoking Hitchcock's *The Birds*. Minutes later, he felt a sharp pain on his vertex, like a beak pecking his skull. When he raised his left arm, he was horrified to see what he believed was a crow on his head, pulling out his hair. In panic, he tried to strike it—only to realize he was fighting with an arm he did not initially recognize as his own (only later did he infer from the red sleeve that it must have been his left arm).
>
> He lost consciousness mid-struggle. When found by another hiker, his hair had been pulled out in clumps. He was airlifted to a hospital. Imaging revealed a recent infarct, as well as scattered ischemic lesions throughout the right hemisphere.
>
> Later, during neuropsychological testing, he drew a striking image (Figure 5a): the invisible crow perched at the top of his head. (The patient explained that he was not talented enough to draw the arm as well.)

Asomatognosia

Asomatognosia literally means the absence of somatognosis. Critchley (1953) defined it as a "feeling of 'nothingness'" echoing Bonnier's (1905) older term "aschematia." This definition remains widely accepted by leading experts in corporeal awareness.

More recently, however, the term has been confused with somatoparaphrenia (as criticized by Saetta et al. 2021). In its classic form, asomatognosia is non-delusional. Most cases are unilateral, affecting the left body half—functionally akin to a sensory alien limb extended to an entire side. Unlike in somatoparaphrenia, the person does not deny ownership; they simply experience the body part as puzzling, spooky, or frightening.

A patient might ask: "How can it be that I see something whose absence I so convincingly feel?" (Saetta et al. 2021).

a

Anarchic limbs may attack the own body, the hostility ascribed to outside forces; see text. (A patient's drawing, here used with permission).

b

In somatoparaphrenia and misoplegia, the "foreign" hemibody may be physically abused.

c

Violent attacks between the self and its heautoscopic double come in four variants (see text, p. 231).

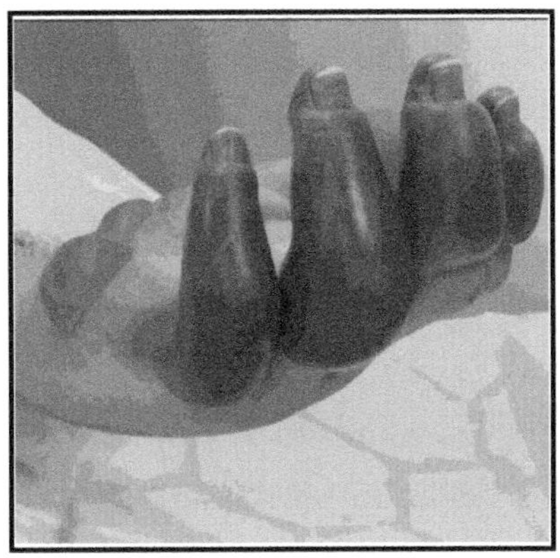

d

The left hand of a patient with BID after immersion in dry ice to enforce amputation (photography by the patient, with permission).

Figure 5: Hostile interactions between body and self. a) A patient's drawing of how a crow pulled out his hair, which, in reality, his anarchic left hand had pulled out. b) A patient with misoplegia in the act of beating her left leg. c) An antagonistic double sneaks in from the left side and harasses the self (from Menninger-Lerchenthal (1946) with permission by Hans-Huber publisher). d) The left hand of a young patient with BID after being held in dry ice for 3 hours. See text for explanation.

Lesions are typically found in the posterior parietal cortex, though the right premotor cortex has also been implicated (Arzy et al. 2006). When an absent visual experience matches an absent feeling, one speaks of (hemi)asomatoscopy (see the section on autoscopic phenomena and Figure 7e).

Anosognosia

Long before Babinski coined the term in 1914, clinicians had noticed that some patients deny or ignore their own illnesses. In anosognosia, a patient exhibits agnosia for their own condition—especially common in cases of paralysis, aphasia, blindness, and cognitive deficits.

Despite over a century of research, the mechanisms remain obscure. A special issue of *Cortex* (Vol. 61) marking the centenary of Babinski's term noted:

"In spite of this century of almost continuous research, a clear understanding of the mechanism, and even the nature of the disorder, remain elusive." (Turnbull et al. 2014).

Turnbull and colleagues emphasized the emotional regulation role of the right hemisphere, suggesting that anosognosia involves more than failed sensory registration—it reflects a broader deficit in self-monitoring and affective integration.

Neuroanatomical correlates include:

* premotor, motor, and somatosensory areas (Berti et al. 2005)
* the right insular cortex (Karnath and Baier 2010)
* the right inferior frontal gyrus (Kortte et al. 2015)

These findings explain the *classical* form of anosognosia (e.g., denial of hemiplegia). But their explanatory power is questionable for other striking forms, such as

* anosognosia for amnesia,
* anosognosia for Wernicke's aphasia (see Chapter 4),
* anosognosia for psychosis (Lehrer and Lorenz 2014), and
* healthy people's "prospective anosognosia," i.e., unrealistic optimism—a denial no less irrational than any clinical symptom of denial (see Chapter 9, p. 201).

? **Can patients with anosognosia lie?**

A provocative question—and surprisingly underexplored.

The answer is "yes"—at least when the lies are unrelated to illness. In sentence verification tasks, patients with anosognosia for hemiplegia or cortical blindness can lie about neutral statements (e.g., "Frogs sometimes eat flies") just like healthy controls. Lying takes longer, but it is cognitively possible.

However, the answer is "we don't know" when the statements are illness-related (e.g., "My left side is paralyzed" or "I can move all my limbs"). Our small number of pilot participants produced ambiguous results: some affirmed their paralysis when instructed to lie, others persisted in their original denial.

We also tested vision-related statements in a patient with Anton's syndrome. Interestingly, he treated "I am blind" as just another neutral sentence, while treating paralysis-related statements as emotionally charged.

This research is not yet published—but may inspire you to explore it yourself. We recommend an auditory sentence verification task. A caveat: lying requires frontal lobe integrity (Yang et al. 2005; Spence, 2005), which may be compromised in anosognosia. Still, correlating lying performance with lesion site could offer deep insight into the neural architecture of symptom denial.[2]

Somatoparaphrenia (Critchley's "Hemiplegic Twin") and Misoplegia

Somatoparaphrenia (a term introduced by Gerstmann, 1942) is a delusion of the bodily self. The alternative label «personification anosognosia» captures the essentials: patients with anosognosia for (typically left-sided) hemiplegia not only deny the paralysis but also disown the paralyzed half of their body. They may ask medical staff to "remove the other person" from their bed. This "other" is often described as a relative, a nurse, a malign stranger, or even a dangerous animal.

MacDonald Critchley (1953) famously referred to the "somatoparaphrenic other" as the hemiplegic twin. A comprehensive review of the literature is provided by Vallar and Ronchi (2009). A recent case study (Saetta et al. 2021) used fMRI to pinpoint the locus of personification to the right inferior frontal gyrus. Moreover, a resting-state fMRI experiment in the same study supported the idea that the confabulatory component of somatoparaphrenia results from a disconnection between right parietotemporal regions and the left hemisphere's language areas (Halligan et al. 1995; Bartolomeo et al., 2017).

Misoplegia is a morbid dislike or hatred of a paralyzed limb. The term was coined by Critchley (1953; 1974). Patients may curse, strike, or attempt to mutilate their hemiplegic limb. Unlike in somatoparaphrenia, the aggression is explicitly directed at a body part that is acknowledged as one's own. Even when the affected limb is given a nickname, it is not attributed to another person.

Loetscher et al. (2006) report a striking case of misoplegia without hemiplegia in a woman with a right temporo-parietal tumor. Her aggression—repeated, excessive beating of her non-paralyzed left leg (see Figure 5b)—may have been linked to a premorbid condition: childhood poliomyelitis had left her left leg 6 cm shorter than the right.

A short video clip documenting the patient's misoplegia can be downloaded as supplemental material from Loetscher et al. (2006).

? **You mention various clinical conditions of phantom limb. What about the experimental induction of phantom limbs in healthy volunteers?**

There is indeed a substantial literature on the experimental evocation of phantom phenomena in healthy individuals. One of the earliest demonstrations used pressure cuff anesthesia of an arm stretched out on a table (Melzack and Bromage 1973). With eyes closed, the volunteer reports that the anesthetized arm is felt in a more proximal position than it actually is. As soon as the eyes are opened, the illusory percept vanishes: the phantom arm fuses almost instantly with the physical one. This elegantly illustrates the role of vision and

[2] To avoid confounds from brain damage, you might adapt this paradigm for use in healthy participants who differ in their degree of unrealistic optimism.

its interaction with proprioception (see also the paragraph on phantom limbs in paraplegia, p. 222).

The rubber hand paradigm can likewise be seen as a method of experimentally inducing a phantom hand. The simultaneous binding of vision, touch, and proprioception causes a person to experience their own hand as shifted in space—toward the rubber one (see Figure 1e above).

Another classic technique involves muscle or tendon vibration, which produces an illusory lengthening of the vibrated muscle and thus a perceptual dissociation between the real limb and its phantom. Lackner (1988) and Kammers et al. (2006) provide a range of such experiments. Figure 1f exemplifies this principle in the induction of a phantom nose, following vibratory stimulation of the biceps.

Finally, the illusion of phantom touch can be induced using a mirror setup: the participant observes both hands being touched in the mirror, and then the touch on one hand is suddenly withdrawn. The continued visual feedback paired with a lack of tactile input creates a ghostly, often startling experience (see Figure 1d).

Whole-Body Phantoms: The Six Variants of Autoscopic Phenomena

Autoscopic phenomena are a heterogeneous class of experiences in which one's own body is duplicated or displaced in space. They may be unimodal—purely somesthetic, visual, or vestibular—or they may involve multisensory integration across these modalities.

Autoscopic phenomena are highly relevant to phantomology because they demonstrate that not just limbs, but the entire body can become a phantom. Moreover, they illustrate the importance of paying close attention to the subtle details in patients' descriptions of hallucinatory experiences.

If a patient ever tells you, "I have seen myself," the paragraphs below offer a guide to the many follow-up questions you should ask in order to specify and classify the experience.

Figure 6 introduces the six recognized variants of autoscopic phenomena.

Extended Legend to Figure 6

The sensed presence (Figure 6a). In some respects, this variant is most closely related to the experience of a phantom limb. The being whose presence is sensed is not seen, but can often be precisely localized in near-extracorporeal space. Unlike a phantom limb, however, the sensed presence is not recognized as a part of oneself. Still, some observations suggest it may be a reduplication of non-visual aspects of one's own body. The sensed presence often imitates one's own movements, particularly in the so-called "third man experience" reported by mountain climbers (Child 1989; Brugger et al. 1999).

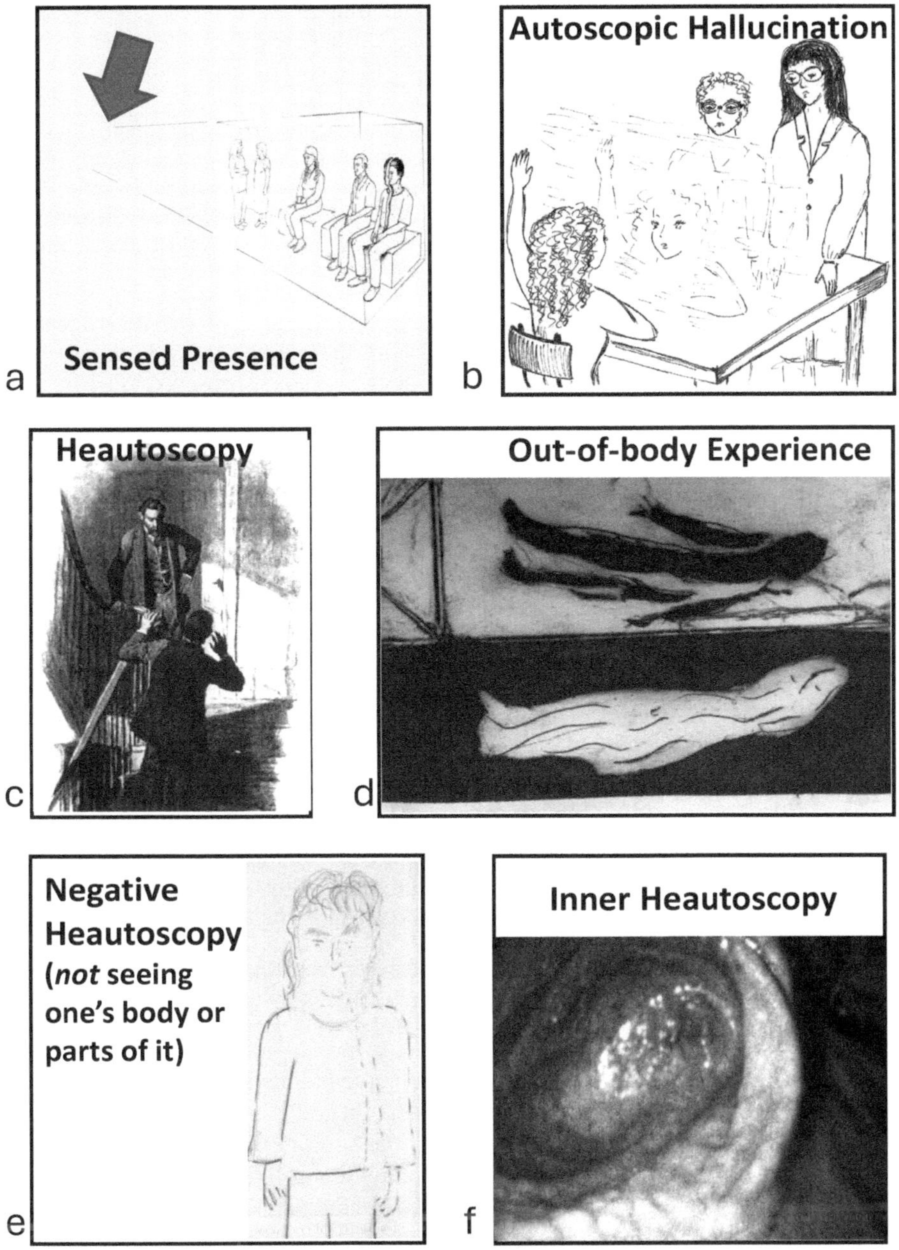

Figure 6: Felt or seen reduplications of one's own body. Six types of autoscopic phenomena (see the text). Sources: a) Own collage, modified from Brugger et al. 2006, Figure 1. b) Reprinted, with permission, from Zamboni et al. 2005, Figure 2. c) Retrieved from Project Gutenberg: http://onlinebooks.library.upenn.edu/webbing/gutbook/lookup?num=6995. d) reprinted with permission by the artist. e) With the permission of the patient. f) Endoscopic image by Stelarc (original in color), with permission by the artist.

> A particularly vivid case was described in Brugger et al. (2006): a 41-year-old right-handed pottery maker, PH (for poliopic heautoscopy), hallucinated an entire *doppelgänger* family (Figure 6a). While the five figures closest to him were both seen and felt, a boy—positioned some 20 meters to the right—was only felt, not seen (see arrow). Remarkably, PH felt that *all* hallucinated figures, including the invisible boy, were aspects of his own self. His experiences combined features of heautoscopy (Figure 6c) and the sensed presence, suggesting that the latter may, in some cases, reflect own-body projections into extracorporeal space. All of PH's doubles, strictly lateralized to the right hemispace, were experienced as consoling and supportive. He died 16 months after excision of a left insular tumor.

Barnby et al. (2023) provide a comprehensive review of the heterogeneous literature on the sensed presence, including work that treats it outside the scope of autoscopic phenomena (see also the monograph by Alderson-Day 2023).

Autoscopic Hallucinations (Figure 6b). This illustration shows a 30-year-old right-handed female patient (foreground) and two examiners (background). Between them is the patient's visually hallucinated double, which partially overlaps with the background. This case, reported by Zamboni et al. (2005), is prototypical for an autoscopic hallucination:

* The double is a mirror image of the patient.
* Only visual features are reduplicated; the observer's perspective remains body-centered.
* The patient's lesions were primarily occipital in location.

Zamboni's patient experienced autoscopy (henceforth used as a shorthand for autoscopic hallucination) after recovering from cortical blindness due to a bilateral occipital stroke. Other cases involve unilateral posterior strokes (more often right-sided), with the autoscopic double appearing in the blind visual field (Bhaskaran et al. 1990). Seizures—especially with occipital or occipito-temporal foci—may also be accompanied by autoscopy (Maximov 1973; Genner 1947; Devinsky et al. 1989).

Heautoscopy (Figure 6c). While autoscopy derives from the Greek αὐτό (self) and σκοπεῖν (to look at), heautoscopy includes the reflexive ἑαυτός (oneself), signaling that the self is seeing another self.

This is the core distinction between autoscopic hallucination and heautoscopy: in the former, the observer is clearly anchored to the body; in the latter, there are two centers of selfhood—one in the patient, the other in the double. Visual similarity is secondary in heautoscopy; the double may look different but is identified by somesthetic features—a bodily feeling projected onto the double, a "shared proprioceptive unit."

Note that in Figure 6b, the autoscopic double waves with its right hand when the patient waves with her left—a mirror reversal. In contrast, no such reversal occurs in heautoscopy (Figure 6c). There are two distinct perspectives, one centered on the patient, the other on the heautoscopic double.

The two perspectives may rapidly alternate within a single heautoscopic episode, especially in patients with seizure disorders. This unsettling experience of being in two places at once can be so disorienting that a patient may attempt suicide "in order to stop the intolerable feeling of being divided in two" (Brugger et al. 1994a).

This phenomenon, termed "heautoscopic suicide" (Brugger 2007), has been categorized into four variants:

(1) the patient's attempt to kill himself as an escape from the split self,

(2) the attempt to kill the double,

(3) the experience of being attacked by the double, and

(4) the observation of the double commiting suicide.

It is this intricate psychological interplay between the self and its reduplication that makes the *doppelgänger* such a compelling motif in literature and film (Dieguez 2013).

Before this explicit distinction (Brugger et al. 1997; Brugger 2002; Blanke et al. 2004, Figure 5), autoscopy and heautoscopy were often treated interchangeably. Yet such a distinction is crucial if one aims to elucidate the neural mechanisms underlying the specific aspects of autoscopic phenomena—visual, somesthetic, and vestibular.

A recent symptom-lesion mapping study in 26 patients by Blondiaux et al. (2021) identified the bilateral temporo-parietal junction (TPJ) as the common core region implicated in autoscopic hallucinations, heautoscopy, and out-of-body experiences (OBEs). Earlier work by Heydrich and Blanke (2013) had suggested distinct but overlapping networks for these phenomena:

* the right occipital lobe for autoscopy,
* the left posterior insula for heautoscopy, and
* the bilateral parieto-temporal cortex for OBEs.

Involvement of the left hemisphere in the sense of presence is also supported by the observation that—when consistently lateralized—the sensed presence is more often felt in the right hemispace (Brugger et al. 1996b).

Out-of-Body Experiences (OBEs; Figure 6d). Before OBEs were recognized as part of the broader family of autoscopic phenomena (Brugger et al. 1997), they were largely confined to the domain of parapsychology (see Chapter 12, footnote 13). Thanks to the collaborative work of neurologist Olaf Blanke and philosopher Thomas Metzinger, OBEs are now acknowledged within mainstream cognitive neuroscience as key phenomena in the empirical study of the bodily self (Blanke and Metzinger 2009; see also Brugger and Mohr 2009).

A seminal moment was their joint report in *Science* (Lenggenhager et al. 2007), published alongside a convergent article by Ehrsson (2007), using virtually identical methodology to induce OBEs in healthy volunteers via virtual reality setups.

Both experimental approaches exploited the principle underlying the rubber hand illusion (Figure 1e), which induces an "out-of-hand" experience by synchronizing visual observation of touch on a fake limb with tactile stimulation of the unseen real hand.

In both Lenggenhager et al. (2007) and Ehrsson (2007), participants wore virtual reality goggles through which they saw themselves being touched by an experimenter standing behind them, while simultaneously feeling touch on a corresponding area of their body (the back in Lenggenhager et al., the chest in Ehrsson).

Participants experienced a dislocation of the first-person perspective, reporting that their observing self was centered behind the physical body—matching the camera's viewpoint, feeding the goggles. In both paradigms, the out-of-body illusion was induced by shifting the self's spatial origin away from the body normally anchoring it.

Of historical interest, a simplified version of such a "virtual reality" illusion—mediated by mirrors rather than cameras—was already employed over a century earlier to elicit OBEs (Stratton 1899).

Figure 6d shows an *intaglio* print by a 75-year-old right-handed woman (a visual artist) with a long history of migraine. During attacks, she feels herself hovering above her physical body, viewing the room—including her own body—from an elevated perspective. Sometimes, as depicted here, the body itself is not seen because her gaze is directed upward and she perceives fantastic scenes. She remains fully conscious and aware of the illusory nature of the experience. Feelings of lightness and floating frequently accompany OBEs, suggesting the involvement of the vestibular system. More artistic depictions of OBEs and related autoscopic experiences can be found in Podoll and Robinson's (1999) collection of migraine art.

The last two variants of autoscopic reduplication—negative and inner (or internal) heautoscopy—appear to be historically bound. Around the turn of the 20th century, case reports and even monographs (e.g., Sollier 1903) were common. Today, especially the latter type, has almost vanished from the medical literature. Nonetheless, we include a brief discussion of both, even though their classification as truly autoscopic phenomena has been questioned (Dening and Berrios 1994).

Negative Heautoscopy (Figure 6e). This phenomenon refers to the visual disappearance of a body part—often an entire hemibody. The term "negative heautoscopy" is synonymous with asomatoscopy, a label preferred by some authors (Magri and Mocchetti 1967; Devinsky et al. 1989). Asomatoscopy may or may not co-occur with asomatognosia. In a case reported by von Stockert (1934), a patient with a thalamic tumor experienced left-sided hemiasomatognosia. The absence of feeling triggered horror—until he looked down and saw that the left side of his body was also visually gone. This bizarre confirmation brought a sense of comfort.

The self-portrait in Figure 6e is by a 23-year-old ambidextrous cabinetmaker referred to a psychiatric outpatient clinic for neuropsychological evaluation. No assessment could be completed because, after completing only a verbal learning test (performed within normal limits), the patient left and never returned. He had reported panic attacks triggered by a gradual vanishing of his left upper hemibody, from the shoulder to the hip. These episodes, initially rare, had increased in frequency to several times a week. Each lasted about 10 minutes, typically accompanied by dizziness and stereotyped upper body movements (as observed by his mother), though without loss of consciousness. In about half the episodes, the visual experience of the affected body part also vanished: "like being wiped out with an eraser." The skin would become black-and-white, and then increasingly transparent, allowing the patient to still see objects or people covered by his left arm. This partial asomatoscopy intensified the panic associated with his complete hemiasomatognosia. EEG showed mild to moderate nonspecific changes and epileptiform potentials over the right centro-parietal region; carbamazepine was prescribed, but no follow-up was possible.

Inner Heautoscopy (Figure 6f). In this rare variant, one sees one's own internal organs outside the body. A synesthetic component seems present—somatic sensations or fearful beliefs are projected visually. Thus, the autoscopic aspect arises from a blending of visual and somesthetic input, with individual variation in which is dominant.

Modern case reports are rare (but see Menninger-Lerchenthal 1935; Rao 1992). Figure 6f symbolically illustrates the phenomenon via Stelarc's endoscopic image, from his 1973 project, *Inside of My Stomach.*

For centuries, phantom limbs and autoscopic phenomena have inspired both ghost stories and philosophical reflection. The concept of the "phantom body" and its various components is essential for understanding seemingly ghostly experiences that—though strange—can be a painful reality for patients with brain damage.

While Stanislaw Lem's phantomology envisioned the technologies of virtual reality, the phantomology introduced in this chapter seeks to deepen our understanding of the virtual realities of the bodily self, in both health and disease. In doing so, it may help bridge the long-standing divide between neurology and psychiatry (Price et al. 2000), with neuropsychology as a mediating discipline (Cunningham et al. 2006).

Incarnation Without Animation: BID as a Neuropsychiatric Condition

We have conceptualized phantom sensations of congenitally absent limbs (see above, p. 218) as a case of animation without incarnation (Hilti and Brugger 2010). A person experiencing congenital phantoms "animates" limbs that have never physically existed.

There is one peculiar condition in which—at least metaphorically—the opposite seems to hold. Some individuals express a persistent desire for the amputation of a fully developed limb that they nevertheless do not experience as part of themselves. When I first encountered such reports in the popular press, I shook my head and could not empathize. I dismissed the phenomenon, not taking it seriously enough to consider it worthy of investigation.

However, one day I remembered the animation/incarnation distinction. It suddenly dawned on me: why not conceive of this non-acceptance of a physically intact limb as the mirror image of congenital phantom limbs? A case of incarnation without animation? Here is a limb that is fully formed, neurologically intact, and physically present but experienced by the individual as no more than a piece of meat. A limb without a soul. Unanimated. In stark contrast to the rest of the body.

That insight, in 2010, marked the beginning of our systematic research into what is now recognized by the World Health Organization (WHO) as a distinct mental disorder. The WHO General Assembly adopted the 11th version of the International Classification of Diseases (ICD-11) in May 2019. It became globally effective on January 1, 2022, the year of my retirement.

The desire for amputation of a healthy limb now has a diagnostic label and a code: Body Integrity Dysphoria (BID), ICD-11 code 6C21. Previously, it had been described as a paraphilic disorder (under the label apotemnophilia) or, more cautiously, as Body Integrity Identity Disorder (BIID), a term suggested by the American psychiatrist Michael First (2005).

Others have proposed the term "xenomelia"—from the Greek *xeno* (foreign) and *melos* (limb)—which I preferred, as it is descriptive and free of the pathologizing overtones of "disorder" or the assumptions of deviant sexuality.

Regardless of terminology, here are some relevant facts. The ICD-11 description of BID reads in part:

> Body Integrity Dysphoria (BID) is characterised by an intense and persistent desire to become physically disabled in a significant way (e.g., major limb amputation, paraplegia, blindness), with onset by early adolescence, accompanied by persistent discomfort...
>
> The desire to become physically disabled results in harmful consequences, as manifested by either the preoccupation with the desire (including time spent pretending to be disabled) significantly interfering with productivity, with leisure activities, or with social functioning (World Health Organization (2019).

Apart from the agreed-upon points listed in Table 1, there is a general consensus that BID is associated with alterations in cortical and subcortical grey and white matter structures. While the earliest neuroimaging studies on the structural correlates of BID implicated relatively focal and circumscribed cortical areas—especially the right superior parietal lobule and the right insula (McGeoch et al. 2011; Hilti et al. 2013)—each new investigation has broadened the picture, adding further extended networks and circuits to the list of affected regions.

Saetta et al. (2020) examined a relatively homogeneous group of 16 men with a specific desire for the amputation of the left leg. They identified a breakdown at two levels of limb ownership: (1) at the level of structural and functional integrity in cortical areas representing the body as a whole (particularly the right superior parietal lobule), and (2) at the level of functional connectivity between the primary sensorimotor representation of the undesired limb and the rest of the brain. The authors emphasize that their findings are correlational. Are the observed neural alterations the cause of BID, or the consequence of a life lived with a limb insufficiently animated and excluded from the bodily self? Decades of such an experience may have profound effects on both brain function and structure.

Noll and Kasten (2014) reported satisfaction in all 14 individuals who had realized their desire for amputation. Although no systematic long-term evaluation of this radical treatment for BID was available, one is left with a provocative thought: are we faced here with a condition in which it is justified to "cut the body to cure the mind" (Schlich 2015)? Alternative treatment options should, of course, be sought, but as of today, no truly convincing proposals have emerged.

Table 1: Characteristic features of BID and of the persons affected (adapted from Brugger et al. 2016, Panel 2).

Generally Agreed	**Controversially Discussed**
First manifestation in childhood or early adolescence.	Emasculation desire part of BID?
Marked suffering.	Paraphilic component frequent? – No (First 2005) – Yes (Blom et al.2017)
Mostly men affected.	Association with non-heterosexual orientation?
Typically, a higher education status.	What key experience is needed as a trigger for the desire?
Largely unremarkable personality profile.	Is psychotherapy a promising treatment option?
Pretending to have the desired disability (either in private or in public) is present in most persons affected.	Is amputation a promising treatment option?
The most common form of desired disability is major limb amputation – legs more than arms, and left-sided more than right-sided limbs.	Ethics of amputation? (see Loriga 2025 for the debate)
Desire for paraplegia is also frequent, and the proportion of women is significantly higher than in the case of amputation desires.	Will BID remain listed in ICD? Will DSM-6 (to be released within the next two years) list BID as a mental disorder?

A Note of Caution on Classification

While the recognition of Body Integrity Dysphoria as a diagnostic entity in ICD-11 may serve to legitimize the suffering of those affected and facilitate research, it may also carry unintended consequences. Diagnoses codify more than symptoms—they influence public discourse, clinical policy, and self-identification. Whether BID, including the desire for paraplegia (Giummarra et al. 2012), will become more prevalent as awareness rises, much as has been discussed in the context of GD, gender dysphoria, is an open question. So is the possibility that future revisions of classification systems may, in hindsight, view the pathologization of such desires as misguided.

History offers sobering precedents. Homosexuality meandered in and out of official diagnostic manuals for decades (Drescher 2008). Might BID, too, come to be seen not as a disorder but as an uncommon and sometimes erotically charged variant of bodily desire? If so, we may look back on its inclusion in ICD-11 as a well-intentioned, but possibly premature, attempt to draw psychiatric boundaries around an experience that challenges our current frameworks of normality (Brugger et al. 2013).

Where Can We Read More?

Alderson-Day, B. Presence. The Strange Science and True Stories of the Unseen Other. Manchester University Press: Manchester, UK, 2023.

Blanke, O., Landis, T., Spinelli, L. and Seeck, M. Out-of-body experience and autoscopy of neurological origin. Brain 2004; 127: 243-58. doi: 10.1093/brain/awh040.

Craig, A.D. How do you feel? Interoception: the sense of the physiological condition of the body. Nat Rev Neurosci. 2002 Aug; 3(8): 655–66. doi: 10.1038/nrn894.

De Vignemont, F. Mind the Body: An Exploration of Bodily Self-Awareness. Oxford University Press: Oxford, UK, 2020.

Longo, M.R. Distortion of mental body representations. Trends Cogn. Sci. 2022 Mar; 26(3): 241–254. doi: 10.1016/j.tics.2021.11.005.

Loriga, L. Body Integrity Dysphoria and the Ethical Dilemma of On-Demand Amputation. Redefining Wholeness: Identity, Autonomy, and the Moral Boundaries of the Human Body. Springer: New York, 2025. https://doi.org/10.1007/978-3-031-82443-2.

Morin, C. Stroke, Body Image, and Self-Representation. Psychoanalytic and Neurological Perspectives. Taylor and Francis: New York, 2017.

Ramachandran, V.S. and Blakeslee, S. Phantoms in the Brain: Probing the Mysteries of the Human Mind. Harper Collins: New York, 1999.

Schilder, P. The Image and Appearance of the Human Body. Kegan Paul: London, 1935.

CHAPTER 11

What Are FUHEMAS?

The acronym FUHEMAS stands for functional hemispheric asymmetries. Few topics in neuropsychology have caused a deeper rift among researchers. While some authors argue that the differing processing styles of the two hemispheres underlie the very foundations of personality psychology, others consider cerebral hemispheric specialization marginal—at best—and certainly irrelevant for understanding everyday behavior. This clash of opinions highlights how far we still are from an integrated understanding of the nature, evolution, and social consequences of hemispheric asymmetries. Worse still, in response to what has often been an emotionally charged debate, many researchers have simply turned away from the topic altogether. Understandably, the left/right dichotomania promoted by pop psychology has had an intimidating effect on serious scientific inquiry. But this pop-culture baggage should not prevent FUHEMAS research from expanding rather than shrinking. If it finds the right balance, we may have more than pop psychology; it would be Popper, who once remarked that "bold ideas, unjustified anticipations and speculative thought are our only means for interpreting nature" (Popper 1959/1980, p. 280).[1] Research on FUHEMAS must learn to throw out the bathwater—while keeping the baby.

? **What is the baby, and what is the bathwater in work on FUHEMAS?**

The bathwater includes all the "bold ideas, unjustified anticipations and speculative thought," as Popper called them, that have either never been put to serious testing or have already been falsified—but still appeal so strongly to the popular imagination that they continue to muddy the waters of pseudoscience. Worse, serious scientific inquiry is affected too: overly cautious researchers shy away from exploring the topic altogether, reluctant to fish in such troubled waters.

Some of the unjustified assumptions date back to the golden age of laterality research in the 1970s. These include ideas like drawing with your left hand unleashing your artistic creativity, or that "right-brain cooking," "right-brain photography," and "right-brain sex" somehow bypass the rational, narrow-

[1] I have already used this play of words (pop – Popper) earlier: Brugger (2012).

DOI: 10.1201/9781003384991-11

minded left brain and activate a broad, "holistic" worldview. Many such claims have been thoroughly debunked—for more examples, see Ornstein (1997).

The baby is harder to define but can be captured in broader terms. Laterality research is not just about making lists of what one hemisphere can do and what the other cannot. As a multidisciplinary endeavor, it points to broader questions about the biological roots of asymmetries—functional and structural, at the levels of cells, brains, and entire organisms. The molecular origins of FUHEMAS are just as important as their potential implications for social life.

From this broader approach, more specific questions arise naturally. For example:

* Do we really need to assume that handedness originates in the cortex—or, as McManus (2002) provocatively suggested, should we "think the impossible" and look elsewhere in the brain?
* How can we move away from the term "dominance" when describing FUHEMAS? Personally, the word makes me queasy. I realize many textbooks still use it, but to me, it belongs in the bathwater, not with the baby.
* Should we abandon the idea that the right hemisphere may play a vital role in creativity? Or can we instead try to define "creativity" more precisely, in ways that generate falsifiable hypotheses?

We need scientists who are not locked into existing prejudices (Fleck 1935), but whose curiosity does not spiral out of control when faced with grand questions about left and right—from atoms to entire universes (McManus 2002).

? I do not understand your active dislike of the term "dominance" in describing task sharing between the two hemispheres. Is this not just a shorthand for saying that one hemisphere does more for certain functions, like language?

That is a fair question—and no, I am not rejecting the empirical fact that certain functions are more strongly represented in one hemisphere than the other. What I object to is the implication built into the word "dominance". It suggests a power hierarchy, as if one hemisphere is in charge and the other merely follows. But that is not how the brain works.

The hemispheres are not rivals in a contest, nor is one the boss. Think of them instead as collaborators with different strengths, perspectives, and strategies. In language, for instance, the left hemisphere typically handles fine-grained syntactic processing, while the right contributes to intonation, emotion, and pragmatic understanding. Calling the left hemisphere "dominant" for language obscures the fact that both are doing essential—but different—parts of the job.

In some cases, the so-called "non-dominant" hemisphere even inhibits or modulates the other hemisphere, suggesting to a dynamic, and reciprocal relationship rather than a one-sided command. This interaction is complex and context-dependent. Referring to it as "dominance" not only oversimplifies but also misleads students into imagining a rigid hierarchy rather than a nuanced exchange.

Yes, the term is still widely used in research papers and textbooks, often for historical reasons. But it is worth asking whether the metaphor still serves us. Scientific language should clarify, not obscure. If we shift toward terms like specialization, complementarity, or task-preferential processing, we offer students—and ourselves—a more accurate and intellectually honest model of brain function.

The Notion of FUHEMAS: Historical Roots

According to current knowledge in the history of science, the earliest detailed theory of functional specialization between the two cerebral hemispheres originates in classical antiquity (Révész 1917; Lokhorst 1982). The Latin text itself dates from the 12th century and contains an unusually explicit description of FUHEMAS, framed in terms of perceptual versus intellectual processing:

> "… the brain provides sensation and intellect. That with which we understand is namely different from that with which we perceive. There are accordingly two brains in the head. The one gives us our intellect, the other provides the faculty of perception. That is to say: the brain on the right side is the one that perceives, whereas the left brain is the one which understands." (Lokhorst 1996)

The reasoning behind this claim of differential processing remains unclear. One possibility is that the attribution of intellectual function to the left hemisphere drew on analogies between the heart and the brain. At the time, the thicker wall of the heart's left ventricle was considered superior to the right side, and a similar hierarchy may have been projected onto the brain. Since the intellect was regarded as superior to perception in Greek thought, it may have been symbolically placed in the left hemisphere, leaving perception to the right (Lokhorst 1996). In any case, the intellect–perception dichotomy is a striking anticipation of the kinds of functional hemispheric distinctions that would reappear in the 19th and 20th centuries.

The most comprehensive, wide-ranging, and engaging analysis of 19th-century theories of FUHEMAS is owed to the American historian Anne Harrington (e.g., Harrington 1987; 1995). She focuses particularly on developments in France, where early laterality research received its most significant impetus. However, she also addresses important contributions from other regions, including the American neurologist John Hughlings Jackson, the British physician Arthur Wigan, and various other European thinkers.

In earlier work, Harrington commented on the fantastic range of functional dichotomies proposed during the 18th and 19th centuries, noting that they often reflected the prevailing social values of the time: "It is interesting that, once one has given the two hemispheres sexual identities, the idea of cerebral dominance becomes a rather apt metaphor for the social and economic domination of men over women in 19th-century Europe" (Harrington 1985).

Before we turn to some of the more exotic identities that have been attributed to the two hemispheres, we begin with perhaps the most obvious of all FUHEMAS and ask:

What Do We Know About Handedness?[2]

Right after conception, every human being begins as a morula—a perfectly symmetrical sphere. Just a few months later, we have developed into a bilaterally symmetrical organism that already shows striking asymmetries in motor preferences. One of the earliest and most visible of these is handedness: the preference for using one hand over the other. This preference has profound implications for everyday life, especially for the minority of left-handed individuals who must navigate a world designed primarily for right-handers. It has also inspired an enormous body of literature—so vast, in fact, that its volume is "inestimable even by a computer with double-symmetrical counting hemispheres" (Oppenheimer 1977).

Handedness refers to an individual's spontaneous preference for using one hand over the other in most daily activities. In neuropsychological practice, this is often assessed simply by asking which hand a person uses to write. But motor preferences emerge much earlier: even in utero, human fetuses show handedness—left-handed, right-handed, or in rare cases, ambidextrous.

? Is it true that human fetuses already demonstrate hand preferences? How are these assessed? What are the implications of fetal FUHEMAS?

Human fetuses exhibit unilateral arm movements as early as the 10th week of gestation (Hepper et al. 1998). Ultrasonographic recordings from hundreds of uncomplicated singleton pregnancies have shown a clear and consistent right-hand preference (94.6%) for thumb-sucking from the 15th week until birth (Hepper et al. 1990). These prenatal hand preferences predict later handedness as assessed by standard inventories, although a small proportion of left-thumb-sucking fetuses—particularly males—eventually develop right-handedness (Hepper et al. 2005).

The significance of these findings is difficult to overstate. Limb movements at such an early stage are not yet under cortical control, which implies that theories of handedness development must take subcortical mechanisms into account—perhaps even peripheral muscle activity. For instance, Pande and Singh (1971) found that in 9 out of 10 fetuses studied, the total muscle and bone mass was greater on the right side. In a small fetal sample, Ocklenburg et al. (2017) identified gene expression asymmetries in the spinal cord, whose epigenetic regulation could represent a foundational mechanism for both handedness and language-related asymmetries.[3]

---

[2] This section is dedicated to my daughter Hazel, in admiration.

[3] Restrictions by the Ethical Review Board limited the impact of this important study. I have early warned from the unfortunate consequences bureaucratic control can have on scientific progress (Christen and Brugger 2015).

If cognitive FUHEMAS are indeed "triggered" by motor-level asymmetries, then their origins may lie much earlier in ontogeny than is usually assumed. With regard to language lateralization, the temporal planes—key regions for linguistic processing that include Wernicke's area—were found to be larger in the left hemisphere of third-trimester fetuses (Chi et al. 1977), just as cortical gyrification begins. Later *in vivo* neuroimaging studies confirmed and expanded on these findings (Kasprian et al. 2011).

An Early Study on "Dextral Pre-Eminence" Anticipating Many Modern Research Aspects

Direct observations of motor preferences in fetuses are absent in the early essay by Ogle (1871). However, the author speculated indirectly about a possible developmental correlate of handedness—namely, structural differences between the two frontal lobes. Citing a colleague's data, he noted that "the convolutions of the left frontal lobe appear earlier in the foetus than the corresponding convolutions of the right." Remarkably, this largely forgotten British physician addressed many aspects of handedness that remain relevant in current research.

Aspect 1: Right-Handedness Across Times and Cultures

Ogle begins his article by referring to ancient records indicating that the human species has been "from time immemorial a right-handed animal" (p. 279). This aligns with the modern view of the universality of right-handedness across cultures and historical periods. In all known societies—including prehistoric contexts—the majority of individuals have preferred the right hand for fine-motor tasks. This does not mean that cultural factors have played no role; indeed, there have been—and in some places still are—social pressures discouraging or limiting the use of the left hand.

? How can we know that the incidence of right-handedness during prehistoric times was comparable to the one of today?

We "know" this by inference—but the sources that support this inference are numerous, and the data surprisingly robust. Marks on stone, bone, and antler tools found at archaeological sites can often be classified as having been made by right- or left-handed individuals. The observed ratio of right-to-left-handed usage is approximately 9 to 1. Similar findings have come from the analysis of weapons such as spears—whether used with one or both hands—or from laterally concentrated crush marks on predatory skulls.

Another important source is the imprint of hands near cave paintings. These impressions typically reflect the supporting hand, which braced the body while the other hand drew. About 80% of these supporting handprints in Paleolithic caves are left hands, suggesting that the drawings were made with the right (Faurie and Raymond 2004).

For the more recent period of the past 5000 years, Coren and Porac (1977) analyzed pictorial representations of individuals engaged in unimanual activities—such as throwing spears, playing games, or writing—and found an overall incidence of left-handedness of around 8%.

For further discussion of this fascinating intersection between neuropsychology, anthropology, and archaeology, see McManus (2009) and Uomini and Ruck (2018). The latter authors also review work suggesting that cognitive asymmetries may have been modest in early human history but grew significantly over time, due to a nonlinear relationship between brain size and changes in functional neuroanatomy (Herculano-Houzel 2009; see Chapter 2, p. 14).

Aspect 2: Biological vs. Sociological Origins of Right-Handedness

Ogle (1871) describes two competing worldviews about the origins of handedness—views that clashed during his time and, to some extent, still clash today. One is the biological stance: motor asymmetries arise from organic factors, whether innate or shaped during embryonic development. The other is the sociological view: handedness results from social conventions and habits, shaped by family practices or broader cultural norms.

Ogle clearly defends the biological view. His essay is structured around five arguments aimed at falsifying the sociological interpretation. While he acknowledges that a person's environment can influence hand preference, he emphasizes that even a potentially powerful cultural habit—such as the direction of reading and writing—does not alter the predominance of right-handedness (see also the question on "chiral cognition" on p. 247 below).

He also presents observational data suggesting that left-handedness tends to run in families. At the same time, he notes exceptions: left-handed individuals occasionally appear in families where both parents and grandparents are right-handed. Based on data from 2,000 patients (1,000 women and 1,000 men, likely inpatients at the hospital where he worked), Ogle found that left-handedness was more common in men than in women—a finding consistent with modern data. In addressing issues of familial versus non-familial left-handedness, Ogle was already anticipating key themes in the ongoing debate about the genetic basis of handedness.

? **Is there a gene for handedness?**

No. Historically, models proposing a single "handedness gene" have been put forward, but the heritability of hand preference is now understood to be far more complex. While left-handedness is indeed slightly more common in men (as already observed by Ogle in 1871), even more revealing are the differing probabilities of left-handed offspring depending on the handedness of the parents. The following figures, drawn from McManus (1991; see also Llaurens et al. 2009), illustrate this:

Parents' handedness (R = right, L= left)		**Left-handedness in the offspring in percent: (numbers in brackets indicate sample size)**	
Mother	Father	Daughters	Sons
R	R	8.5 (26,020)	10.4 (30,268)
L	R	21.7 (1,688)	22.1 (1,815)
R	L	15.3 (2,100)	18.2 (2,308)
L	L	21.4 (168)	27.0 (215)

This familial effect could, in theory, be due to either genetic or environmental influences. However, studies involving foster parents make an environmental explanation unlikely. One notable finding is the so-called "maternal effect": in mixed-handed couples, a left-handed mother is more likely to have a left-handed child than a left-handed father is.

Moreover, monozygotic twins are often discordant in handedness—one twin may be right-handed, the other left-handed—which argues against a purely genetic explanation. One proposed mechanism involves the intrauterine environment: the spatial constraints of close proximity may restrict one arm more than the other, typically the left for one twin and the right for the other.

Large-scale twin studies estimate that direct genetic effects account for only about 25% of the variance in handedness (Medland et al. 2009). In short, while hand preference is heritable, its expression is shaped by a complex interplay of genetic, hormonal, perinatal, developmental, and cultural influences. For a comprehensive review, see Güntürkün et al. (2020).

Aspect 3: Motor Lateral Preferences in Non-Human Animals

The issue of lateral preferences in animals deserves its own treatment, as in Ogle's time there was no established literature on FUHEMAS in the animal kingdom. Only much later did researchers begin to appreciate the broader significance of laterality across species—whether in the form of adaptive traits unique to a single species, or in comparative studies of lateralized behavior more generally.

As an introduction to this field, the edited volumes by Rogers et al. (2013) and Hori and Takahashi (2022) are highly recommended. Also worth exploring is the journal *Laterality*, founded in 1996 with the explicit aim of bringing together research on "*Asymmetries of Brain, Behaviour, and Cognition*" (as its subtitle states). From the beginning, studies on non-human animals have featured prominently in its pages. A forthcoming special issue will be devoted to "*Translational Research on Laterality: From Animals to Humans and Humans to Animals*."

Ogle (1871) was far ahead of his time when he began his discussion with observations of monkeys. He claimed that, as a rule, they were right-handed. From today's perspective, this statement is not particularly informative: he only observed captive individuals and did not specify which species he studied. We now know that primate handedness often depends on the type of task. In general, a right-hand preference becomes more apparent in tasks requiring fine motor control.

Interestingly, Ogle anticipated what would later become the standard test for paw preference in mammals. He described a basic paradigm in which an animal is required to reach for food placed equidistantly from its two forelimbs. As he put it:

"If, standing close to a monkey, one offers it a nut or apple, the monkey takes it with the nearest, and so the most convenient, hand, be this the right or the left [...]. But if, instead of standing close to the monkey's cage, one stands, bait in hand, at some distance—that is, at such a distance that right and left hand are equally distant from the tempting morsel—the monkey will stretch out one of its arms as far as possible through the bars of the cage."

Ogle also reported on footedness in parrots. Of the 86 birds he observed, the majority (63) preferred to stand on the right foot, using the left claw to handle the food offered through the bars. He acknowledged that this could be interpreted as sinistral pre-eminence—but he preferred to frame his results in favor of right-sidedness. In any case, later research confirmed a left-claw preference in most parrot species (Harris 1989).

In his study of other bird species, Ogle (1871) also anticipated what is now known as the distinction between individual-level and population-level asymmetry (see the question below). In some species, individual animals show a consistent preference for one limb or the other, but with no clear population-wide trend. In others, nearly all individuals show the same directional preference, resulting in a species-typical asymmetry.

? What about today's views about lateral preferences in animals?

Pet owners may have noticed it, but for a long time, an animal's preference for one paw over the other was not a topic of academic interest. It was considered a uniquely human trait. However, in the second half of the 20th century, research into limb preferences in animals began to flourish. It became clear that such studies could offer crucial insights into the development and genetics of human handedness—and, more broadly, into the biological basis of FUHEMAS.

The picture that emerged was complex. In some species, individuals show a consistent preference for one limb, but about half prefer the left and half the right—this is called individual bias. In other species, there is a *population-level* bias: most individuals use the same limb, a pattern similar to that seen in human handedness (Ströckens et al. 2013, for an overview).

In mice, for instance, researchers explored whether pawedness might be genetically determined. As reviewed by Sun and Walsh (2006), inbreeding experiments failed to produce mouse strains with a consistent lateral bias. However, they succeeded in breeding for strength of limb preference—from absent to strongly lateralized.

In many species, the type of task matters. Great apes, for example, tend to use the right hand for fine-object manipulation but show no strong preference in social tasks such as grooming (Forrester et al. 2012). Such task-specific limb preferences suggest that hemispheric asymmetries go well beyond simple motor control—and deserve to be studied in ecologically meaningful contexts.

What about animals that primarily use medial action organs? The elephant's trunk, for instance, is a far more important tool than either forelimb. Lefeuvre et al. (2022) found a right-side bias in exploratory trunk movements but also reviewed studies suggesting task-specificity—e.g., pinching versus grooming—similar to what has been observed in primates.

In dogs, about 70% show a consistent paw preference, though no population-level bias exists. Interestingly, paw preference in dogs has been linked to emotional reactivity—whether on the aggressive or the anxious side. Simon et al. (2022) review these findings but also point to methodological inconsistencies across studies, which may partly explain the variability in results.

For dog owners: several standardized assessment tasks are available to test paw preference (see references in Simon et al. 2022). The "Give Paw" test simply records which paw the dog offers most readily. Other tests assess which paw is used to remove a blanket placed over the head or adhesive tape stuck to the snout. Spontaneous food-grabbing is another method, with the "Kong test" being the most standardized version.

Some More Aspects Discussed by Ogle

Discussing all the variables Ogle (1871) considered in his reflections on the preeminence of right-handedness would be an exhaustive task. Here are just a few of his more minor discussion points:

Situs inversus. Ogle speculated (pp. 299–300) about the condition of *situs inversus*, in which the positions of the internal organs are mirror-reversed. Although cautious due to limited data, he predicted that based on vascular asymmetries—and a societal bias toward right-handedness in ambiguous cases—there would be only a slight increase in left-handedness among individuals with situs inversus. McManus et al. (2004) would have pleased him: in their study of 46 such individuals, the incidence of left-handedness was only modestly elevated (15.2%) and not statistically significant. Interestingly, earlier research (Kennedy et al. 1999), based on only three individuals, found a reversal of certain cerebral landmarks—but all three were right-handed and showed typical left-hemisphere language dominance. A recent single-case report of a right-handed man with complete situs inversus (Zhao et al. 2024) is open to varied interpretations: the man developed aphasia following a stroke in the *right* parietal lobe.

Observations on "Footedness". Ogle's (1871) observations on footedness are worth noting, as they reveal his early recognition that asymmetries might extend beyond the hands. His sample was too small and his conclusions were premature—he assumed that handedness and footedness always align. Modern studies show that correlations between laterality measures—such as handedness, footedness, eye dominance, and ear dominance—are generally weak. However, thinking beyond hands is important: footedness has been shown to be a better predictor of some cognitive FUHEMAS than handedness. Broader assessments of "sidedness" (Coren 1993) have led to the concept of inconsistent laterality or fluctuating asymmetry—when an individual prefers one side for certain limbs or sensory organs and the opposite side for others. This variability has clinical relevance: it is elevated in individuals with schizophrenia, schizotypal personality disorder, and in healthy individuals with strong paranormal beliefs (Mohr et al. 2003c).

Handedness in Sports. Ogle's casual observations of left-handedness in certain sports, such as cricket (Ogle 1871, p. 291), turn out to be surprisingly prescient. Modern research shows that left-handers are overrepresented in interactive sports, that require rapid, competitive responses (Loffing 2017). This finding is potentially relevant to resolving the so-called Darwinian paradox of handedness (see Chapter 8, p. 178): if left-handedness carries health-related fitness costs, its persistence in human populations might be explained by its advantages in competitive situations—a classic case of frequency-dependent selection.

One Last Aspect From Ogle (1871): Ontogenetic Development

I have saved what may be Ogle's most important idea for last, as it offers a natural transition from lateralized motor preferences to FUHEMAS in cognition and emotion. In brief, Ogle suggested that the motor dominance of the left hemisphere might stem from superior vascularization during development. He did not cite the diameter of the carotid arteries, but instead emphasized differences in tortuosity. The left carotid artery branches directly from the aorta, delivering blood unimpeded to the left hemisphere. By contrast, the right carotid must detour around the brachiocephalic artery (which has no left-side equivalent). As Ogle wrote: "The left side of the head will thus, as a rule, receive a more abundant flow of blood than will the right" (p. 298).

van Vuuren et al. (2022) provide a methodologically sophisticated analysis confirming a leftward asymmetry in middle cerebral artery perfusion in neonates. In contrast, Burlakoti et al. (2019) dismissed vascular asymmetries by measuring arterial diameters at the skull base and finding no left-right differences. Unfortunately, their paper—already hostile to theories of hemispheric asymmetry in its introduction—jumped to the conclusion that FUHEMAS do not exist at all. It is disheartening to see such modern-day naïveté clash with the broad, integrative thinking of an assistant physician in the 19th century.

Limbs and Language

The sensorimotor system is intimately linked to language. Even passive reading of the words LICK, PICK, and KICK activates areas of the brain responsible for face, hand, and foot movements, respectively (Hauk et al. 2004; see Chapter 4, p. 67). The link between upper-limb movement and linguistic processing is especially strong. Reading words—even those referring to objects that cannot be grasped (e.g., HOUSE)—increases excitability in the hand area of the motor cortex in the language-dominant hemisphere (Meister et al. 2007).

These findings support theories that propose a common evolutionary origin for hand motor actions and language. Most prominent among these is the long-standing idea, dating back to the 18th century, that vocal language evolved from gestures—a path "from hand to mouth" (Corballis 2002). However, according to Corballis (2003), the emergence of handedness followed the opposite route: a path "from mouth to hand." Once vocal language had developed, he argued, it liberated the hands from communicative duties, allowing specialization of the right hand for fine motor control. For critical perspectives on this proposal, see the peer commentaries accompanying his target article in Behavioral and Brain Sciences (Corballis 2003).

? What about those people born with only one arm? Are they not "extreme handers" and thus interesting with respect to their hemispheric competence for language?[4]

I agree—people with unilateral hand amelia (absence of one hand from birth) are an intriguing group for neuropsychology. One surprising observation emerged during recruitment: individuals with left-sided amelia were far more common than those missing the right hand. This asymmetry has long been known, but no clear explanation has been offered. Interestingly, most embryologists I spoke to were more puzzled by my interest in this asymmetry than by the asymmetry itself. "Come on, a hand is a hand!" one remarked—and even my question about his own handedness didn't prompt recognition of the potential significance.

We eventually recruited 11 individuals with unilateral hand amelia of unknown origin (excluding thalidomide embryopathy), all of whom were otherwise physically and mentally healthy. Of the six participants missing the left hand, all showed left-hemisphere superiority for lexical decisions in a divided-visual-field task. Among the five with right-hand amelia, only one showed a typical left-hemisphere advantage, one showed a reversed asymmetry, and in three there was no clear lateralization.

These preliminary findings are consistent with the idea that intrauterine activity of the right arm may play a causal role in the development of individual handedness. Of course, they cannot rule out additional influences—genetic or early embryological.

? What is "chiral cognition," and what should we think about the concept?

The term "chiral" describes objects that are not superimposable on their mirror image. It derives from the Greek word for hand (*χείρ*), and indeed, the left and right hands are canonical examples of chiral forms. In chemistry, chirality typically refers to two molecules with identical atomic composition but differing in three-dimensional structure—so-called enantiomers. These can have dramatically different effects on living organisms: one may promote health, while its mirror image can be toxic.

[4] Asked by a student born without a left forearm.

Chiral cognition refers to purported fundamental differences in how left-handers and right-handers think. Hardyck and Petrinovich (1977) observed that:

"The teasingly suggestive nature of this relationship [between handedness and cognition] is rather striking in its persistence through generations of research workers. There is usually just enough of a relationship to suggest a possible link and never enough of one to establish firmly a solid correlation."

Martin and Jones (1999a, 1999b) found systematic differences between left- and right-handers in memory for horizontally asymmetric objects or events. However, they remained cautious about how truly "cognitive" these differences were, or whether they ultimately reflected motor asymmetries. This caveat applies to other examples of chiral cognition as well, including the processing of action verbs (Willems et al. 2010), mental rotation of body parts (Ionta and Blanke 2009), and—perhaps to a lesser extent—differences in local versus global visual perception (Mevorach et al. 2005).

The notion of "chiral personalities" has faced further complications. While personality differences between left- and right-handers have often been reported (Furnham 1983; Hicks and Pellegrini 1978), a potential modulator is whether an individual grows up mostly among same- or opposite-handed peers. This asymmetry in social exposure could shape personality traits and cognitive styles.

Distinguishing between personality-based and cognition-based differences is also challenging in studies of belief in unscientific or "magical" phenomena. Nevertheless, Barnett and Corballis (2002) found that "magical ideation is related in an orderly fashion to variations in handedness." In their sample, ambidextrous students scored highest on the Magical Ideation Scale (Eckblad and Chapman 1983). The authors interpreted this through the lens of a genetic theory of laterality, suggesting that the allele associated with ambidexterity might also promote cognitive polymorphism—adding "lateral thinking" to a more linear, left-hemisphere-dominated processing style (see also Leonhard and Brugger (1998) for a similar but less handedness-dependent view).

One final note concerns emotional chiral cognition. In most cultures, right is associated with correctness, desirability, and goodness, while left connotes error, sinisterness, and negativity. For left-handers, this spatial-valence mapping is typically reversed. Casasanto (2009) demonstrated this reversal, and it was later confirmed in a multi-lab replication (Yamada et al. 2024). These effects are best understood as body-specific influences on space-valence associations. However, they do not sit comfortably alongside findings from the animal literature, which links the left side of space with threat, aggression, and danger (see Rogers et al. 2013 and the question on p. 265).

Perhaps we should not abandon the concept of chiral cognition too hastily. Instead, we need experiments capable of disentangling the cultural, motor, and intrinsic cognitive factors that may shape how left- and right-handers perceive, feel, and think (Mohr et al. 2003b,c).

To sum up, what we know for certain is that right-handedness is statistically associated with a superiority of the left hemisphere for speech and language functions. This relationship was first established through clinical observation and has since been confirmed by the majority of studies using the methodologies listed in Table 1 (257–260). Before we explore methods and findings related to non-motor FUHEMAS, we will first turn to the famous "split-brain" experiments, which allowed for the investigation of the two hemispheres in relative isolation.

The Impact of the "Split-Brain" Studies

"Split-brain" is a semi-popular term used to describe a neurosurgical procedure that became famous in the 1960s. The more technical terms are commissurotomy or callosotomy. The operation involves a longitudinal cut through the corpus callosum—often only partially—the largest commissure connecting the two hemispheres of the brain. The procedure was carried out to prevent seizure activity from spreading across the brain in patients with severe epilepsy. By interrupting interhemispheric communication, generalization of seizures could be avoided, offering relief to individuals whose epilepsy was resistant to pharmacological treatment.

The first human callosotomies for epilepsy were performed in 1939 by the American neurosurgeon William P. van Wagenen (1897–1961) (Mathews et al.

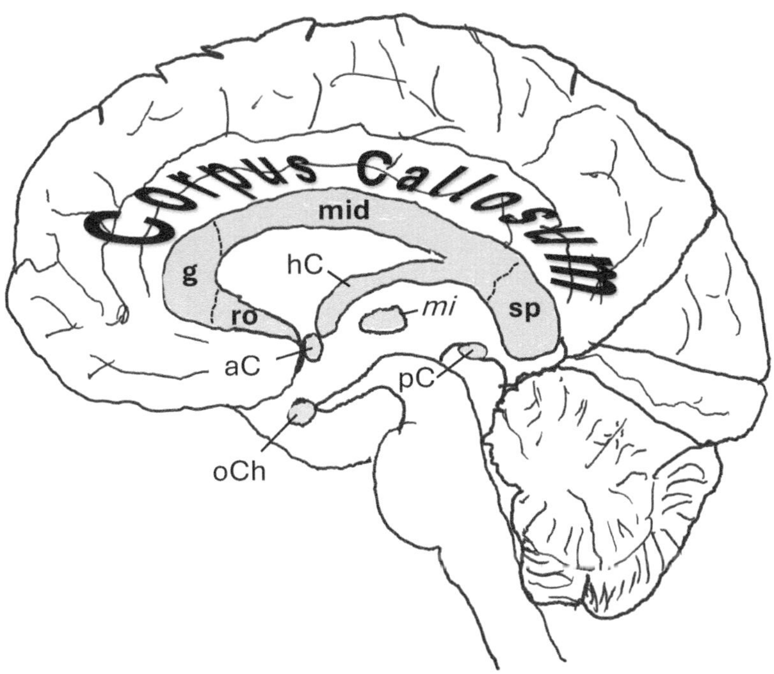

Figure 1: A medial view of the brain showing the corpus callosum and other major commissures connecting the two hemispheres in light grey. Selected single parts of the corpus callosum, from anterior to posterior: g: genu, ro: rostrum, mid: midbody, and sp: splenium. aC: anterior commissure; hC: hippocampal commissure; *mi*: *massa intermedia*, a fiber bundle connecting the two thalami; oCh: optic chiasm; pC: posterior commissure.

2008). In the early 1940s, he published on postsurgical symptoms, including apraxia, in collaboration with his colleague Andrew J. Akelaitis (Akelaitis et al. 1942). Akelaitis, a neurologist and neuropsychiatrist, later reported further motor, visual, and language peculiarities in callosotomy patients, although he did not attribute these directly to the severing of the corpus callosum.

It was the life's work of American psychologist Roger W. Sperry to replicate and refine the clinical observations first made by Akelaitis, and to draw the decisive conclusions. For this work, Sperry was awarded the Nobel Prize in Physiology or Medicine in 1981. Alongside his name, those of Michael S. Gazzaniga, Joseph E. Bogen, Joseph LeDoux, and several other neurologists, neurosurgeons, and psychologists are closely associated with the groundbreaking discoveries of split-brain research.

Before we describe the most telling tests—some of which still have clinical relevance today (see also Chapter 12)—let us take a brief look at the structure and function of the corpus callosum.

A Brief Look at the Corpus Callosum

The location of the corpus callosum—the "highest integrating organ in the human body" (Sperry 1964)—is shown in Figure 1.

? Is the function of the corpus callosum excitatory or inhibitory?

What a dispute this question has generated in the literature! Some researchers, especially those focusing on neurotransmitters, argue that the corpus callosum is primarily excitatory, as most of its axons are glutamatergic (van der Knaap and van der Ham 2011). However, many of these axons synapse onto inhibitory interneurons, meaning that the net effect from one hemisphere to the other may often be inhibitory.

Researchers who favor behavioral and noninvasive stimulation methods tend to frame the question in terms of inhibition and release. For example, TMS-induced "transient lesions" over the left parietal lobe can alleviate left-sided neglect following right parietal damage (see Chapter 9, p. 204). This is interpreted as the removal of interhemispheric inhibition from left to right, thereby functionally releasing the homotopic region in the right hemisphere (see the references cited in Chapter 9).

Whether the callosal effect is excitatory or inhibitory likely depends on several factors: the types of axon fibers involved, the segment of the corpus callosum through which they pass, the origin and target regions in each hemisphere, and the task at hand. The relative activation state of the two homotopic areas also plays a role.

In motor tasks, for example, excitatory and inhibitory phases may alternate during bimanual coordination to optimize performance (Duque et al. 2007). Importantly, a similar dynamic switching between excitation and inhibition may also underpin higher cognitive processes, such as creativity (Bogen and Bogen 1969; 1988; see the final paragraph of this chapter).

> One neuropsychological task has generated a particularly large body of research: a simple reaction time test designed to estimate interhemispheric transfer time. In this paradigm, participants respond as quickly as possible with a single hand to brief light flashes presented either in the left or right visual field, in pseudorandom order. Two separate blocks are recorded—one for left-hand responses and one for right-hand responses.

The trials are categorized into two types: uncrossed and crossed. In uncrossed trials, the hemisphere receiving the visual input also controls the motor response (e.g., right-hand response to right visual field stimuli, or left-hand response to left field). These require no interhemispheric transfer. In crossed trials, however, the hemisphere that perceives the stimulus must communicate with the opposite hemisphere to trigger the motor response (e.g., a left-hand response to a right visual field flash). The average difference in reaction time between these two types—calculated over many trials—provides an estimate of the time it takes for information to cross from one hemisphere to the other.

This paradigm was introduced by American psychologist Albert Theodore Poffenberger (1885–1977). In a seminal 1912 paper, he described his chronometric studies of callosal transfer in healthy individuals and estimated the transfer time at around 5 milliseconds (Poffenberger 1912). Modern techniques have confirmed the accuracy of this early estimate.

? How could Albert Poffenberger determine an interhemispheric transfer time of around 5 milliseconds at his time, that is, around the turn of the 19th to the 20th century?

This is a sensitive question. When introducing the Poffenberger paradigm in my lectures, students rarely ask how such accuracy could have been achieved with the chronometric devices available at the time, which regularly prompts me to interrupt the lecture and point out that some curiosity here would be quite appropriate.

Poffenberger used a fast-rotating drum blackened with soot. When he triggered the stimulus light using a telegraph key, it marked the drum at that exact moment. When the participant indicated awareness of the stimulus by pressing another telegraph key, a second mark was made. The distance between the two marks served as a time stamp. Knowing the speed of the drum's rotation, Poffenberger could translate this distance into milliseconds. By averaging the differences between crossed and uncrossed trials across many repetitions, he arrived at an estimate of 5 ms.

The fact that interhemispheric transfer times in individuals with congenital absence of the corpus callosum are only about three times longer than in healthy individuals suggests that alternative pathways must support communication between the hemispheres. Figure 1 (above) shows some of these cortical and subcortical commissures in cross-section.

Even split-brain patients show crossed–uncrossed differences in Poffenberger paradigms (Westerhausen 2023 for a meta-analysis). These differences are approximately ten times greater than in healthy controls. Still, no callosal disconnection is ever so complete as to eliminate all awareness of stimuli presented in the visual field ipsilateral to the responding hand.

Nonetheless, the surgical disconnection is profound enough to give rise to some of the most astonishing behavioral phenomena ever observed in neuropsychology.

The Key Disconnection Symptoms in Split-Brain Patients

These are:

(1) tactile anomia of the left hand,
(2) agraphia of the left hand (and visual-constructive deficits specific to the right hand),
(3) apraxia of the left hand, and
(4) left hemialexia and the inability to verbally describe pictures briefly presented to the left visual field (LVF).

We will describe these four symptoms in turn.

Tactile anomia of the left hand[5]

Objects palpated with the left hand cannot be named. This is because the tactile information reaches the somatosensory cortex of the right hemisphere but cannot be transferred to the left hemisphere—where language resides—for naming. To ensure that naming failures are not due to visual identification, the objects must be shielded from view.

Agraphia of the Left Hand (See Footnote 5)

The left hand is unable to write words to dictation. However, it can copy visual forms or draw objects from memory (e.g., a flower or a house). The right hand, in contrast, can write dictated words fluently but tends to produce clumsy copies or freehand drawings, especially of 3D forms—indicative of constructional dyspraxia.

Apraxia of the Left Hand (See Footnote 5)

When asked to pantomime using tools, the left hand is unable to comply. This apraxia is typically restricted to verbal commands, highlighting the failure of language-based instruction to reach the motor systems of the right hemisphere.

Left Hemialexia and Inaccessibility of LVF Pictorial Material to Verbal Report

Words briefly flashed to the left visual field—which projects to the right hemisphere—cannot be read aloud, nor can pictures be named. Naming and reading aloud require the transfer of visual information to the language-dominant left hemisphere via the corpus callosum, which is severed in split-brain patients.

[5] Strictly, one should speak of "a person's anomia [agraphia, apraxia] for objects explored by touch with the left hand." Hands cannot have anomia, agraphia, or apraxia.

Together, these four symptoms form the classic callosal disconnection syndrome—a functional split between verbal and nonverbal access to the world. This dissociation has generated an enormous body of neuropsychological research, most of it in neurologically intact participants. Divided visual field tasks, which allow for isolated stimulation of either hemisphere, remain one of the most widely used paradigms in behavioral laterality research (see Figure 2 and Table 1 below). In clinical neuropsychology, they are used to assess hemispheric competencies for language processing.

? **Is not the impetus for consciousness research by far the most important contribution of split-brain research?**

This question splits my mind. I am inclined to say yes when I consider the sheer volume of split-brain–inspired publications in the field of "consciousness research"—a domain whose boundaries stretch from fertile philosophical inquiry on one side to a dark esoteric abyss on the other.

But I am also inclined to say no when I reflect on the confusion that split-brain research has introduced into the study of consciousness. The dissociation between patients' lack of verbal awareness of stimuli presented to the left visual field and the measurable influence of those same stimuli on nonverbal behavior (see below) has led some to suggest that the neurosurgeon's scalpel did more than sever the hemispheres—it also split the patient's once-unified mind.

Yet the community of philosophers and neuroscientists has not reached consensus, and the debate continues. I see here a baby that should not be thrown out with the bathwater (see above, p. 237). So let us devote a short paragraph to the question of the unity or disunity of consciousness, as a way to respond to your question in a suitably ambiguous manner.

How Many Minds in a Split-Brain?

Nearly 100 years before the first split-brain operation was performed in a human being, the idea of a duality of mind was already being proposed, based on the simple observation that humans possess two cerebral hemispheres. British general practitioner Arthur L. Wigan (1785–1847) was primarily concerned with the psychiatric implications of this duality in his book *The Duality of the Mind* (Wigan, 1844). He believed that each hemisphere harbored its own mind, and that mental illness resulted from an imbalance between the two. One striking example of how far such an imbalance could go is illustrated by Wigan's report of a man suffering from heautoscopy—a reduplication of the self (see Chapter 10, p. 230):

> "He became gradually convinced that he was haunted by himself. Or (to violate grammar for the sake of clearly expressing his idea) by his *self*. This other self would argue with him pertinaciously, and to his great mortification sometimes refute him..." (Wigan 1844)

> One day, the man shot himself in an attempt to escape the intolerable persecution by his second self. Similar motivations for suicide have been documented in more recent case reports (Ames 1984; Brugger et al. 1994a), and the phenomenon of *heautoscopic suicide* has been explored in both literary and neuropsychiatric contexts (Brugger 2007). Regardless of these dramatic cases, doppelgänger experiences have long been linked—though only moderately convincingly—to the duality of the cerebral hemispheres (Grotstein 1983).

While many of Wigan's contemporaries agreed that a single person might harbor two minds, the German philosopher and founder of psychophysics, Gustav T. Fechner, held the opposite view. About 20 years after Wigan, he anticipated—through a *Gedankenexperiment*—the surgical severing of the corpus callosum, and predicted that such a cut would create two minds from what had previously been a unified one. For Fechner (1860), the corpus callosum guaranteed the unity of a healthy mind; only its severance would allow each hemisphere to develop an independent consciousness.

One might think that the advent of split-brain studies would have settled the question of unity vs. duality of consciousness. In fact, the opposite has happened. Most commentators on Sperry's work interpreted it as a confirmation of Fechner's view: split-brain patients have two minds, while normal individuals have one (see Marks 1980, for references). But some saw Wigan's view confirmed instead—arguing that even normal people have two minds (e.g., Puccetti 1973).

Schechter (2018), in her book *Self-Consciousness and 'Split' Brains*, offers a more nuanced take by distinguishing between mind and personhood. While not denying that a split-brain patient may possess two minds, she argues that the patient remains a single person. In exploring this paradox—dual minds within one person—Schechter examines two extraordinary individuals: the Hensel twins. These conjoined sisters, A and B, have two heads but share one body from the chest down. Each has her own mind, ID card, and driver's license. One is married; the other is single. Each of them, moreover, has two cerebral hemispheres—and their mental experiences are just as heterogeneous as those of any other person.

Schechter captures the "one person, multiple minds" idea in the subtitle of her book: *The Minds' I*. The apostrophe placement is no accident.

The difficulty in counting the minds of a split-brain patient (Lokhorst 1996) is much the same as in counting the minds of any "normal" person. Careful introspection quickly reveals that we are not unitary beings. No surgical commissurotomy is needed to change our mind on a whim, to desire something we wish we did not, or to act on a gut feeling despite contrary advice from reason.

Mind, consciousness, and awareness are properties of whole persons—not of brain halves, connected or disconnected. Future research should take this into account, and recognize that terms such as mind, consciousness, awareness, and personhood require clearer definitions than they are often granted in the researcher's own mind.

? Why has "split-brain" research become unpopular these days?

Commissurotomy as a method for preventing seizure propagation across the hemispheres has become rare. This is largely due to significant improvements in pharmacological seizure control. In addition, alternative neurosurgical procedures have been developed to remove or inactivate the seizure focus, such as amygdalohippocampectomy, highly selective resective surgeries, or deep brain stimulation.

In rare cases of intractable epilepsy, commissurotomy is still performed. However, these patients are often of lower intellectual functioning and not easily testable. On the other hand, new experiments involving split-brain patients whose surgery took place decades ago are difficult to interpret and cannot be directly compared to earlier findings (e.g., Pinto et al. 2017). This is most likely due to substantial functional reorganization of their brains and interhemispheric communication over time.

FUHEMAS: From Dichotomies to Complementary Specializations

Long before the era of split-brain research, the two hemispheres were attributed exclusive properties thought to represent fundamentally different styles of thinking. In the 19th century, these took the form of opposing pairs such as civilized vs. uncivilized, intellectual vs. emotional, male vs. female—dichotomies deeply shaped by the cultural and ideological *Zeitgeist* of that period in Europe (Harrington 1987; 1995).

These absolute dichotomies—or polarities—were inspired by clinical observations (e.g., aphasia following left-hemisphere damage; speech arrest during seizures originating in the left hemisphere), by clinical-experimental findings after commissurotomy (see above), and by a wide array of empirical studies using diverse methodologies (see Table 1). Participants in these studies were typically healthy volunteers—mostly students, more often male than female, and almost always right-handed —WEIRD people in the world of laterality research.

? What are WEIRD people in laterality research?

Henrich et al. (2010) criticized psychological research broadly for its bias in studying almost exclusively WEIRD participants—volunteers from Western, Educated, Industrialized, Rich, and Democratic societies. In laterality research, this acronym might even need some adjustment. I propose WASP: (almost) Exclusively male, Individualistic, Rational, and Dextral.

The term "WASP," more than Western, points to White Anglo-Saxon Protestants—a demographic that has dominated the participant pools in U.S.-based studies. This carries cultural implications beyond geography, including the influence of left-to-right reading direction (Friedrich and Elias 2014), racial homogeneity, and monotheistic religious backgrounds. All of these may be relevant to neuropsychological traits in general, and to FUHEMAS in particular (Han and Northoff 2008; Persinger 1987; McGilchrist 2009; Prete and Tommasi 2019).

Especially during the second half of the 20th century, most research subjects in FUHEMAS studies were male. This bias reflected more than just the male-dominated nature of science at the time. Functional hemispheric asymmetries are influenced by sex hormones and the menstrual cycle (see Hausmann (2017) for a review), meaning that including women increases intra-group variability. This may have made researchers reluctant to recruit them, but the cost was a narrowed perspective on laterality.

Cultural factors matter as well: individuals from individualistic versus collectivistic societies differ in various cognitive domains, some of which involve FUHEMAS (Han and Northoff 2008). Even the degree of rationality—or so-called irrational thinking—has been linked to hemispheric processing styles as a trait (Brugger and Mohr 2008).

Finally, the dominance of dextral participants in laterality research is both a practical and methodological choice: right-handers are easier to recruit, and including only right-handers reduces variability. But this choice neglects 10% of the population. As Willems et al. (2014) argue, excluding left-handers may lead researchers to miss valuable neuroscientific insights by failing to study the minority brain.

The most widely accepted absolute dichotomy in hemispheric processing is between verbal and visual (or visual-spatial) functions. It is the only dichotomy discussed in some detail in this chapter. Many of the distinctions developed later (see Figure 2) show considerable overlap with this verbal–visual divide. More importantly, even this well-established type of FUHEMAS is not truly absolute: it reflects a relative hemispheric preference for specific aspects of verbal processing. And "verbal processing" is as multifaceted as language itself (see Chapter 4, p. 76).

We will also focus on one of the many methodologies used to establish hemispheric dichotomies or to refine particular pairs of complementary specializations (see Figure 3). The divided visual field (DVF) paradigm has been chosen for three reasons. First, it was the most frequently used method in the lateralized testing of split-brain patients. Second, it has an advantage over auditory paradigms in that visual information is more completely crossed between hemispheres than auditory input. And third, the DVF paradigm offers maximal flexibility: words, numbers, faces, and scenes can all be presented laterally to either patients or healthy research participants. DVF methods should be familiar—and ideally mastered—by clinical neuropsychologists, as they enable both quantitative and qualitative assessment of a patient's hemispheric superiority for language processing.

The DVF paradigm rose to prominence during the heyday of split-brain research. It remains important enough to merit its own figure. Figure 2 shows the visual pathways from the retina via the optic chiasm and the thalamus to the primary visual cortices, and highlights key features of lateralized visual stimulation. The central fixation point must be maintained throughout stimulus presentation. Once a stimulus is directly looked at (foveated), it is represented in both hemispheres. In fact, a small

Table 1: Methods to investigate FUHEMAS. Note that some of the methodologies are included primarily for historical context. For instance, the Wada test—long considered the gold standard for determining hemispheric "dominance" in cognitive functions (particularly language)—has largely been replaced by less invasive techniques. Like electrocortical stimulation and unihemispheric convulsive treatment, it is now used exclusively in clinical contexts, such as for patients with epilepsy or severe depression.

Method	**Key Reference(s)**	**Comments and References to Specific Aspects**
(a) BEHAVIORAL		
Visual:		
Divided Visual Field (DVF) paradigm	Ocklenburg 2025; Parker et al. 2021; Vingerhoets et al. 2023; Motz et al. 2012, for a didactically valuable early online version of a tachistoscope (no longer functional).	Historically, the DVF paradigm required a tachistoscope (Benschop 1998); software-based versions are now available (Peirce et al. 2019). There are numerous variants involving different stimulus types, eccentricities, tasks, and response modes. See Bourne (2006) for a comprehensive review. Central fixation must be strictly maintained (Jordan et al. 1998), and exposure times should not exceed 150 ms to prevent saccadic eye movements. Dependent measures can include accuracy, reaction time, or derived laterality indices. For applications to lateralized lexical decisions and semantic priming, see the main text.
Divided output paradigm	Root et al. 2006; Kong 2013	Pictorial stimuli are presented centrally, accompanied by specific task instructions and a unimanual response mode. Reaction times of the left or right hand serve as indirect indicators of right- or left-hemisphere task superiority. This method is especially useful when lateralized stimulus presentation is not feasible—for instance, when sequential information is involved (Anagnostopoulos et al. 2013).
Auditory:		
Dichotic listening (DL) paradigm	Kimura 1961; Bryden 1988; Ocklenburg 2025	Simultaneous presentation of different auditory stimuli to each ear. The main variable is the ear from which the participant reports. A right-ear advantage (REA) is commonly observed in individuals with left-hemisphere superiority for phonological or lexical processing. The paradigm has been adapted for lexical decision tasks with manual responses (Schüeli et al. 1999). A variant of the free-report method involves selective listening (forced-attention) procedures. For a comprehensive review of factors influencing dichotic listening (DL), see Hiscock and Kinsbourne (2011). A smartphone-based version of the task is presented in Bless et al. (2015).

Table 1 contd. ...

...Table 1 contd.

Method	**Key Reference(s)**	**Comments and References to Specific Aspects**
Motor-Expressive:		
Mouth asymmetry (MA)	Graves et al. 1982; Code et al. 2005	Among the indicators of facial asymmetry that can serve as markers of FUHEMAS in cognition and emotion, one particularly informative measure is mouth asymmetry (MA). This method assesses the asymmetric opening of the corners of the mouth during speech or nonlinguistic vocalizations. In right-handed populations, earlier and wider opening of the right side is the typical pattern. This asymmetry is reduced—or even reversed—when speaking metaphorically (Argyriou et al. 2015), and similarly altered when singing compared to speaking (Graves et al. 1985). Differential MA has also been observed in nonhuman primates: fearful vocalizations tend to produce a left hemimouth effect, while approach-related social vocalizations show a bias toward right-sided opening (Hook-Costigan and Rogers 1998).
Motor-Cognitive		
Motor-cognitive interference	Kinsbourne and Cook 1971; Medland et al. 2002	A unimanual motor task (e.g., speeded tapping, maintaining a constant tapping rate) is performed once under baseline conditions and once concurrently with a cognitive task. The performance decline relative to baseline is calculated separately for each hand. The greater decline typically occurs in the hand controlled by the hemisphere that is dominant for the concurrent cognitive task. As a general rule, right-hand performance declines more during verbal-manual tasks (e.g., spontaneous speech, word fluency, verbal recall), whereas left-hand performance declines more during tasks involving visuospatial processing, such as solving block design problems (Hellige and Longstreth 1981).
(b) PHYSIOLOGICAL/ NEUROIMAGING		
"Wada test" (intracarotid amobarbital procedure, IAP)	Wada and Rasmussen 1960; Strauss et al. 1985	Named after Jun Wada, who introduced the procedure in the second half of the 20th century, this highly invasive "test" involves anesthetizing one hemisphere by injecting sodium amobarbital into either the left or right internal carotid artery. It is used in the presurgical evaluation of patients with epilepsy to determine language lateralization. Speech arrest following anesthesia of one hemisphere indicates that it is language-dominant. While the Wada test has largely been replaced by less invasive techniques such as fMRI, it still retains relevance in certain clinical contexts (see Noorizadeh et al. 2024).

Electrocortical stimulation	Penfield, 1972	Electrical stimulation of cortical tissue can either transiently disrupt a function or, particularly when applied to mesial temporal regions, elicit experiential phenomena suggestive of underlying functional specializations. Today, this method is most commonly used in the context of awake brain surgery, where it helps the neurosurgeon preserve tissue critical to language and other cognitive functions.
Unilateral Electroconvulsive Treatment (ECT)	Fleminger et al. 1970; Pratt and Warrington 1972	This method functionally "knocks out" an entire hemisphere, similar to the Wada procedure, but uses electrical shock rather than pharmacological agents. Deficits in language or verbal memory are observed almost exclusively after electrode placement over the left hemisphere—typically in the temporoparietal region—and primarily in right-handed individuals (Fleminger and Bunce 1975). Geffen et al. (1978) validated the approach by comparing it with both the Wada test and a dichotic listening procedure.
Tympanic thermometry	Swift 1991; Cherbuin and Brinkman 2004	Tympanic temperature tends to decrease on the side of the hemisphere that is more actively engaged in a cognitive task.
Functional transcranial Doppler ultrasonography (fTCD)	Knecht et al. 2000; Stroobant and Vingerhoets and Luppens 2001	This method measures task-induced changes in cerebral blood flow velocity relative to baseline levels. Sampling is typically done via the temporal bone, targeting the anterior, middle, or posterior cerebral arteries. Tasks used may involve verbal processing or complex visuospatial functions.
Functional near-infrared spectroscopy (fNIRS)	Villringer and Chance 1997; Colier et al. 2001	An optical imaging method based on the differential light absorption properties of oxyhemoglobin and deoxyhemoglobin. Task-induced changes in hemoglobin concentration are measured using multichannel sensors placed over specific scalp regions. One major advantage is its relative insensitivity to motion artifacts. Different aspects of linguistic processing can be lateralized even in one-day-old neonates (Vannasing et al. 2016). The method can also be easily combined with other techniques, such as EEG (Pinti et al. 2021).
Repetitive Transcranial Magnetic Stimulation (rTMS) and transcranial Direct Current Stimulation (tDCS)	Duecker and Sack 2015; Duecker et al. 2013	Both methods—repetitive transcranial magnetic stimulation (rTMS) and transcranial direct current stimulation (tDCS)—can disrupt or facilitate task performance depending on pulse frequency (rTMS) or electrode placement (tDCS). Comparing stimulation-induced performance changes between left and right hemispheres can provide insight into FUHEMAS associated with specific brain regions or cognitive functions.

Table 1 contd. ...

...Table 1 contd.

Method	Key Reference(s)	Comments and References to Specific Aspects
Electroencephalography (EEG), Event-Related Potentials (ERPs), and Spectral Analysis of the EEG	Eichele et al. 2005; Weng and Lee 2020; Nix et al. 2024	Surface EEG records the brain's electrical activity via electrodes placed on the scalp. Functionally meaningful information can be derived from different frequency bands. Event-related potentials (ERPs) are obtained by repeatedly presenting a specific stimulus and averaging the resulting brain responses. Differences in EEG frequency bands or ERP components across hemispheres can reveal which hemisphere is specialized for particular aspects of a cognitive task.
Magnetoencephalography (MEG) /Magnetic Source Imaging (MSI)	Passaro et al. 2011	MEG measures the magnetic fields generated by neural electrical activity. Magnetic source imaging (MSI) combines MEG with MRI to localize brain functions more precisely. For applications in presurgical determination of the language-dominant hemisphere, see Doss et al. (2009).
Positron Emission Tomography (PET)	Belin et al. 1998	Radioactively labeled glucose accumulates in brain regions that are particularly active during a task. In language tasks, this allows identification of the hemisphere more strongly involved in processing.
Functional Magnetic Resonance Imaging (fMRI)	Abbott et al. 2010	Measures the level of blood oxygenation (the BOLD signal), which increases in brain regions actively engaged during a specific task.
Diffuse correlation spectroscopy (DCS) and speckle contrast optical spectroscopy (SCOS)	Valdes et al. 2014; Kim et al. 2023a	Both DCS and SCOS use near-infrared light to measure blood flow in neural tissue. DCS can be considered a form of "dynamic NIRS," offering significantly better spatial resolution than traditional NIRS. SCOS measures deep-tissue blood flow non-invasively by analyzing speckle contrast. Wide-field acquisition with thousands of speckles in a small region enables high-density tomographic imaging.
Functional Ultrasound Imaging (fUSI)	Deffieux et al. 2021	fUSI, like fMRI and optical imaging, is based on neurovascular coupling. Its ease of use, portability, and high temporal and spatial resolution make it a promising candidate among emerging neuroimaging techniques.
Ultrafast ultrasound localization microscopy (uULM)	Demené et al. 2021	uULM using intravenously injected microbubbles offers an unprecedented combination of deep tissue penetration and high spatial resolution. This method has the potential to revolutionize the assessment of blood flow dynamics in response to cognitive activation.

strip to the left and right of the fovea projects bilaterally, which is why the inner edge of any lateralized stimulus should be placed at least 2–3° of visual angle away from the center. To prevent eye movements toward the stimulus, presentation times should not exceed 150–200 ms.

Figure 2 also illustrates selected types of lateralized stimulation, demonstrating the richness of research questions that can be addressed using the DVF paradigm—as well as its clinical potential.

Two key task variants involving word stimuli should be highlighted: the lexical decision paradigm and the semantic priming paradigm.

Lateralized Lexical Decisions

Words or pseudowords are presented either to the left or right of a central fixation cross. The participant presses a key whenever a meaningful word appears (see Table 2 and Figure 2a). Responses can be made with the hand ipsilateral to the visual field of stimulus presentation, or bimanually to keep motor demands equal across hemispheres. Some studies use bilateral presentation, which enhances hemispheric asymmetries (Boles 1990). In this version, two pseudowords appear simultaneously on 50% of trials (no response required), while on the remaining 50% of trials, one of the letter strings is a real word—25% on the left, 25% on the right. As a group, right-handers respond faster and more accurately to targets in the right visual field (RVF), which projects to the left hemisphere (LH). Most left-handers show the same pattern, but their asymmetries tend to be smaller—i.e., the RVF/LH advantage is less pronounced.

Lateralized Semantic Priming

In this task, a target word (or pseudoword) is flashed to either the LVF or RVF. Shortly beforehand (with a variable stimulus onset asynchrony), a centrally presented "prime word" appears. Some authors use lateralized presentation for both prime and target. A key press is required only for meaningful target words. Primes can be semantically related (closely or distantly) or unrelated to the target. Lexical decision latencies following unrelated primes serve as a baseline for calculating priming effects.

Trial	Prime word	Prime-target interval [ms]	Target word	Associative relatedness	Reaction time [ms]	Semantic Priming (SP) type
1.	BOOK	20 or 200	NOIL	-	-	-
2.	BOOK	20 or 200	LION	none[6]	x ms	no SP
3.	MANE	20 or 200	LION	"close"	x-50 ms	direct SP
4.	STRIPES	20 or 200	LION	"distant"	x-30 ms	indirect SP

[6] No two words have zero associative relatedness. "None" is taken from collected ratings, and the mean associative relatedness between BOOK and LION is negligible, compared to the other types of relatedness.

The basic trial types are illustrated in the table above. While the numerical values are arbitrary, the timing between prime and target is critical: short intervals promote automatic processing, whereas long intervals engage controlled processing. These two processing types may interact with visual field presentation—automatic priming favoring the LVF/right hemisphere, and controlled priming favors the RVF/left hemisphere (Chiarello 1985).

Trial 1 illustrates a nogo trial: "noil" is a pseudoword and does not require a manual response. In Trial 2, a key press is required, as "lion" is a meaningful English word. Trial 3 features a prime that is closely related to the target, resulting in a faster reaction time. Crucially, in Trial 4, reaction time is also faster than baseline—even though *striped lions* do not exist. In this case, either tiger or zebra serve as mediators of a semantic connection between the prime and the target.

Semantic priming effects are reportedly larger in unmedicated psychiatric patients with delusions (Spitzer et al. 1993), with mediated semantic priming effects being disproportionately enhanced—especially following left visual field stimulation (Weisbrod et al. 1998). Similar effects have been observed in healthy students who report having experienced or believing in "paranormal" phenomena (Pizzagalli et al. 2001).

Extended Legend to Figure 2 (b to g)

Figure 2b:

This panel depicts a left visual field (LVF) trial from a task requiring a key press (using both hands simultaneously to balance motor activation) when the two letters presented are either nominally identical (e.g., "A A") or, in another run, match in graphical form (e.g., "e a" or "E A"). Nominal matches are generally more accurate and faster following right visual field (RVF)/left hemisphere (LH) stimulation, whereas graphical matches tend to show a LVF/right hemisphere (RH) advantage (McCarthy and Beaumont 1983).

Figure 2c:

This panel shows a trial from a lexical decision task using shorthand script (Regard et al. 1985). The ideographic stenography system allowed secretaries to transcribe meetings verbatim. Unlike standard print, shorthand carries both spatial and phonological meaning. In the figure, the word "ABER" ("but") is flashed to the LVF, and a pseudoword to the RVF. The authors used double simultaneous stimulation, which is known to enhance hemispheric asymmetries (Boles 1990), and required bilateral manual responses. With 50 ms exposure, an RH advantage emerged for shorthand words. In contrast, the same words presented in standard print (Experiment 2) yielded the typical LH superiority. This elegant demonstration underscores the relativity of "dominance." A similar pattern has been observed for Japanese: Kanji (logographic) lacks the LH dominance typically seen for Kana (phonetic) script (Hatta 1977; Nakamura et al. 2005). Attenuated LH involvement has also been reported for cursive handwriting relative to print (Hellige and Adamson 2007).

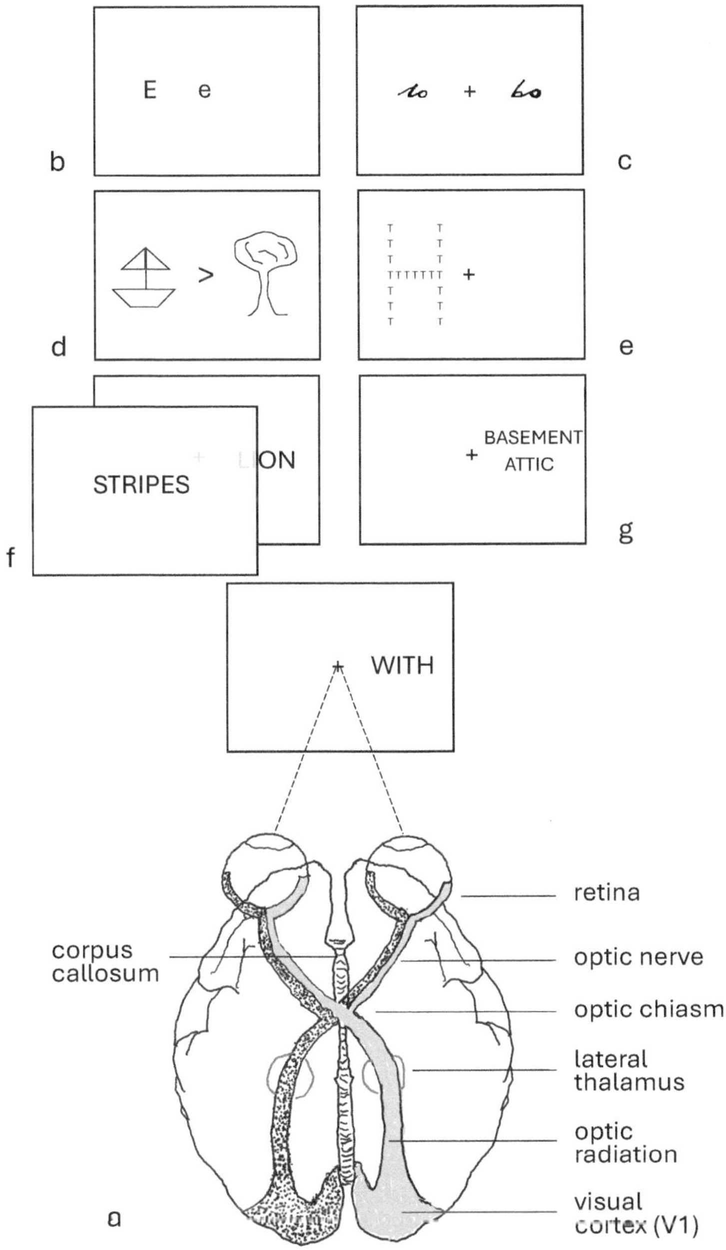

Figure 2: Crossing of the information from the two visual half-fields to the brain.
a: On central fixation, stimuli in the right visual field (RVF) are projected to the left-sided half of each eye's retina and, vice versa, left visual field (LVF) information is projected to the right hemiretinae. As the nasal halves of each retina cross in the optic chiasm but the temporal halves do not, the LVF is linked to the right hemisphere (RH) and the RVF to the left hemisphere (LH). The letter string WITH is recognized as a meaningful English word as it is received by the LH, which is superior to the RH in making lexical decisions.
b–g: sample paradigms of Divided Visual Field Studies (DVFs). See text for explanations.

Figure 2d:

Here, a rarely used DVF paradigm is illustrated (Hunter and Brysbaert 2008). Pictorial stimuli are presented bilaterally—two recognizable objects, one on each side of fixation. A central fixation cross briefly turns into an arrow, indicating which object is to be named aloud (naming latencies recorded). In this trial, the correct response is "tree," not "boat." The paradigm has been validated with fMRI (mental word generation). A general RVF/LH advantage for correct naming was observed. This task shows promise for clinical and research use. If the typical RVF/LH naming advantage is attenuated—e.g., by emotionally charged or spatially encoded pictures (see Figure 2g)—it may reveal nuanced patterns of hemispheric interaction.

Figure 2e:

A local–global chimera is shown. The large letter "H" is composed of small "T"s. The local-global dichotomy (Figure 3) suggests LH/RVF advantage for local elements and RH/LVF advantage for global structure. This idea has been debated (e.g., Sergent 1982), who attributed the effects to spatial frequency differences—high for the LH, low for the RH. Nevertheless, the DVF paradigm using verbal stimuli for local–global contrast still offers opportunities to explore hemispheric cooperation. For a critical discussion, see Bardi et al. (2013). For hormonal modulation of global–local processing asymmetries, see Hausinger and Pletzer (2021).

Figure 2f:

This trial from a semantic priming task consists of a central prime (e.g., "MANE") followed by a lateralized target (e.g., "LION"). Participants must decide whether the target is a meaningful word. Direct semantic associations (as in this trial) typically produce faster responses after RVF/LH stimulation. However, for indirect associations (e.g., "STRIPES" → "LION" via "TIGER" or "ZEBRA"), LVF/RH advantage is observed (Beeman et al. 1994). This RH bias for distant associations links right-hemisphere function to creativity and, in independent research, to paranormal beliefs and psychosis-related thought processes (Pizzagalli et al. 2001; Taylor et al. 2002).

Figure 2g:

This paradigm presents word pairs vertically aligned, and participants judge their semantic relatedness (Zwaan and Yaxley 2003). For example, "BASEMENT" and "SALAD" require a "NO" response, while "BASEMENT" and "ATTIC" require a "YES." However, when "ATTIC" appears below "BASEMENT" (as in the figure), the word arrangement contradicts their real-world spatial relationship. Participants were unaware of this design variable. While an overall RVF/LH advantage was found for the task, a striking interaction emerged: only the RH was sensitive to the spatial congruence or incongruence of word referents. A similar RH sensitivity was found with vertically arranged body parts (e.g., "EYES" above vs. below "FEET"; van Elk and Blanke 2011).

Summary of Figure 2:

The DVF paradigm has produced a wide range of findings, often giving rise to dichotomous models of perception, cognition, and emotion. But even its most established polarity—the LH "dominance" for verbal and the RH "dominance" for nonverbal material—represents, at best, a complementary specialization. More nuanced views suggest that the principal differences lie in processing style rather than in strict content-based divisions. For example, the LH may specialize in high spatial frequencies and brief integration windows (as in printed words), while the RH may favor low spatial frequencies (as in the smooth brightness gradients of a face). Canadian neuropsychologist Justine Sergent (1982) was an early proponent of this view. Related is the "asymmetric sampling in time" (AST) model proposed by Poeppel (2003), which links hemispheric specialization to differences in temporal resolution and spectral processing. These frameworks are compatible with—and may even subsume—many of the classical FUHEMAS dichotomies outlined in Figure 3.

? Opposite cultural habits, like the direction of horizontal script, *must* have some influence on FUHEMAS? By the way, why are more languages with a horizontal script written from left to right than the other way around?

People from cultures with right-to-left (R-to-L) scripts—such as Arabic, Hebrew, or Urdu—perform differently in tasks involving scanning and manual exploration, including line bisection and greyscale judgments (Chokron et al. 1998; Friedrich and Elias 2014). Some lateral biases in numerical processing are also influenced by script direction (Fischer et al. 2010). However, most cognitive FUHEMAS—such as the left hemisphere's superiority for language functions—are biologically grounded and largely unaffected by reading and writing habits. Similarly, emotional associations with "left" and "right" tend to be culture-invariant: across diverse societies, "left" carries more negative, and "right" more positive, connotations, regardless of script direction (Hertz 1909; Tamagni et al. 2009).

Interestingly, even many animal species exhibit left-to-right scanning behaviors—despite having no culture at all. Ants, for instance, show right-side traffic patterns on foraging trails (Heuts and Brunt 2005), and chicks "read" horizontal displays from left to right (Rugani et al. 2015). Politically delicate or not, the default direction for horizontal writing seems to be from left to right.

? What do we know about emotional FUHEMAS in animals?

Animal studies have greatly advanced our understanding of emotional processing in humans. A broad overview can be found in Rogers et al. (2013). To pick just one aspect: birds have a simpler visual-to-hemisphere mapping than mammals—the left eye projects to the right hemisphere, and the right eye to the left. (But see Vallortigara et al. (2001) for additional functional specializations within the avian visual system.)

This simplicity allows clear behavioral interpretation in open-field studies (see illustration below) and facilitates experimental manipulation, such as patching one eye to probe hemispheric involvement.

So, finally, to your question: chicks show left-eye (right-hemisphere) dominance for spotting distant threats, and right-eye (left-hemisphere) dominance for detecting food items. The right hemisphere is broadly more sensitive to threat detection across nonhuman species, while the left hemisphere is more involved in courtship and other approach-related behaviors. This avoidance–approach distinction has largely replaced older models that cast the right hemisphere as "the seat of emotions" (see Figure 3).

For dog owners: next time your dog wags its tail, notice whether it wags more to the left or to the right. It turns out this depends on the dog's emotional state (Quaranta et al. 2007).

Left-eye/right hemisphere specialization for the detection of danger in far space, right-eye/left hemisphere dominance for the detection of food grains in the chick (adapted from Rogers et al. 2004 and Güntürkün et al. 2020).

? **The digital-analog polarity listed in Figure 3 intuitively makes sense to me. Aren't there rational arguments to defend it? Apart from this, is our brain working in a digital or analog mode, quite irrespective of hemispheric processing?**

The only rational argument for the digital–analog polarity may be... that it intuitively makes sense. "Analog" resonates with stereotypes about right-hemisphere processing: pictures, metaphors, intuitions (see also the discussion of clockface agnosia in Chapter 7, p. 148). "Digital," in contrast, seems to fit with the more literal, verbal, and rational character attributed to the left hemisphere.

So yes—your right hemisphere can intuitively grasp the digital–analog metaphor, while your left hemisphere keeps searching for a precise, testable rationale and comes up empty-handed.

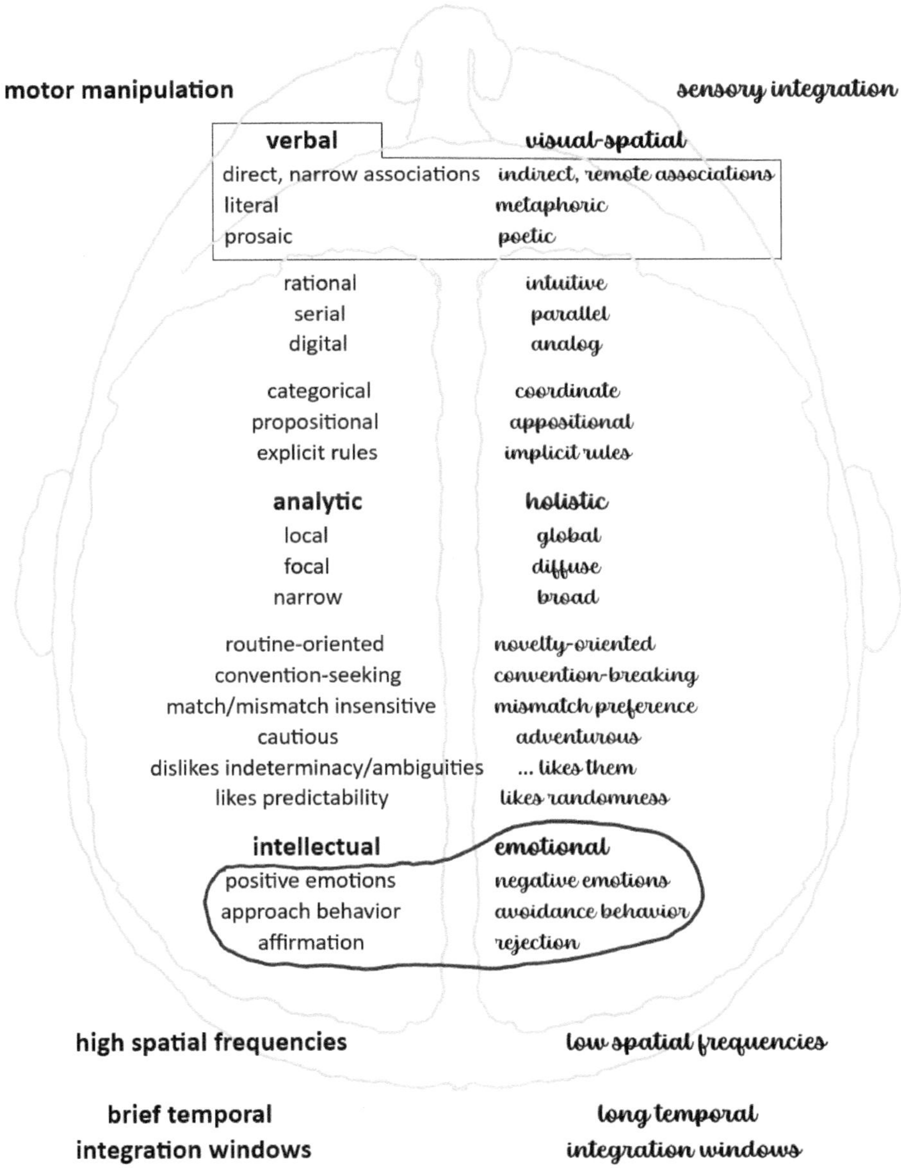

Figure 3: Two takes of the world. Dichotomies proposed in the literature to capture essential differences between the two hemispheres in cognitive and emotional processing

Like many of the dichotomies in Figure 3, this one is both a helpful metaphor and an oversimplified binary. You need both hemispheres to understand complex matters. Metaphors and intuitions offer a rapid, global "hunch" (sometimes accompanied by that forest-like feeling of being surrounded by insight). Careful sequential thought, on the other hand, lets you zoom in and grasp a single tree in detail—possibly with its own beauty.

As for your second question: biologically speaking, the brain probably operates primarily in an analog fashion. There are digital elements—for instance, a neuron either fires or it does not, which is binary. But even synaptic transmission blends digital and analog processes (Alle and Geiger 2006). So, like for many things in neuroscience, the answer is: both.

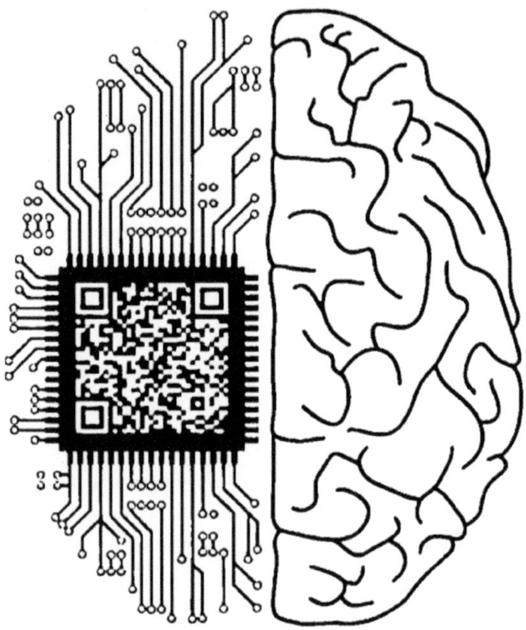

A digital left and an analog right hemisphere: an artistically nice but neuroscientifically simplistic analogy.
Reprinted, with permission, from *Hochschule für Life Sciences FHNW.*

A rare threesome of hemispheric specialization—including the corpus callosum—was proposed in relation to creativity by Bogen and Bogen (1969; 1988). Joe Bogen, one of the neurosurgeons involved in the earliest split-brain operations, and his wife Glenda, suggested that creative insight depends on a dual function of the corpus callosum: the ability to alternate between interhemispheric inhibition and facilitation, depending on the stage of the creative process.

Their model draws on Wallas's (1926) classic phases of "The Art of Thought." Inhibition, they argued, is beneficial during the incubation phase, allowing each hemisphere to process ideas independently—essentially, "two minds" viewing a problem from different angles. This kind of dual exploration could offer evolutionary advantages. Later, during the refinement phase, facilitation becomes crucial to integrate the two perspectives into a coherent whole.

This elegant hypothesis has never been fully appreciated. Instead, public discourse has simplified the matter by placing all creativity in the right hemisphere. Equally unhelpful have been efforts to debunk the "right-brain creativity" myth without acknowledging the more sophisticated interplay between the left and right, and the mediating middle (e.g., Corballis 2007).

I predict that future empirical research on this LH–RH-corpus callosum triad will not only enrich the science of creativity but also offer new insights into the long-standing debate about the connection between genius and madness. The complementarity between distant and close associations, or between divergent and convergent thinking, is a theme that transcends neuropsychiatry. It is foundational to natural language (Jung-Beeman 2005), to our understanding of metaphor and poetry (Pobric et al. 2008), and perhaps—ultimately—to the evolution of FUHEMAS itself.

Where Can We Read More?

Güntürkün, O., Ströckens, F. and Ocklenburg, S. Brain lateralization: A comparative perspective. Physiol. Rev. 2020; 100: 1019–1063. doi: 10.1152/physrev.00006.2019.

McManus, I.C. Right hand, left hand. The Origins of Asymmetry in Brains, Bodies, Atoms and Cultures. Weidenfeld and Nicolson: London, 2002.

McGilchrist, I. The Master and His Emissary: The Divided Brain and the Making of the Western World. Yale university Press: New haven (CT), 2009.

McGilchrist, I. Reciprocal organization of the cerebral hemispheres. Dialogues Clin Neurosci 2010; 12: 503–15. doi: 10.31887/DCNS.2010.12.4/imcgilchrist.

Springer, S.P. and Deutsch, G. Left Brain, Right Brain: Perspectives from Cognitive Neuroscience (5th edition, originally published 1985). W. H. Freeman and Co.: New York, 1998.

CHAPTER 12

How to Assess The Untestable?

At the end of the past millennium, the Houston Conference Guidelines on Specialty Education and Training in Clinical Neuropsychology defined who qualifies as a clinical neuropsychologist. They list the type of university training needed, the areas of core knowledge to be acquired, and many other standards. For an update, see Smith and members of the Clinical Neuropsychology Synarchy (Smith and CNS 2019). van Patten et al. (2022) expand on this by describing the roles of podcasts, webinars, and social media in both the formation and ongoing education of clinical neuropsychologists.

Fascination with popular books on the brain, the spark of curiosity ignited by puzzles of brain–behavior relationships—solved or still unsolved—and passion for working with individuals affected by brain damage are rarely mentioned in these formal documents. Yet fascination, enthusiasm, and passion are central to the profession (as they are to finding joy in any profession). They are the foundation on which a career in clinical neuropsychology must be built—and from which every patient will benefit. Merely performing the duties of a psychometrician, by contrast, shortchanges your patients.

Sadly, today's training culture is saturated with terms such as validity, standardization, and test batteries (see Chapter 8, p. 173 for a reflection on the latter). In the context of assessing brain damage, this often leads to a flattening of clinical neuropsychology into the narrower domain of psychometric testing. That is not meant cynically: there *is* an art to psychometric assessment, and we should not dismiss it. But it is not the art that this book—and especially this chapter—is about.

There are already enough textbooks that take inspiration from the business world, where standardization and validation are seen as guarantors of progress. In that world, the lack of validation or standardization slows innovation by leaving customers uncertain and open to alternatives. But clinical neuropsychology is not, in the first instance, a business.

It is the craft of understanding individual behavior—behavior shaped not only by disease, but by a complex mix of personality, life history, coping styles, and circumstances. Before quantifying any deviation—be it sensorimotor, cognitive, or emotional—we must pause to appreciate the possible sources of its determination.

This chapter explores selected approaches to navigating the broad and nuanced palette of behaviors observed in patients, whether at the bedside or when a more

DOI: 10.1201/9781003384991-12

formal neuropsychological assessment is not conceivable. Its purpose is to support the pre-assessment phase, before one reaches for a test manual.

Depending on why a given patient may be untestable—or untestable by conventional means—the following suggestions are grouped into three broad categories. First, testing may be limited by the circumstances in which it takes place. Second, there may be patient-specific variables that render standard procedures ineffective. And third, the examiner may need to confront their own conceptual or professional limitations.

Ways to Circumvent Situational Limitations

The most frequent limitation to a comprehensive neuropsychological evaluation is a lack of time. Clinical neuropsychologists working in country practices, seeing one patient a day, are the exception. If you are part of this fortunate minority, you may skip the rest of this section. Everyone else will likely wonder how to save time in one part of the assessment in order to reallocate it to another. Of all situational limitations, time pressure is the only one addressed here.

Parsimonious Assessment of Seeing, Naming and Doing

Figure 1 shows the contours of simple objects which, while individually familiar, are not trivial to recognize when presented in overlapping form (cf. Chapter 7, Figure 5). Some of the objects are tools and invite you to ask for pantomime use (see Chapter 5). Show the figure to patients and ask them to name what they see (Chapter 4).

There is no need to provide a formal instruction. Something like "I'm going to show you some objects that are a bit hard to recognize—what do you see?" will suffice. Most patients will name them effortlessly, and your preface creates a sense of success rather than failure.

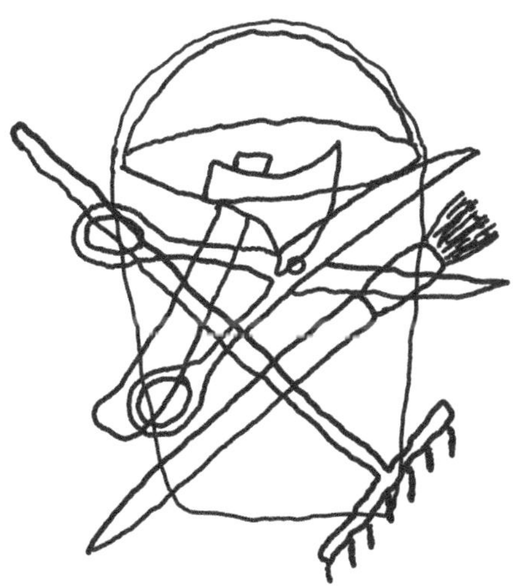

Figure 1: Overlapping object images allow to check for visual perception, naming and praxia. See text.

Some may not immediately name the scissors, but offer a descriptive workaround: "Something to cut with." Others might struggle to find the word for "paintbrush" and instead say, "You use it for colors... to draw something... no, to paint... it's not a toothbrush..." In such cases, you should follow up with more targeted probes to better characterize the anomia.

When asked to pantomime using an axe, rake, or paintbrush, most patients will demonstrate the correct action, with both right and left hands. The same typically holds for the scissors—but here, some may symbolically open and close their fingers to represent the blades. That is a body-part-as-tools error (see Chapter 5, p. 85), and you should gently point this out. This alone, however, is no indication to proceed with a full standardized praxis battery.

If a patient frowns at your request to pantomime during a serious neuropsychological examination, they will usually relax upon hearing the brief but authoritative explanation: "The ability to pantomime is a higher brain function that we need to assess!"

What about patients who do not recognize overlapping objects? Take a blank sheet of paper, lay it over the stimulus sheet, and trace one of the contours with a thick felt pen. Or draw any familiar object from memory. Does she recognize it now? If yes, continue with five or six more non-overlapping figures and assess naming and pantomime. If no, proceed to evaluate visual perception. Or—if it suddenly dawns on you that the patient's companion dog might be more than a friend—you might want to switch to the auditory modality.

Parsimonious testing is not a workaround for patients who are untestable. It is a thoughtful, targeted response to the time pressures that clinical neuropsychologists increasingly face. You should never hesitate to be concise in your report—ideally written in a structured yet flexible format:

"No apraxia. Patient pantomimes well the use of an axe, a rake, scissors, etc." If you test full sequences (e.g., "brewing coffee and serving it to guests"), you might add: "Also, entire sequences of actions are performed without difficulty."

If sticking with the objects from Figure 1, you could ask the patient to describe a full sequence like cutting small branches with the axe, raking up the pieces, and filling them into a bucket.

The time you save by omitting lengthy standardized assessments like the VOSP (Visual Object and Space Perception Battery), the TULIA-R for praxis, or the Boston Naming Test can be redirected—perhaps to explore (hemi)spatial neglect more thoroughly, or to provide a richer evaluation of calculation abilities.

None of this is a call for "quick and dirty testing." Rather, it is encouragement to avoid test-machine behavior and, as a pleasant side effect, turn the assessment into a more enjoyable experience for the patient.

? This sounds like a first screening, before you initiate serious testing. Why not use valid and reliable instruments like the MMSE (Mini-Mental State Examination), which has high sensitivity and specificity?

The MMSE was introduced by Folstein et al. (1975) in a good spirit: they sought to offer "a practical method for grading the cognitive state for the clinician" (as stated in the original paper). I suspect the authors were themselves surprised by how quickly their instrument gained worldwide fame. Over the past half-century, the MMSE has become the most widely used screening tool for cognitive decline in the elderly.

This is despite debates over sensitivity and specificity, which range from over 80% to under 50% across reviews. The MMSE is highly training-dependent. When used by neurology residents, the range of scores given to a fake patient simulating a score of 18 can vary by nearly 20 points (n ≈ 600 students, collected over six years; unpublished data).

More critically, the MMSE does not assess executive or prefrontal functions—yet a patient's total score depends heavily on attention, concentration, interference control, and other executive processes. For detecting mild cognitive impairment or dementia—especially of frontotemporal origin—the MMSE cannot be recommended (Mitchell 2009).

To paint a bold picture: the MMSE's value is akin to that of a fictional *Mini* Motor *Status Examination*, scoring between 0 and 4.

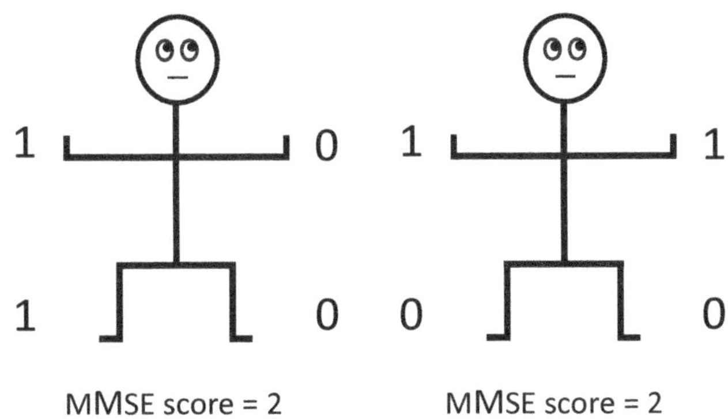

Scoring is 1 for preserved functionality of an extremity and 0 for the loss of it. Both patients depicted above have a score of 2. According to the *Mini Motor Status Examination*, hemiplegia is thus identical to paraplegia.

The MMSE is often cited, but the original publication is rarely read. This is especially true in research contexts, where it is included in study protocols simply because others have included it before. It reminds me of a poster I had as a teenager: Frank Zappa on a toilet, with the caption: "EAT SH** — TEN BILLION FLIES CAN'T BE WRONG!" Pardon the vulgarity. The problem is not with the MMSE itself, but with researchers who uncritically consume what is already on the table.

Once again, in 1975 the MMSE was a welcome addition to the limited repertoire of brief cognitive instruments. These days, however, we are seeing a gratifying increase in the use and citation of the MoCA—the Montreal Cognitive Assessment (Nasreddine et al. 2005). This trend is likely to support the gradual decline of the MMSE in the years ahead.

Parsimonious Assessment of Seeing, Comprehension and Perspective Taking

Returning to quick—but not careless—methods for assessing selected cognitive domains: while Figure 1 targeted lower-level perceptual functions, naming, and (ideomotor) praxis, Figure 2 builds on the same principle of parsimony while shifting the focus to scene comprehension, command following, and aspects of perspective-taking. Be prepared to accept the trade-off: you gain flexibility at the expense of standardization.

Figure 2: A picnic scene to assess perceptual integration, language comprehension, and left-right discrimination. See text.

Show Figure 2 to the patient and ask for a title. Respecting the person's educational background and general cognitive status—as you intuitively do when speaking to people in everyday life—you will find that both the choice of title and the search for it reveal much about intellectual functioning and current mood.

Most healthy individuals, and many patients, will offer a clear and coherent answer: "A picnic," "Outdoor scene," or some semantically related variant. By contrast, patients who struggle to integrate the visual scene—such as those with Balint's syndrome (Chapter 7, p. 150)—will name only isolated elements, depending on where their gaze happens to land: "A bottle." "A house." "A woman." "A dog." They will not arrive at a global title such as "Picnic scene."

Even less severe integration problems become apparent in the patient's verbal description. Subtle forms of simultanagnosia (Chapter 7, p. 150) and various degrees of impaired eye movement control can be detected this way.

Next, explain why you are asking the patient to follow simple and increasingly complex commands. Keep the atmosphere light, even playful. Remember: language comprehension is not a "pure" function. It depends heavily on attention span and working memory. By gradually increasing the number of steps and the grammatical complexity of your instructions, you can gain a nuanced impression of where problems may arise—particularly the kinds of misunderstandings that affect real-life communication.

This mirrors our own experience: short, direct sentences are easy to understand. Longer sentences, especially those that include subordinate clauses irrelevant to the main point, become harder. What really increases difficulty, however, is the nesting of clauses—when one is embedded inside another. Depending on whether this embedded content is essential or not, comprehension may fail even in otherwise functional language users.

You might begin with instructions like the following, while pointing to Figure 2:

* Please touch the kite.
* … one of the dogs.
* … the girl in the foreground.

Then gradually add complexity:

* Please touch the house first, and then the boy's father.
* … first one of the clouds, then the sailing boat, and finally the basket.
* … first the tree on the other side of the lake, then the blanket on which the family sits, and finally the cup in the boy's hand.
* After you touch the lake, please move your finger from the right cloud to the left—without touching the kite.
* Please put this paper clip [assuming you have one at hand] over the two chasing dogs. Then move it to the mother in the foreground after knocking on the table three times.

Continue as needed—not for an hour, of course, but until you feel confident reporting whether the patient can follow three-step commands, understands spatial and temporal conjunctions, and, on the expressive side, speaks with grammatically structured sentences and appropriately modulated intonation.

To probe left–right discrimination, body schema, and perspective-taking, you might ask:

* Is the ball to the right or left of the dog? (from the patient's perspective)
* How many left-handed people are shown in the picture?
* Is the boy sitting to the left or right of his mother?

You can increase the challenge by rotating the image upside down and repeating the same questions.

Figures 1 and 2 are not substitutes for a full aphasia examination, nor are they intended to cover the full range of perceptual, cognitive, and body schema functions. But they offer efficient shortcuts—shortcuts that allow you to reallocate time and energy to areas where deeper examination is more urgently needed.

Parsimonious Assessment of Hemispheric Language Competency: The "Luria+" Assessment

This paragraph is a tribute to the Russian psychoanalyst-neurologist[1] Alexander Romanovich Luria (1902–1977; see Cole et al. 2005, for a biography, and the special issue of *Cortex* titled *Luria's Legacy in the Era of Cognitive Neuroscience*, Vol. 169, December 2023). Luria founded the science of neuropsychology in the Soviet Union and gained international recognition through his engaging case studies—often in a style reminiscent of Oliver Sacks.

One such book is *The Mind of a Mnemonist* (Luria 1968/1972), which portrays a man with extraordinary memory and synaesthesia. Another is *The Man with a Shattered World* (1971/1972), which chronicles the life and rehabilitation of a brain-injured soldier, with a trace of literary stylization.

Luria's most technical contribution is the volume *Higher Cortical Functions in Man* (Luria 1962/1980), where he details the kinds of sequencing deficits—both cognitive and motor—that can result from frontal lobe damage.

A now-classic element of his assessment approach is the Fist-Edge-Palm task. The patient is asked to observe the examiner perform a three-step sequence of hand postures, then imitate it while the examiner repeats it, and finally execute the sequence independently.

Luria's original sequence was **fist → edge → palm**, though others have adopted different versions (e.g., palm–fist–edge in Cummings and Mega 2003; fist–palm–edge in Scott and Schoenberg 2011).

Figure 3 shows the sequence as originally described by Luria (1980).

[1] What a neat combination! Note that this was quite a while before *neuropsychoanalysis* was founded in the Western world—see the International Society and its *Journal* of the same name.

The Fist-Edge-Palm task is also part of the Frontal Assessment Battery (FAB; Dubois et al. 2000), a two-page screening instrument freely available online, with normative data stratified by age and education. It is a highly recommended tool for rapid clinical use.

I am shocked: *you* recommend a test battery (see your mocking words in Chapter 8, p. 173)?! 😉

You caught me! I remain reluctant to use the word "battery" in connection with (neuro)psychological testing. But the content of the FAB makes a lot of sense. It provides a parsimonious selection of tasks that tap into the diversity of frontal lobe–mediated functions, and its psychometric properties are surprisingly good. The FAB offers a properly standardized screening: beyond motor and grasping behaviors, it includes verbal fluency and conceptualization tasks, as well as assessments of imitation, inhibition and cognitive flexibility.

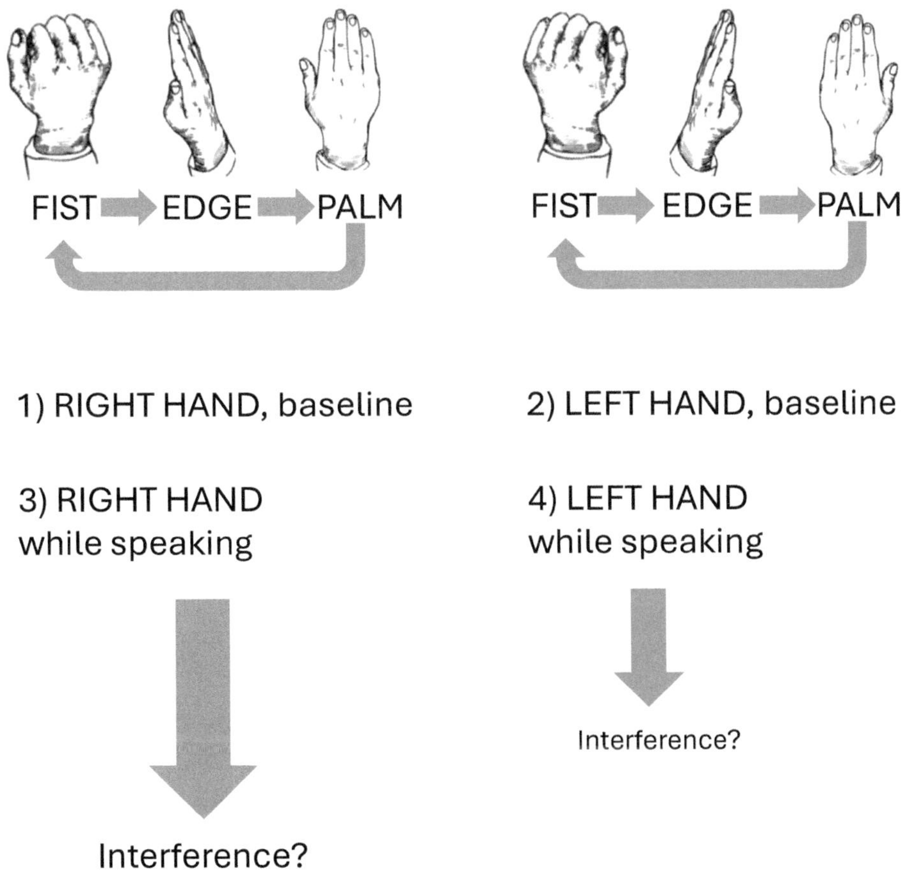

Figure 3: Luria's "Fist-Edge-Palm" Test, here executed with both hands in succession and, in addition, once again in a dual task condition. See text (pictures of hands from the FAB webpage, originally drawn by A. R. Luria).

The *Luria+* assessment, as we call it, begins with the original *Fist–Edge–Palm* task, performed twice: once with the dominant hand, and once with the non-dominant hand. Clinically, differences in manual skill do not produce major sequencing effects.

The twist lies in the next two runs—one per hand—during which the patient[2] is asked to continue the now partially automatized sequence while speaking. They might tell you what they usually eat for breakfast, what they are planning for the weekend, or anything else they like. No standardization is needed here; the key is that the patient keeps talking while continuing the hand sequence.

Avoid chunked execution (e.g., fist–edge–palm, pause, repeat). Instead, the transition from one hand posture to the next should be smooth and continuous, so that a bystander could not tell where the cycle begins or ends. This smooth rhythm should already be encouraged in the baseline condition. The target speed is about three hand positions every two seconds. More importantly, however, the pace remains stable across baseline and dual-task conditions.

Video recording is strongly recommended for later offline analysis.

The interference between manual sequencing and simultaneous speech varies widely between individuals. What is of primary interest, however, is the intraindividual asymmetry—the difference in interference when the task is performed with the right versus the left hand.

Based on the model of functional cerebral distance (Kinsbourne and Hicks 1978), interference should be stronger for the hand whose cortical representation lies closer to Broca's area. In other words, right-hand sequencing should be more disrupted by simultaneous speech in individuals with left-hemisphere language dominance—and vice versa. In people with right-hemisphere language dominance, the interference should be greater when they use their left hand during speech.

> An elegant demonstration of the principle of functional cerebral distance is found in the seminal study by Kinsbourne and Cook (1971). Healthy participants practiced balancing a rod on the index finger of each hand. After training, they were asked to repeat the balancing task while either speaking or remaining silent. During the "speaking" condition, balancing time was reduced only for the right hand. The authors concluded that interference was greater for right-sided motor control because both speech and right-hand movement are programmed by overlapping areas in the left hemisphere—an example of "shared functional space".

? This sounds like an interesting experiment. However, it is an isolated observation. Or have there been follow-up studies?

There were not just a few follow-up studies—the work by Kinsbourne and Cook (1971) sparked an entire subfield of laterality research focused on concurrent manual and verbal tasks. In most studies, "manual" refers to simple finger tapping, which pairs reliably with a wide range of verbal activities (Clark et al. 1985).

[2] Only patients without upper limb paresis were tested.

> Hellige and Longstreth (1981) reported asymmetric effects of *reading* on tapping speed. Specifically, for the right hand, instructions to memorize a text produced more interference than instructions to read merely. In a complementary manual–nonverbal task, they found the inverse pattern: solving a block design task interfered more with the left hand than with the right.
>
> Crucially, these effects were soon shown to be related to hemispheric differences in speech processing—that is, to language dominance. Patients with left-hemisphere speech control, as determined by the Wada technique (see Chapter 11, Table 1), showed larger right-hand performance decrements during speaking. Those with right-hemisphere speech control showed the opposite: larger left-hand decrements (Kosaka et al. 1993).
>
> A thorough review of studies up to the turn of the millennium is provided in Medland et al. (2002).
>
> Why has this elegant research tradition lost momentum? Likely because simple behavioral studies on FUHEMAS (Chapter 11) have fallen out of fashion—even when some of them were both brilliant and influential. This is even though many open questions in this domain would still be best addressed through carefully designed behavioral paradigms.

The interpretation of hand differences in verbal–manual interference using Luria's *Fist*–Edge–Palm test does not require advanced expertise. It remains a crude but clinically valuable observation. The procedure saves time compared to more elaborate methods for assessing hemispheric language asymmetries, such as dichotic listening or divided visual field paradigms.

However, if one of the referral questions explicitly concerns hemispheric superiority for language functions, these more formal, lateralized tasks become indispensable.

Still, the Luria+ approach offers a rare opportunity: it allows for data collection even in clinical settings, especially once a more standardized protocol is developed. For example, incorporating structured reading material—as in Hellige and Longstreth (1981)—would permit a more objective evaluation of speech rate, fluency, and possible irregularities under dual-task conditions. Likewise, the manual sequencing component could be scored more quantitatively.

Yet developing such methodological refinements would go beyond the spirit of parsimony emphasized throughout this chapter. The strength of the Luria+ procedure lies precisely in its clinical accessibility: it can be implemented quickly, with minimal equipment, and still yield meaningful, patient-specific insights.

Ways to Circumvent Limitations by Patient Characteristics

The limitations discussed under this heading arise from the *severity* or *nature* of a patient's impairment. I have chosen two illustrations: *transient global amnesia* and patients who are "behaviorally difficult" for a variety of reasons.

In terms of the principle of testing parsimony, this section and the previous one are as different as chalk and cheese. What you gain from working with the types of patients described here is extraordinary insights—into how the brain functions and into the ways traditional neuropsychological assessment might be transcended. But there is a price: you must be prepared to invest considerable time—often a large portion of your personal time. In the case of TGA (see below), you cannot even plan when or how to spend that time. You will need to be at the hospital emergency unit when it happens. And you might wish to stay there for a couple of hours, both in your own and the patient's interest.

Transient Global Amnesia (TGA)

As noted in Chapter 6 (p. 120), transient global amnesia is a sudden-onset amnesia that occurs without trauma, seizures, or focal neurological signs. The episode typically lasts around 24 hours, during which the amnesia gradually recedes. The anterograde component is most severely affected. If you ask the patient to remember three words and prompt recall after 30 seconds, you will likely hear: "Which three words?" Retrograde memory can also be impaired, to varying degrees, ranging from minutes to many years.

TGA patients are typically seen at the emergency department of a larger hospital. Experienced neurologists make the diagnosis based on behavior: repeated questioning about why they are in the emergency unit, what happened, and—again—why they are in the emergency unit. A partner or friend often accompanies patients. The third-party history usually confirms the diagnosis: sudden onset without head injury, possibly triggered by a cold shower, sexual activity, physical exertion, or strong emotion. The patient reportedly began asking questions such as "Where are we?" and "What are we doing here?"

Standard medical workup includes EEG (to rule out epileptic activity and a transient epileptic amnesia, Chapter 6, p. 121) and MRI. Neuropsychology is typically *not* involved. The case is clear. Any experienced neurologist can diagnose a TGA within five minutes of the patient's arrival.

The problem is that neurologists do not have the time to explore amnesia in the depth that makes it interesting. Ideally, senior neuropsychologists would hand this case to a motivated assistant. Yet in practice, many seniors—especially those specialized in psychometrics—feel overwhelmed by a patient who interrupts the reading of a 15-word list after every two words to ask what happened and why they are where they are.

> Importantly, consciousness is never impaired in TGA, and patients remain oriented to person and identity. However, they do notice that "something is wrong." It's helpful to explain that the areas of the brain responsible for memory experienced a brief, mild lack of oxygen. To the accompanying person—usually a partner or close relative—you should also explain that this is *not* a stroke or a case of dementia. But avoid these terms when speaking directly to the patient, unless the patient specifically asks.
>
> Do not feel awkward about repeating yourself: you *will* be asked the same questions over and over again, and responding patiently is part of the clinical process. Semantic memory is preserved, but prospective memory is virtually absent. Orientation to time is

lost. Don't miss the opportunity to repeatedly ask the patient what day, month, and year it is. Write down the *exact clock time* of each response (and any other specific answer—see Figure 4). Video documentation of the full assessment is strongly recommended; keeping detailed protocol notes while managing spontaneous interactions will be difficult.

Assessing mental status in a patient with TGA is both a clinical service and a rare learning opportunity. It teaches you, as the examiner, a great deal about the components of memory. It is also a test of your flexibility. Finally, your time and attention will likely be most appreciated by the accompanying person—often more desperate and frightened than the patient herself.

In what follows, the term "episode of TGA" refers to an amnesia that began only a few hours ago—or at least one that has not yet resolved to the extent that regular neuropsychological testing would be possible.

Insist on the partner's presence during the examination. They will be a crucial collaborator throughout. Ask them to prepare a list of key autobiographical episodes using simple keywords.

Differentiate among the three temporal layers:

* The very recent past (minutes and hours before onset):
  Where was the patient? What had been planned for the day? Was there an emotionally arousing situation? Sudden immersion in cold water?[3] What physical activity occurred? Do not hesitate to ask about sexual activity—this is a well-known trigger of TGA.[4]
* The past week:
  What happened during the last weekend? Was there any special event? A concert, museum visit, sports event, or social gathering? Any recent accident? Which relatives did the patient meet?
* The more distant past:
  Where was Christmas celebrated? How many days of vacation have been taken this year, and where? Have there been trips out of town, abroad? With whom? Can the patient recall meaningful details?

Ask these sets of questions repeatedly and always in the same order. The partner may supplement with additional prompts, as appropriate.

This approach is instructive for two reasons. First, the autobiographical memory timeline does not perfectly overlap with the timeline for public events (see below). Second, recovery of autobiographical memory gives you a sense of "where in time" the patient is situated. This mental position gradually approaches the present. In contrast, anterograde memory deficits often persist for another hour or two.

[3] The original term for TGA was «amnesia by the seaside» because it was observed in people jumping into the ocean instead of allowing slow immersion (Tubridy et al. 1999).

[4] Readers should not worry about this association. There are more serious neurological sequelae of sex (Grasland et al. 2021 for overview). On the other hand, frequency of intercourse appears to be associated with improved verbal recognition memory in healthy young women (Maunder et al. 2017).

I recall the case of a photographer in her early 30s who was brought to the emergency unit by her husband (she had not yet seen her GP). She was a heavy smoker, so I asked her husband to write down the brands she had smoked over the years. He noted four distinct periods, each defined by an exclusive preference for a single brand.

Over approximately six hours of observation, the patient named these brands in sequence. In the first two hours, she repeatedly answered "Brand 1" when asked what she smoked. Then came a brief "Brand 2" period. When I asked again, she paused: "Didn't you ask me that before? I think I said [Brand 3]... wait... did I say that? That was a while ago. It's been [Brand 4] since last fall, I think." She recalled the brands in correct order—but according to her husband, she had switched to the final one more than two years ago.

Two additional details stuck with me:

(1) Recovery of autobiographical memory preceded her recovery of public event knowledge (e.g., current politicians, major accidents, local news).

(2) When I asked the smoking question again in hour four, her husband quietly handed me a note: "I can't remember having seen her without craving for such a long period of time!"

"Memory" for one's addiction seems impaired in TGA as well![5]

For public events, a list of questions pertaining to the periods described above ("the past few weeks" and "the more remote past") should be prepared in advance. The structure may follow the Remote Memory Questionnaire used by Bergin et al. (2000). In contrast, questions about public events immediately preceding the onset of TGA will need to be improvised in the moment.

TGA is an enormously rich source for learning how to assess brain functions that escape standardized testing. In every patient, you can demonstrate dissociation between procedural memory and content (or declarative) memory (see Chapter 6, p. 123).

One effective way to illustrate this is to use the "Spot the Differences" task depicted in Figure 4. Present the figure at least six times, ideally at regular intervals. Patients are typically familiar with the general nature of such tasks, but you should still explain them during the first presentation. Say, for example: "There are five differences between the top and bottom images. All of the differences are additions to the bottom image."

It may surprise you when, after two or three exposures, the patient asks: "There are five differences, right?" —despite firmly denying that they have seen the image before or that you ever explained the task.

Figure 4 includes a tabular overview of the gradual decrease in time required to spot all five differences, despite the patient's unawareness that they had solved the task several times.

[5] In at least three other cases, we observed that memory for pain—such as chronic back pain that typically forces patients to walk with a guarded posture—can be similarly compromised.

> The patient was a 46-year-old right-handed general practitioner whose TGA episode began abruptly over lunchtime. There was *no retrograde amnesia.* He had spent the morning on his private sailing boat, had sexual intercourse, and then went for a swim in the lake (water temperature approx. 18°C). Already in a state of amnesia, he managed to return to shore in a jolly boat, drove his (secret) girlfriend to the train station (a 5-minute drive), and then proceeded to his medical practice (20 minutes). There, he asked his assistant to take him to the emergency unit of a nearby university hospital (another 20-minute drive). Neuropsychological assessments began approximately two hours after onset.
>
> The accompanying table shows the patient's consistent lack of awareness that he had seen the spot-the-differences task before. On the fifth exposure, he hesitantly suggested that he "might have seen it once in the newspaper, a long time ago." Despite this persistent unawareness, his time to detect the five differences decreased steadily across trials.
>
> Due to situational constraints, more than seven exposures were not feasible. We remained with the patient until late evening, at which point he was still not testable in any conventional sense (e.g., he could recall only the final two items from a 15-word list). By the following morning, he appeared back to normal. A formal neuropsychological assessment could not be conducted for logistical reasons, but based on clinical experience, subtle deficits in anterograde memory and interference control are likely to have persisted below normative levels.

The tendency to confabulate can be provoked by asking verbal riddles and conundrums such as:

1. "Who can jump higher than a house?" (Everyone. Houses can't jump.)
2. "What has a neck but no head?" (A bottle.)

You can also use more figurative prompts. The riddle in Figure 5 provides an example, designed for a protocol sheet. The solution is zebra. The image shows the letter C followed by a bra. Together, they form the pun "c-bra"—phonetically zebra in an animal context.

You may choose to reveal the solution right after the first exposure, or only after several unsuccessful guessing attempts. It is improbable that the patient has encountered this riddle before.[6]

Once the solution is given, a "creative spark" may occur on a later trial: the stripes on the bra are now suddenly recognized as zebra-like. A paradoxical response might be: "Well… it's not a zebra, but…" —indicating implicit recognition without full mnestic awareness.

When the correct solution is finally retrieved, ask the patient how they arrived at it, you will hear every kind of rationalization imaginable—except the simple truth: "Because you told me before."

[6] It was invented for this chapter.

Exposures 1-7 (minutes since previous exposure)	Exp. 1 -	Exp. 2 (4)	Exp. 3 (9)	Exp. 4 (12)	Exp. 5 (12)	Exp. 6 (16)	Exp. 7 (11)
«Have you seen this picture before?»	No	No	No	No	Not here	No	No
Time to find the 5 differences [s]	74	69	49	52	32	27	24

Figure 4: A "Spot-the-differences" task. Can you find the five differences between the top and the bottom part of the picture? See the text, also for an explanation of the table.

This phenomenon illustrates a point made in one of the most-cited articles in cognitive science: Nisbett and Wilson (1977).[7]

You will witness many dissociations between what is reported and what is known when examining patients with TGA. Covert knowledge, confusion about both the content and source of memories, and the paradox of verbally rejecting a solution that has clearly been retained somewhere—these all reflect what might be called paradoxical knowledge: the solution is present, but inaccessible to conscious acknowledgment.

What animal hides here?

Patient's answers (verbatim!):

Exposure 1

time: ______

Have I asked you before?

Exposure 2

time: ______

Have I asked you before?

Exposure 3

time: ______

Have I asked you before?

Exposure 4

time: ______

Have I asked you before?

Figure 5: A riddle to be repeatedly exposed to a patient during an episode of TGA. This page may be used as a scoring sheet (use one per riddle). See the text.

[7] After my earlier disrespectful comments about citation frequencies (see p. 273 above), I feel I must emphasize: this one is a must-read for every neuropsychologist!

? TGA seems to have had a lasting impression on you. Is it not sad to gain insights into the workings of the brain by investigating people with such severe impairments, like the loss of memory? And, if I may add one more question, what is the biggest paradox you have stumbled over in your work with patients with a TGA?

To your first question: working with patients with TGA is not sad at all! You know with near certainty that memory will recover by the next day. This knowledge helps you maintain a cheerful mood during testing, which in turn has a calming effect on both the patient and the accompanying person.

To the second question: for me, the greatest paradox lies in a small, seemingly trivial observation. We once collected data using the "Mental Dice Task," (Brugger et al. 1996a), which requires patients to generate a random number from 1 to 6 every 30 seconds. The examiner interrupts the ongoing assessment by saying, "Dice!"—and the patient is supposed to respond with a random number.

Not one single patient ever asked, "What do you mean by 'dice'?"—they simply named a number between 1 and 6. This included patients who, just seconds earlier, had asked for the second or third time what had happened or where they were.

The only explanation I can offer—though I am not fully convinced by it—is that this behavior reflects a kind of conditioning. Perhaps the word "dice" carries with it an implicit association: it automatically activates the expectation of a number between 1 and 6. That is not full-blown memory. But it is not absence of memory either.

And that, I think, is precisely what makes it so paradoxical.

How to Cope With "Difficult" Patients

"Difficult" is difficult to define in a universally accepted way. Perhaps it is best circumscribed as a challenge to our own limitations—that is, behaviors we struggle to deal with because our university training has not adequately prepared us for such encounters.

In that sense, many behaviors of otherwise "regular" patients can be challenging. Consider the Parkinsonian patient whose tremor is so pronounced that graphomotor testing seems unthinkable. Or the patient with advanced Huntington's disease and prominent choreiform movements: have them write large letters in the air, and you will avoid writing in your report the embarrassing line, "Writing could not be assessed due to motor symptoms."

Or take the patient with Balint syndrome (see Chapter 7, p. 150). Who has been taught how to assess writing in such a case? Ask her to close her eyes and write her name. Dictate a full sentence. What may look like severe agraphia with eyes open may simply resemble what happens when you or I write blindfolded. Under visual control, her hand movements are chaotic; with vision eliminated, they normalize.

Even in the domain of oral language, our neuropsychological training often leaves us ill-prepared. How does one hold a conversation with a patient with expressive aphasia?

This list could go on—but the "difficulty" that is hardest to define deserves special attention. Below, we will attempt to organize it under a few descriptive headings.

? Sorry to interrupt the flow, but have you just meant to say that I can have a conversation with a patient with Broca aphasia?

Absolutely yes. One of the first patients I ever examined with a classic Broca aphasia (after a left MCA stroke) had a vocabulary of just three expressions: "Yes," "No," and "My god!"

With him, I learned that "yes" doesn't always mean yes, and "no" is not always a negation. He happened to be my former botany professor, specializing in the taxonomy of algae and lichens. Through increasingly well-informed questioning—and by paying close attention to his prosody—I was able to reconstruct the key differences between the two.

He used "yes" and "no" interchangeably,[8] while "my god!" typically signaled dissatisfaction with the proposal embedded in my question. So, if I asked, "Isn't immersion in water the principal difference between them?" he might reply with a mix of those three utterances. But the prosody—his tone and facial expression—made it clear that I had missed the point.

In the course of our "conversation," he gestured that algae are a single organism, while lichens are always a symbiosis—using his thumb alone versus thumb and index finger together. By miming sun rays beaming down to Earth, he reminded me that only algae can perform photosynthesis.

Of course, you cannot expect to discuss a topic that you are not familiar with. But through a strategy of increasingly educated guesses and attention to nuance, you can have a surprisingly technical conversation with someone with Broca aphasia.

This is the opposite of parsimonious, time-saving testing. But it is deeply encouraging for the patient and highly instructive for the examiner. It sharpens your ability to interpret nonverbal signs, prosody, and gesture.

In fact, all successful "mediumistic" performers—consciously or not—use a similar form of *cold reading* (Dutton 1988) to impress their naïve clients. So in the end, you might cease being a neuropsychologist and start a more lucrative career as a spirit medium!

The behavioral elements described below may not be caused exclusively by brain damage. In many cases, they likely reflect a mixture of neuropsychological dysfunction and premorbid personality traits. That said, details from the personal

[8] Alice in Wonderland once complained that her cat would always purr or mew, but that neither sound consistently meant "yes" or "no." "But how can you talk with a person if they always say the same thing?" (Chapter XII of Through the Looking-Glass; Carroll, 1862/1984).

history of the first patient—referred to here as RB (for Rule Breaking)—make premorbid factors seem highly unlikely.

Rule Breaking Behavior

RB is a 68-year-old right-handed woman, a retired real estate manager with 18 years of education. She was referred for a preoperative neuropsychological evaluation before resection of a left temporal lobe tumor that extended into the orbital aspects of the frontal lobe.

Her initial appearance contrasted strikingly with her subsequent behavior. She was well-dressed—elegant, even—despite being an inpatient in the neurosurgery department, where most patients wear standard hospital clothing. Her makeup was discreet and tasteful, perfectly matched to her light blouse and elegant skirt. She carried a valuable handbag on her lap and insisted on holding it throughout the entire examination.

Her verbal behavior, however, was unusual from the start. She was so impolite that the examiner—ordinarily a patient and empathic listener—called me in to help. I couldn't. RB was equally offensive toward me. (No "white coat effect" here!)

When asked about her medical history, she responded obliquely: "You'd do well to read the charts," or "I've stopped counting how many times I've been asked this—[followed by a clipped, telegram-style summary]."

She did comply with test instructions—but always after a dismissive preamble such as:

"What a silly test!" or "This is the stupidest request I've ever heard!" Then she would follow the instructions for a few seconds (demonstrating adequate auditory comprehension), only to deliberately and provocatively violate the task.

In the Trail Making Test A (TMT-A), she connected 1 to 2 to 3 to 4, found 5 and 6, and so on quickly—and then violently erased all numbers. She copied a semi-complex ten-element figure by drawing a spiral, shouting, "Look, I've done it!" and tossing the pencil onto the table.

Later, during an individually adapted apraxia task, she scribbled on a blank sheet in front of her. Alongside abstract round and curved shapes, she drew a miniature but accurate copy of the ten-element figure, which still lay on the table.

When asked to demonstrate how to use a hammer, she responded, "Why would I ever use a hammer?"—and then added, meeting my gaze, "Well, I could hit it on your head?!" When asked to pantomime the action, she executed an entirely appropriate movement. (Left-hand apraxia was not tested.)

This brief vignette—rough as its summary may be—should serve as encouragement: rule-breaking behavior, at its most pronounced, can offer rich diagnostic and therapeutic opportunities. RB gave written informed consent for segments of the assessment to be videotaped. The resulting footage has become a valuable record of how temporo-frontal lesions can alter behavior and personality—echoing classic observations in the context of traumatic brain injury (Stuss 2011).

Third-party history later provided by RB's sister confirmed that the patient had changed "as if turned inside out like a glove" following her diagnosis, made three weeks earlier. (She had been diagnosed with glioblastoma, later resected with apparent success, although she refused to undergo postoperative follow-up.)

The central message of this case: it pays to remain patient and understanding when confronted with pronounced hostility in a brain-damaged individual. Sometimes, beneath the provocation lies a deeper—and deeply altered—neurological reality.

Individual Quirks

Now and then, a patient peculiarly approaches a task. These peculiarities are not mere idiosyncrasies of personality or reflections of a general mannerism. Nor are they linked to an increased tendency toward rule-breaking behavior. They often seem independent of the difficulty the patient experiences with the task itself.

We have not followed these patients as part of any prospective study. What follows are simply observations—one study that arose from a chance discovery of such a "quirk," and two exploratory assessment procedures that were inspired by similar behaviors. Both methods await validation.

Perhaps a reader of this book will take these ideas further?

Keep Smiling. A question for the clinically active neuropsychologists among you: How often have you noticed that a patient misremembers element no. 11 of the Rey–Osterrieth Complex Figure (Rey 1941; Osterrieth 1944)—the small circle with three internal dots—as a smiley?

According to Regard and Landis (1994), this occurs in slightly more than one out of every hundred patients. They retrospectively examined the Rey figure reproductions of 2,400 consecutively admitted patients over four years. Thirty-seven of them (23 women) drew a smiley—either in the copy condition, the recall condition, or both.

This investigation began with the incidental observation of a 47-year-old right-handed woman with a large right frontal meningioma. The tumor was both neurologically and neuropsychologically silent. Her performance on all standard tests of frontal lobe function—quantitatively and qualitatively—was within normal limits.

However, there were longstanding neuropsychiatric symptoms, described as "mildly maniform, hyperverbal, unconcerned, and irritable" (Regard and Landis 1994, p. 304). And in her recall of the Rey figure, element no. 11 appeared as a smiley.

Regard and Landis considered a range of explanations for this quirk—from mere chance, to disinhibition, rule-breaking, or even deliberate humor. Ultimately, they concluded that the most plausible explanation was what they called "a true mnemonic transformation of an ambiguous graphic sign by a pathologically elated mood" (p. 304).

A closer look at the 37 smiley drawers revealed a consistent pattern:

* 33 of them showed structural abnormalities on CT or MRI.
* In most cases, the abnormalities were pre-Rolandic, more than twice as often in the right hemisphere.
* Corresponding neuropsychological deficits were typically in the nonverbal domain: impaired nonverbal fluency, concept formation, learning, and recall.

* 83% of these patients had affective or mood disturbances, described as "distractable, affectively unstable, fluctuating from elation and inconcern to irritability and sometimes aggressive behavior" (p. 303).

What began as a single curious case—a neurologically silent patient with a bright mood and a drawn smiley—led the authors to spend many hours combing through archived records. The combination of an icon of happiness and an elevated affective state proved too intriguing to ignore.

With the development of automatic scoring systems for the Rey figure (Langer et al. 2024; Simfukwe et al. 2025), clinicians will save substantial time during both copy and recall assessments. Perhaps this extra time, gained through automation, will help neuropsychologists become more attuned to subtle individual quirks—and to what these quirks might teach us about the human brain.

Golden Bisections. This addition to the long list of variants of the line bisection task (see Chapter 9, p. 189) was actually invented by one of my patients. GS (for "Golden Section") was a 74-year-old right-handed man who had suffered a large anterior cerebral/middle cerebral (ACM) insult of the right hemisphere about three months prior. This event left him with left hemiparesis (without hemianopia) and marked neglect, which he typically did not even recognize.

When I instructed him to "bisect" a line presented on a sheet (line length 22 cm), he responded with amusement: "What do you mean—you want me to bisect it in the Golden Ratio?"

As an architect, GS was familiar with the concept of bisecting a line according to the properties of the irrational number phi (1.61803…). Such a bisection occurs at the point where the ratio of the smaller segment to the larger is the same as that of the larger segment to the entire line.

> The Golden Section—often dubbed the "divine proportion"—has long been heralded as the gold standard for maximizing aesthetic pleasure, whether one is viewing proportions or listening to musical intervals. Artists and architects have employed it to achieve harmony and balance in their work. The Golden Ratio appears in structures such as the pyramids of Giza, in Chinese pottery, and in animal cognition and human behavior. The human body itself reflects divine proportions in the morphology of the skull, face, trunk, and limb kinematics (Verrelli et al., 2025), as well as in the aesthetics of the female genitalia (Qiang et al., 2020). Phi, like e and pi, is irrational—so much so that most attempts to "see" phi in nature (beyond clearly geometric contexts) remain approximative. Livio (2002) devotes an entire book to the wonders and irrationalities of this number, while Iosa et al. (2018) urge us not to throw out the baby with the bathwater by either mystifying or dismissing all such occurrences a priori..

GS made his remark with a twinkle in his eye—half in jest. I explicitly asked him to bisect the line at its midpoint. True to his neglect, he attempted to do so but failed in a manner reminiscent of spatial neglect. In his first set of trials, he bisected nine lines (all 22 cm long, with three lines each presented on the left, center, and right of the sheet). His mean displacement was M = 5.8 cm to the right of the objective midpoint (SD = 1.4 cm). A sample of these typical bisections is shown in Figure 6a.

I then presented the patient with a second, identical set of lines and asked him for "Golden Bisections" as he had suggested. He smiled, took the pencil, and bisected the lines—this time with a mean displacement of only M = 0.8 cm to the right of the objective midpoint (SD = 0.7 cm; only 1 out of 9 lines was bisected to the left). Although his placements did not match the true Golden Ratio bisection (which would be about 2.6 cm to the left or right of the midpoint, depending on orientation), it was remarkable how close his bisections were to the objective midpoint when he was given the "Golden" instruction (Figure 6b presents a redrawn sample of these trials).

Unfortunately, I missed the opportunity to collect judgments on the pre-bisected lines under both the regular and Golden conditions, and the patient was not available for further testing. He later died from causes unrelated to his neurological condition.

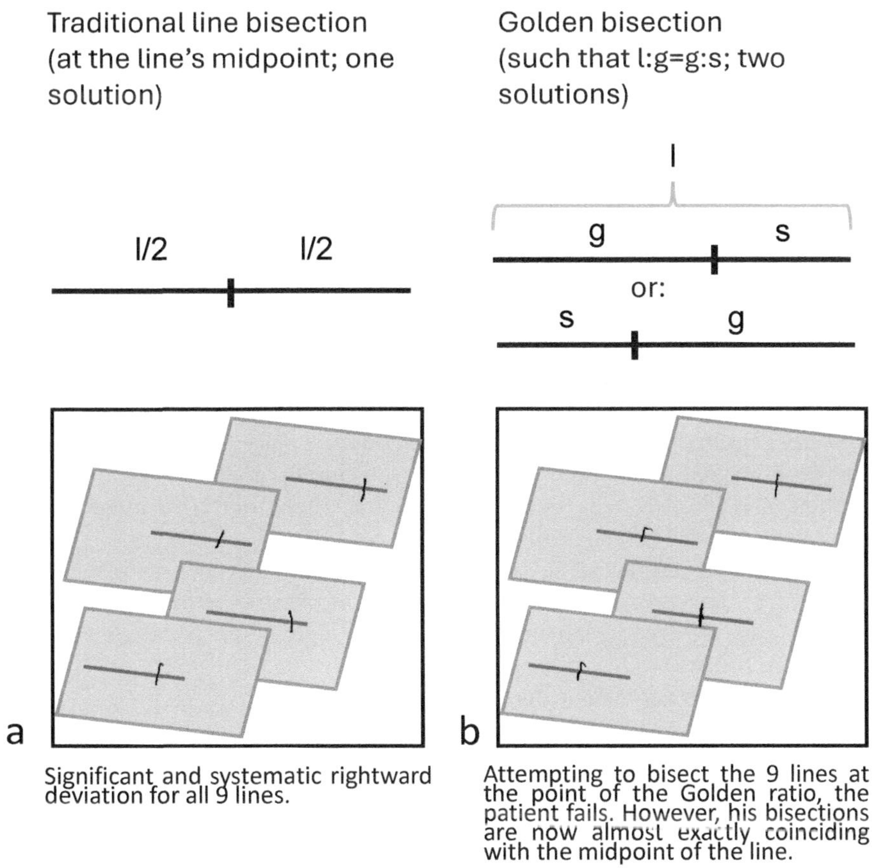

Figure 6: Your next little research project: ask for Golden Bisections (b) instead of bisections at the midpoint of a line (a). The point of the Golden Section is such that the ratio of total line length (l) to the greater segment (g) equals the ratio of *g* to the smaller segment (*s*). The bisections shown are from a patient with left-sided hemineglect (a sample of 4 of 9 lines, redrawn by the author for improved visibility). See text for explanation.

GS's offhand idea, though fleeting, suggests a worthwhile avenue for research—a project in which patients are presented with their own bisections and asked to judge their accuracy. Such an approach could further clarify whether aesthetic or cognitive factors influence spatial bisection, as suggested by Figure 4b of Chapter 9.

Manieristic Bisections. It feels almost embarrassing to add yet another variation to the already vast literature on line bisection procedures. But the method invented by MLB (for Manieristic Line Bisection) is so unusual—and so stylistically marked—that it deserves mention.

> MLB is a 30-year-old ambidextrous man with 10 years of formal education. He worked as a flat-painter and now lives on a disability pension. He was diagnosed with paranoid schizophrenia four years ago. His psychiatrist initiated the current referral to our unit due to renewed signs of psychosis: visual hallucinations, disorganized thought, and ideas of reference.
>
> MRI was unremarkable; EEG showed non-specific alterations in the left parietal lobe.
>
> Neuropsychological testing revealed deficits across all domains of attention and executive function (more pronounced for verbal than nonverbal tasks). Both short-term memory and learning capacity were reduced, especially for verbal material. There was no aphasia (though spontaneous speech was verbose and copious), no apraxia, and no agnosia. Testing showed no hemispheric superiority for language (tachistoscopic methods). MLB scored extremely high on the *Magical Ideation* scale (Eckblad and Chapman 1983).

When asked to bisect a line at its midpoint, MLB consistently used a highly idiosyncratic strategy. He began by carefully drawing a large dot—either on the left or right end of the line, or directly over its midpoint.[9] Then he would draw a bisecting line from this dot toward what he perceived as the center.

Across approximately 40 lines, the initial dot was placed left, right, or centrally with roughly equal frequency. Regardless of the starting point, however, the actual bisection marks consistently deviated to the left of the objective midpoint. This deviation was statistically significant.

The leftward displacement—already evident when the dot was placed at the actual center—is clearly greater than what is typically observed in healthy individuals. It even exceeds the average leftward bias shown by high scorers on schizotypy scales (Mohr et al. 2003a).

This pronounced pseudoneglect, seen here alongside active psychotic symptoms, is in line with findings by Bracha et al. (1993). It also echoes the absence of left-hemisphere superiority for lexical decisions reported in patients with schizophrenia or schizotypy (Leonhard and Brugger 1998)—although this feature is also observed in healthy ambidextrous individuals.

The key point here is the immediate clinical usefulness of the patient's manieristic bisection habit (Figure 7)—if appropriately standardized. Although

---

[9] In a few cases, MLB also drew the dot below the line, always at a distance of roughly one finger's width.

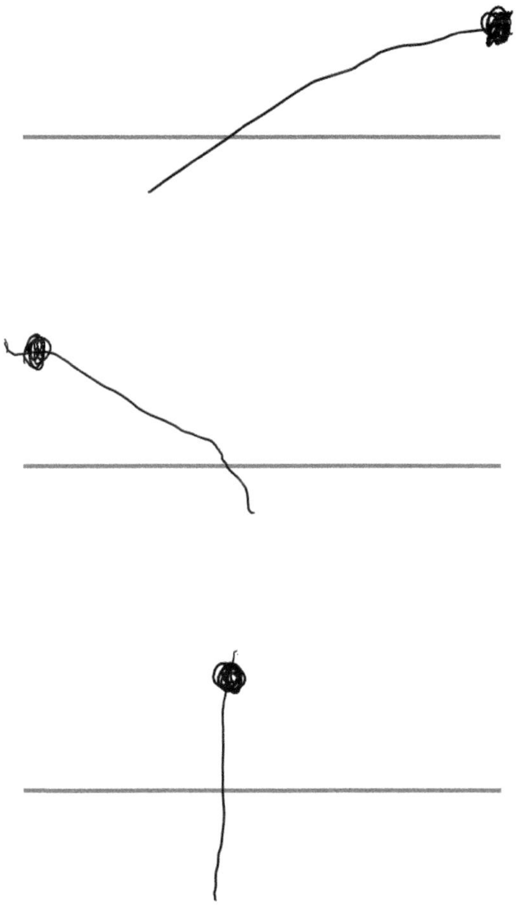

Figure 7: Patient MLB's manieristic line bisections (3 representative trials). Each bisecting line was drawn from a large initial dot, which he placed arbitrarily at the left end, right end, or assumed midpoint of the line. A consistent leftward displacement of the bisection mark was observed—even when the dot was placed near the line's true center. This is interpreted as pseudoneglect, known to be exaggerated in psychotic disorders. The (incidental) value of this bisection method may lie in its potential to help differentiate motor from sensory components of neglect. See text for discussion.

we did not pursue a formal standardization ourselves, here are some preliminary considerations:

We prepared a new set of line bisection sheets, each featuring a conspicuous dot placed 1.5 cm above either the left or right end, or the true midpoint of the line. We administered this modified set to a small group of patients with left-sided neglect. For each, we had already determined the relative contribution of sensory versus motor components using several published methods (see also Chapter 9, p. 186), some of which are quite time-consuming (for an overview, see Husain et al. 2000).

In patients with a predominantly motor component of neglect, placing the dot at the left end of the line led to a clear reduction of neglect. Midline dots also reduced neglect, but in both motor and sensory types. Right-end dot placements had no appreciable effect in either group.

Given the simplicity of this procedure, its potential utility in differentiating the *motor* and sensory components of neglect deserves further exploration.

To summarize the three examples of patient "quirks" discussed above: we should take patients' idiosyncrasies seriously. A behavior that appears unique at first glance may, on closer inspection, be part of a broader and recurring pattern. Once recognized, such patterns can be studied systematically—and their causes, or at least their clinical implications, better understood.

This leads us to consider one final source of limitation in the assessment of the untestable.

Ways to Circumvent Own Limitations

One's own limitations are, naturally, the least often considered. Yet they lurk in the background of every clinician's practice. The more routine you develop in testing, the better: you would not be easily distracted by difficult situations or difficult patients.

At this stage, you have left behind the status of a mere "tech." You are now a senior tech—approaching the level of a testing machine, increasingly unresponsive to the subtle cues that might threaten the smooth progression of standard neuropsychological routines.

Much of what follows has already been emphasized earlier in this chapter. It is about our initiative: our willingness to take shortcuts through standard procedures. These shortcuts are not careless—they aim to create time. Time to examine "extraordinary" behaviors in greater depth.

To do so, we may have to break rules now and then. Of the many rules that surround us, I want to highlight just two.

The first concerns the location of testing (see the next section). The second is location-independent: it involves the preparedness to violate test instructions.[10] Put less provocatively: clinical neuropsychologists should be curious enough to occasionally try, alongside standard versions of a task, slightly altered instructions, to gain an impression of what they might one day recommend to their experimentally working colleagues. Here, we will restrict this discussion to the administration of fluency tasks (see the section on p. 299).

The Haunted Armchair

This case vignette illustrates that, to assess certain behaviors, the examination must take place on-site. Parapsychology differentiates between location-based and person-based hauntings. The former refer to disincarnated ghosts bustling around a particular place—often an old building—independent of any living witness. The latter makes

[10] This formulation is deliberately provocative. I will never attempt to seduce you that test instructions would not have to be adhered to. Without doing so, any validation procedure was in vain, and a patient's performance could no longer be evaluated quantitatively.

more sense from a scientific perspective. They refer to startling events that are paranormal in the sense of not being readily explainable by established sciences.

? Throughout this book, you occasionally refer to "paranormal" topics. What is parapsychology's place in the sciences, and what relationships are there with neuropsychology?

I see parapsychology's role as that of a devil's advocate. Its focus is on phenomena that current scientific models cannot readily explain—but that nonetheless occur. The term "parapsychology" was coined in the 1930s by the American botanist Joseph Banks Rhine (1895–1980), who regarded it as a branch of abnormal psychology.

Unbeknownst to many, Rhine's early work was devoted to debunking phenomena, such as levitation and the fraudulent "materializations" staged in spiritualist séances. His major contribution was to bring some of these unexplained phenomena into the laboratory, where he applied statistical methods to the analysis of tedious guessing experiments. While he interpreted above-chance guessing as evidence for extra-sensory perception (ESP), others joked that ESP stood for Error Some Place (Wolfle 1956).

Still, the mainstream sciences' exclusive focus on detecting errors in a "fringe field" makes them blind to what might actually expand the horizons of the cognitive sciences—if only esoteric interpretations were set aside.[11] Another way of reading ESP is as an Effect of Subjective Probability (Brugger and Taylor 2003). Both belief in an extrasensory causation of everyday coincidences and the allegedly "above-chance" matching in guessing experiments are deeply connected to the concept of randomness (Brown 1953, 1957). And this leads us back to neuropsychology. Guessing is nothing but an implicit attempt to produce a random sequence of responses.

The study of human random generation—long known to cognitive psychology (see Brugger (1997) for an overview)—opens new avenues for neuropsychology. What is the brain's internal random generator? Why do we fail so miserably at producing true randomness? And to bring us back to this chapter: why are we so prone to see "meaningful" patterns against a randomly structured background?

In neuropsychology and neuropsychiatry, one often encounters a mixture of location-based and person-based hauntings. Hallucination-like experiences that clearly originate in a person's mind may still be confined to particular places or contexts.

---

[11] The study of out-of-body experiences is a good example. Menninger-Lerchenthal (1935) recommended these phenomena as ideally suited to bridge the gap between parapsychology and neuropsychology. Long taken seriously in parapsychology, they only began to receive sustained scientific attention when their potential to elucidate the brain mechanisms of body–self integration was recognized (Blanke et al. 2002; see Chapter 10). Other underexplored examples include (neur)astrology (Brugger 2021) and possible neural mechanisms underlying so-called dermo-optical vision (Brugger and Weiss 2008).

Pareidolia offers a beautiful illustration (see Chapter 7, p. 160). It refers to the perception of meaningful forms in an unstructured background, or of "messages" emerging from auditory noise. Leonardo da Vinci encouraged his disciples to sit before stained, weathered walls and imagine battles, gestures, strange costumes, and infinite other scenes. As he wrote in his *Trattato della Pittura* (1519), they should look for "battles and figures with lively gestures and strange faces and costumes and an infinity of things" (translated and cited in Brugger et al. 1993). Pareidolia lies somewhere between pathological hallucination and creative pattern recognition.

Figure 8 presents a pictorial case report of an elderly woman—referred to here as HA (Haunted Armchair)—who experienced pareidolic phenomena confined to a very specific location in her apartment. One day, she discovered the face of her deceased father in the upper part of her armchair. This occurred roughly six months before she was referred for an initial neuropsychological assessment following a recent diagnosis of Parkinson's disease.

Neurologically, her symptoms were mild—essentially limited to reduced arm swing and a shuffling gait. Neuropsychologically, the 78-year-old right-handed woman performed within normal limits for her age and educational background. A former kindergarten teacher, she was an only child, had never married, and had always lived alone. Despite this, she maintained an active social life: she regularly went hiking with a group of friends and remained in contact with former colleagues from the kindergarten where she had worked for over 40 years.

On first contact, HA told me about some "peculiar experiences" she had when passing the armchair in her living room—particularly during the night. She would see, with great clarity, her deceased father's face. It always appeared in the top part of the pillow at the backrest, consistently on one side of the otherwise symmetrical chair.

> Seeing dead people is an under-investigated feature of anomalous perceptual phenomena. It is surprisingly frequent, with both cultural influences (Peltier et al. 2017) and associations with personal concerns—particularly fears about one's own mortality (Taylor et al. 2005).

HA did not believe in an afterlife and was certain her father was "not really there." Still, she had noticed that the vividness of the experiences had increased over recent weeks. (Note: this was before the prescription of dopaminergic medication.)

The onset of the father's appearance was abrupt. Some months earlier, she had purchased the armchair—an expensive acquisition. Just a few days later, she first noticed the face. It was pale and motionless. She became intrigued and found herself deliberately passing by the armchair when returning to her bedroom after a nightly glass of milk. After about a month, she began speaking to her father—"Just saying hello, you know." He remained mute, with no change in facial expression.

Some weeks later, the figure gained a life of its own. Her father now occasionally smiled—"exactly as he had smiled when still alive." She began talking to him more, sharing news from family and friends. Despite recognizing her father's presence as a kind of comforting companionship, HA also began to feel increasingly uneasy. As much as she had loved him, his silent visits became a source of suffering.

a

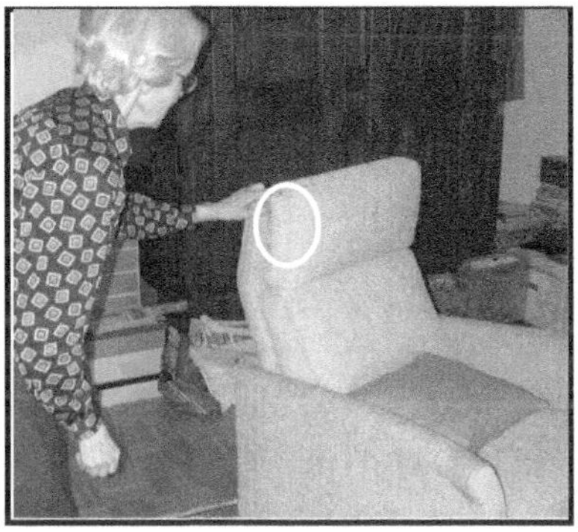

HA demonstrates where in the armchair her deceased father always appears.

b

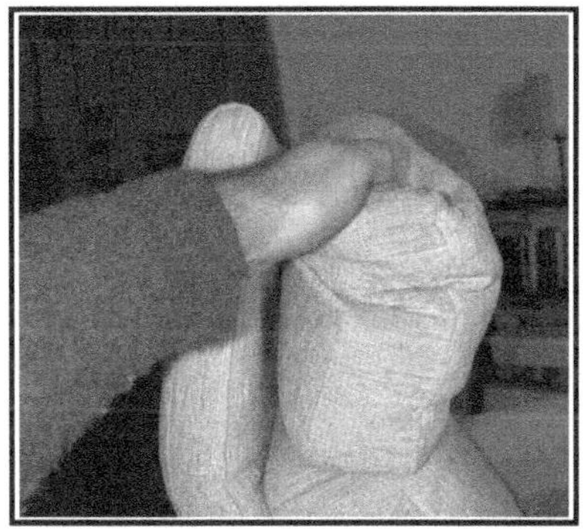

Deforming the pareidolia-critical parts of the pillow also deforms father's face.

c

HA provides a photograph of father's face resembling most accurately its appearance in the armchair.

d

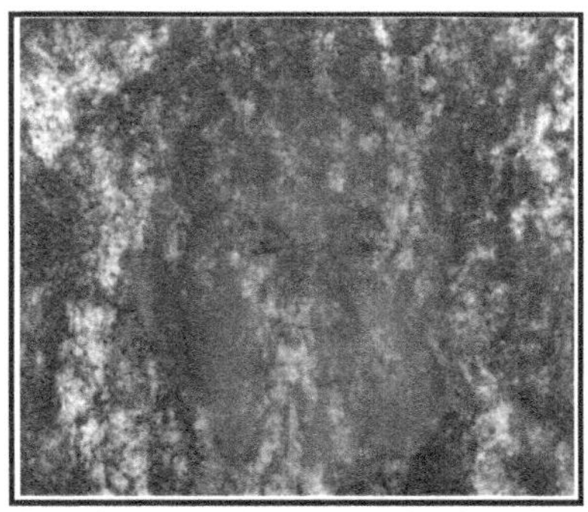

The photograph is morphed with an unstructured background (33%, 50%, 66% face; here a sample compound of 33% face).

e

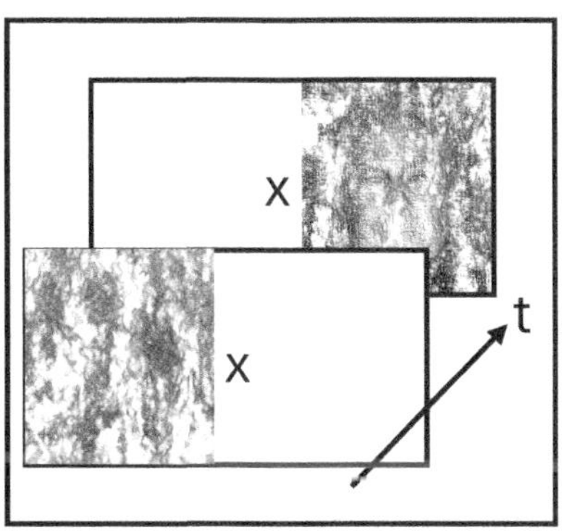

These morphs are presented in a face detection task to HA's left or right visual field (LVF or RVF; shown here is a no-go stimulus to the LVF followed by a 50% face to the RVF).

f

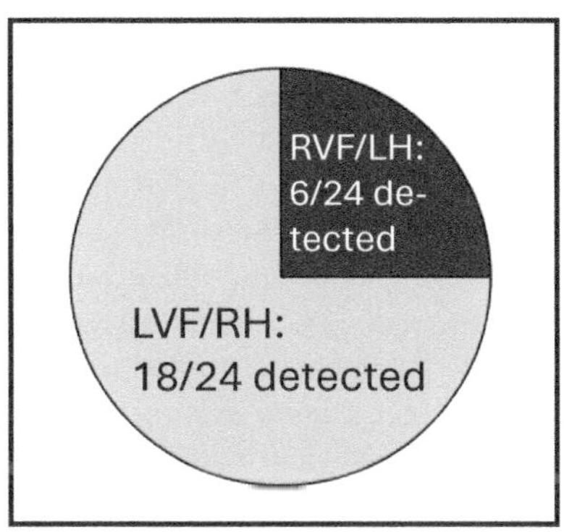

From the 24 stimulations per visual field with 33% face HA detects 3 times as many faces in the LVF/right hemisphere than in the RVF/left hemisphere.

Figure 8: Patient HA's demonstration of her pareidolia on-site and subsequent divided-visual field assessments simulating her illusory perception. See text.

During her regular appointments with her attending neurologist, HA often stopped by to give me updates on her experiences. One day, she insisted I visit her at home—so she could "introduce me to father."

Extended Legend to Figure 8

I accepted her invitation and received permission to videotape the session on-site. HA was able to precisely indicate where the face appeared (the white oval in Figure 8a). At first, she could not introduce me to him, because it was daylight, and she rarely saw the face under bright conditions. We darkened the room slightly—enough that her father finally "appeared" (to HA, at least…). His face remained motionless.

When I deformed the pillow (Figure 8b), the face visibly distorted—but did not vanish. HA sincerely asked me to stop the manipulation. Though she repeatedly affirmed her awareness of the illusory nature of the image, she found it distressing to see him "suffer." This emotional reaction contributed to her later refusal of several well-intentioned suggestions.

For instance, she firmly rejected my proposal to sell the armchair—it had cost a fortune. Likewise, the simple idea of covering the top with an unstructured white towel was unacceptable. Although she did try it once—and the face vanished immediately—she removed the towel. The thought that her father "could not get enough air" was unbearable.

She did allow me to carefully turn the armchair upside down. The face neither rotated nor remained upright—it vanished entirely. Within seconds of restoring the armchair to its normal position, the face reappeared.

HA provided several photographs of her father posing frontally. According to her, the expression in the armchair matched the neutral expression in one of them. We selected the best-matching image (Figure 8c). I then morphed this face onto an unstructured background to create stimuli with 33%, 50%, or 66% facial content (with the complementary percentages as background noise; Figure 8d).

These 72 face-background images, along with 72 pure-noise images (each a different background), were presented in a divided visual field task (see Chapter 11). A total of 144 trials were shown, half in the left visual field (projecting to the right hemisphere) and half in the right visual field (left hemisphere). The trials were pseudorandomized and presented for 50 ms via a Gerbrands two-field tachistoscope. HA was instructed to press a response key with both index fingers whenever she saw her father's face. She completed two runs on different days (data collapsed; see Figure 8e).

From this rich dataset (which showed main effects for face degradation, interactions between hemifield and degradation level, and effects on reaction time), one finding stands out: among the most degraded faces (33% face, 66% noise), detection was three times higher when stimuli were presented to the right hemisphere (left visual field; Figure 8f). This is consistent with findings from right hemisphere stroke patients who show a disproportionately reduced tendency to perceive pareidolic images (Camenzind 2024). It also aligns with results from healthy volunteers, who more often spot meaningful patterns when images are flashed to the left visual field (Brugger et al. 1993).

Whether dopaminergic medication contributed to the enhanced right-hemisphere sensitivity in HA remains unclear. A selective right-hemisphere hyperdopaminergia has been proposed as a mechanism underlying some psychotic experiences (Bracha 1989), and L-DOPA administration in healthy volunteers has been shown to enhance right-hemisphere sensitivity for face detection (Krummenacher et al. 2010b). However, in HA's case, dopaminergic medication did not elicit the pareidolias—they clearly began before treatment.

Conclusion

Assessments of perceptually anomalous experiences are best performed on-site. The observation that the pareidolic image was deformed when the background structure was manipulated could not have been made solely from patient interviews. It highlights low-level, possibly occipito-temporal, contributions to the percept.

What was even more striking was HA's emotional response to these manipulations. She was horrified—genuinely distressed—by the idea that her father might suffer. This dissociation between preserved insight during clinical interviews and diminished insight when directly confronted with the phenomenon is typical of many anomalous perceptions.

Pareidolia, which lends itself to precise experimental manipulation, exemplifies the need for close collaboration between clinical and experimental neuropsychologists.

The Preparedness to Expand Test Instructions

Once again (see Footnote 12): for clinical purposes, test instructions must, of course, be strictly followed. However, as concerns grow that certain procedures lack the sensitivity to differentiate between patient groups—whether by etiology or by underlying neural circuits—it may be time to "break rules" and begin developing better methods. Ideally, this is done in collaboration with the test makers themselves: people who know their field and are often quite keen to explore new niches for test application.

A case in point is fluency tasks. These evolved from the classical Word Association Tests (WATs), once widely used in psychiatry. In such tests, a patient hears a series of words spoken by the examiner and responds with the first word that comes to mind. The resulting associations are then interpreted as reflections of thought patterns, personality features, or psychological concerns.

From these early WATs came the more structured COWATs—Controlled Oral Word Association Tests. Today, this term is largely synonymous with verbal fluency.

In the modern version, a patient is given either a letter (e.g., S) or a category (e.g., animals) and asked to generate as many exemplars as possible—either words (of any grammatical word class) begainning with that letter (excluding proper names and morphological variations), or members of the category. Generation time is usually one, two, or three minutes. A common U.S. variant of letter fluency is one minute each for F, A, and S words.

Figure 9 (top) shows the output of two fictitious individuals for each type of fluency task—"phonemic" (letter) and "semantic" (category).

Scoring of rule breaks, particularly in category fluency, is not always consistent. Depending on the test instructions, listing subordinate items (naming "poodle" after "dog") may or may not be permitted. And, unbelievably but true: in some American protocols, birds and insects are not considered animals! Dragons, on the other hand, do not count either—mythical creatures are not valid, even if not explicitly excluded in most instructions.

The bottom part of Figure 9 presents two more fictitious examples for phonemic fluency. Both individuals produce the same number of correct responses—but a glance reveals that (brain) worlds separate them.

Sample Person 3 shows no clear semantic clustering. Sample Person 4, however, delivers a coherent burst of contextually related items, evoking a runny nose and flu—wide enough as a semantic net. This illustrates that phonemic fluency is *not*

Sample person 1: Phonemic fluency (letter S)		Sample person 2: Semantic fluency (category «animals»)
sea		cat
see		dog
Suzy	← rule break! rule break? →	poodle
shy		dolphin
ship		tiger
sea	← repetition! →	cat
slippery		woodlouse
lascivious	← rule break! →	stable
soul	rule break! →	dragon

Sample person 3 (phonemic)	Sample person 4 (phonemic):
spoon	sneeze
sabotage	sniffle
slippery	sick
satellite	sore
stork	swallow
subway	syrup
synthetic	sleep
salute	shiver
sulfur	sluggish

Figure 9: Top: correct words, repetitions, and rule breaks in letter and category fluency. Bottom: quantitatively identical but qualitatively different performances in phonemic word fluency. See text for more details.

driven exclusively by phonemic retrieval. Some people scaffold their phonemic output through semantic memory; others do not (Schwartz et al. 2003; Sung et al. 2013).

> The widespread belief that letter fluency taps phonemic and category fluency taps semantic processes is a clinical oversimplification. And not a trivial one. In some degenerative diseases, a breakdown of *semantic* memory is key. Category fluency is more sensitive to these early impairments (Monsch et al. 1994, for Alzheimer's disease). Accordingly, *temporal lobe* circuits are more involved in tasks relying on semantic strategies (Martin et al. 1996). In contrast, frontal lobe circuits support phonemic retrieval, which is more affected in conditions historically labeled "subcortical dementias" (e.g., Huntington's disease; Monsch et al. 1994). It would therefore seem essential to capture the strategic differences illustrated in Figure 9 in routine clinical evaluations.

Fluency tasks are among the most creative tools in neuropsychology. Which makes it all the more disappointing how little information we usually extract from them.

Most test manuals offer three main scores with normative data:

* total number of correct responses,
* number of repetitions, and
* number of rule breaks.

What an embarrassing waste of information! Let us perform a little category fluency task tailored to neuropsychologists. Instructions:

Please provide, in the next 60 seconds, as many measures as you can think of that could enhance the clinical utility of verbal fluency tasks (phonemic or semantic).

The following ideas stem from my own brainstorming sessions—admittedly spread out over several years, not seconds. I present them under two headings:

Same Instructions, Novel Procedural Details and/or Measures

* *Tape the output.* The neglect of *mental chronometry* in fluency is astounding. Pauses during speech reflect cognitive planning, and have well-mapped neural correlates (Kircher et al. 2004). Recent work even links the prediction of forthcoming words to richer models of language (Caucheteux et al. 2023). Especially informative: pauses between semantically related words (Schwartz et al. 2003).
* *Extend or Vary Timing.* Do not limit fluency tasks to fixed durations. The slope of word production across time was recognized early as diagnostically useful. So is the serial recall order of produced items, which has clear neural correlates (De Marco and Venneri 2022).
* *Analyze Clustering and Switching.* These now semi-standard metrics can be further enriched when combined with temporal information. Design fluency also benefits from this approach (Resch et al. 2024).

* *Apply Response Commonality Analyses* (Duchêne et al. 1998). How typical is a patient's output compared to a target group (e.g., with similar lesion types or localizations)? Is there a deviation from Marbe's law?[12]
* *Linguistic Profiling.* Categorize each word by grammatical class, frequency, imageability, etc. Nouns and verbs are treated differently in the brain (Bastian 1897; Arévalo et al., 2007). Even among verbs, action-related and "still" verbs diverge—and may have special relevance in some disorders (Péran et al. 2009).

In our own retrospective analyses of native German speakers' fluency output, we looked at the percentage of compound words[13] (e.g., snowflake instead of snow; Katumba et al. unpublished). Patients (n = 138) had focal cortical lesions in one of four quadrants: left/right, anterior/posterior (Goldberg et al. 1994). Results showed a selectively increased rate of compound use in patients with right anterior lesions. Might this reflect a disinhibited left-anterior mechanism? Or a form of verbal creativity (Liederman et al. 1983)? Or is it a subtle rule-breaking behavior?

Novel Instructions

* *Omit Specifying Task Duration.* Then stop after a predetermined time and ask the patient to estimate how long they were speaking. This opens a novel clinical window onto *time perception.* With support from research colleagues, such assessments could link to lesion sites, aging effects, and fluency type (verbal vs. design). Time estimation in the range of seconds to minutes remains vastly underexplored in neuropsychology—despite well-known space–time processing overlaps (Bueti and Walsh 2009; Zäch and Brugger 2008), and despite numerous patient reports of time distortions (Blom et al. 2021).
* *Offer Novel Semantic Categories.* Try affective topics (Gawda 2019) or temporal orientation ("past episodes" vs. "future plans" by Regard et al. (No future in TGA, unpublished manuscript) found that patients in transient global amnesia (TGA) typically underproduce future-oriented responses.
* *Use Structured Multiple-Choice Fluency.* Rather than defining clusters post hoc, assign patients to generate responses from two or more *simultaneous* categories. This allows for more objective measurement of clustering and switching (Schwartz et al. 2003).
* *Revisit Random Generation Tasks* as a variant of multiple-choice fluency (Baddeley et al. 1998). Categories can include numbers, letters, or spatial positions. Range limits may be binary or open-ended (Banks and Hill 1974; Myachykov et al. 2020). Brugger (1997b) reviewed some 300 studies on the variables involved.

---

[12] Marbe's law (linked to the prototype preference effect) describes the likelihood of retrieving a category exemplar based on its typicality. For neuropsychological extensions, see Stilson et al. (1972).

[13] German allows far more compounds than English. Anomia can selectively affect compound production (Blanken 2000).

I am fully aware that the suggestions presented here run counter to any principle of parsimony. Yet the potential gain from these expanded instructions and analyses—for the neuropsychology of the future—cannot be overstated. This holds whether the focus lies on executive functions, associative processing, or divergent creativity.

There is already a growing demand for novel techniques and analytical methods in fluency research (Gaudet et al. 2024), and lists of variables relevant to word generation have been assembled (Henderson et al. 2023; Thiele et al. 2016). Henderson and colleagues (2023) affirmed the general utility of fluency tasks for distinguishing individuals with and without brain damage but criticized their lack of differential sensitivity. In their study, even when a wide range of metrics was applied, the classic one-minute category fluency task failed to offer meaningful diagnostic utility across patients with diverse neurodegenerative conditions. Their conclusion was sobering: verbal fluency may be an insufficient task for diagnostic purposes because it taps such a wide range of brain functions.

That is a frustrating conclusion. But seen through more rose-tinted glasses, the diversity of cognitive operations involved in fluency tasks should not be seen as a limitation, but as a promise. We simply need to establish better measures—ones that capture the relevant features for differentiating the specific patient groups under investigation.

This, however, requires something else: the readiness of clinicians to test their ideas. And the courage to gently break rules when the existing rules do not serve the needs of the individual patient.

A final piece of advice to conclude this book: stay curious. Cultivate your urge to diverge—but do not forget that convergence of ideas will be indispensable for developing truly useful new tools. One more reason to continue crossing the borders between clinical and experimental neuropsychology.

Where Can We Read More?

Anderson, R.M. Behavioral Observations. In: Practitioner's Guide to Clinical Neuropsychology. Critical Issues in Neuropsychology. Springer: Boston, MA, 1994. https://doi.org/10.1007/978-1-4615-2480-9_13.

Roebuck-Spencer, T.M., Glen, T., Puente, A.E., Denney, R.L., Ruff, R.M., Hostetter, G. et al. Cognitive screening tests versus comprehensive neuropsychological test batteries: A National Academy of Neuropsychology education paper. Archives of Clinical Neuropsychology 2017; 32: 491–498, https://doi.org/10.1093/arclin/acx021.

References

Abad, D. and Zwaan, R. Pareidolia as a psychotic and normal mechanism: A systematic review. Interdisciplinaria 2025; 42: 114–141. DOI 10.16888/interd.2025.42.2.6.

Abbott, D.F., Waites, A.B., Lillywhite, L.M. and Jackson, G.D. fMRI assessment of language lateralization: an objective approach. Neuroimage 2010 May 1; 50(4): 1446–1455. doi: 10.1016/j.neuroimage.2010.01.059.

Acharya, A.B., Lui, F. and Maani, C.V. Conduction Aphasia. *In:* StatPearls [Internet]. StatPearls Publishing: Treasure Island (FL), 2025.

Adelson, E.H. Checkershadow Illusion. Perceptual Science Group. MIT, 1995/2005. Last retrieved 2025-04-10.

Adeyemo, B.O. and Nesathurai, S. The original description of Anton syndrome. PM R. 2013 Jan; 5(1): 74. doi: 10.1016/j.pmrj.2012.09.580.

Adolfi, F., Couto, B., Richter, F., Decety, J., Lopez, J., Sigman, M. et al. Convergence of interoception, emotion, and social cognition: A twofold fMRI meta-analysis and lesion approach. Cortex 2017; 88: 124–142. doi: 10.1016/j.cortex.2016.12.019.

Aggleton, J.P., Nelson, A.J.D. and O'Mara, S.M. Time to retire the serial Papez circuit: Implications for space, memory, and attention. Neurosci. Biobehav. Rev. 2022 Sep: 140: 104813. doi: 10.1016/j.neubiorev.2022.104813.

Ahn, S., Jung, K.I., Yoo, W.K., Kang, G.Y. and Ohn, S.H. Dementia due to meningovascular syphilis in medial temporal lobe and cognitive rehabilitation. Ann. Rehabil. Med. 2012 Jun; 36(3): 423–7. doi: 10.5535/arm.2012.36.3.423.

Ajina, S. and Bridge, H. Blindsight and unconscious vision: What they teach us about the human visual system. Neuroscientist. 2016; 23: 529–541.

Akelaitis, A.J., Risteen, W.A., Herren, R.Y. and Van Wagenen, W.P. Studies on the corpus callosum. III. A contribution to the study of dyspraxia and apraxia following partial and complete section of the corpus callosum. Archives of Neurology and Psychiatry 1942; 47: 971–1008.

Alderson-Day, B. Presence. The Strange Science and True Stories of the Unseen Other. Manchester University Press: Manchester, UK, 2023.

Alexander, J.M., Hedrick, T. and Stark, B.C. Inner speech in the daily lives of people with aphasia. Frontiers In Psychology 2024; 15: 1335425 (pp. 1–22). doi: 10.3389/fpsyg.2024.1335425.

Alkadhi, H., Brugger, P., Boendermaker, S.H., Crelier, G., Curt, A., Hepp-Reymond, M.C. et al. What disconnection tells about motor imagery: evidence from paraplegic patients. Cereb Cortex. 2005 Feb; 15(2): 131–40. doi: 10.1093/cercor/bhh116.

Alle, H. and Geiger, J.R.P. Combined analog and action potential coding in hippocampal mossy fibers. Science 2006; 311: 1290–1293. Doi: 10.1126/science.1119055.

Ames, D. Self shooting of a phantom head. Br. J. Psychiatry 1984; 145: 193–194. doi: 10.1192/bjp.145.2.193.

Amoruso, E., Terhune, D.B., Kromm, M., Kirker, S., Muret, D. and Makin, T.R. Reassessing referral of touch following peripheral deafferentation: The role of contextual bias. Cortex 2023 Oct; 167: 167–177. doi: 10.1016/j.cortex.2023.04.019.

Anagnostopoulos, A., Spiegel, R., Palmer, J. and Brugger, P. A left-hand superiority for the implicit detection of a rule. Cortex. 2013 Feb; 49(2): 582–90. doi: 10.1016/j.cortex.2011.12.006.

Anastasopoulos, G.K. Aphasic disorders and verbal hallucinations. J. Neurol. Sci. 1966; 4: 83–93. doi: 10.1016/0022–510x(67)90059–7.

Andermatt, M. Epilepsie als Metapher. Zum Bild der Epilepsie in der Literatur. Epileptologie 2007; 24: 106–112.
Anderson, B. and Winawer, J. Image segmentation and lightness perception. Nature 2005; 434: 79–83. https: //doi.org/10.1038/nature03271.
Anderson, R. M. Behavioral Observations. *In*: Practitioner's Guide to Clinical Neuropsychology. Critical Issues in Neuropsychology. Springer: Boston, MA, 1994. https://doi.org/10.1007/978-1-4615-2480-9_13.
Andoh, J., Milde, C., Tsao, J.W. and Flor, H. Cortical plasticity as a basis of phantom limb pain: Fact or fiction? Neuroscience. 2018 Sep. 1; 387: 85–91. doi: 10.1016/j.neuroscience.2017.11.015.
Antinori, A., Arendt, G., Becker, J.T., Brew, B.J., Byrd, D.A., Cherner, M. et al. Updated research nosology for HIV-associated neurocognitive disorders. Neurology. 2007 Oct. 30; 69(18): 1789–99. doi: 10.1212/01.WNL.0000287431.88658.8b.
Anton, G. Ueber Herderkrankungen des Gehirnes, welche vom Patienten selbst nicht wahrgenommen werden. Wien Klin Wochenschr 1898; 11(10): 227–229.
Antoniello, D., Kluger, B.M., Sahlein, D.H. and Heilman, K.M. Phantom limb after stroke: an underreported phenomenon. Cortex 2010; 46: 1114–22. doi: 10.1016/j.cortex.2009.10.003.
Ardila, A. On the evolutionary origins of executive functions. Brain Cogn. 2008 Oct; 68(1): 92–93. doi: 10.1016/j.bandc.2008.03.003.
Arévalo, A., Perani, D., Cappa, S.F., Butler, A., Bates, E. and Dronkers, N. Action and object processing in aphasia: from nouns and verbs to the effect of manipulability. Brain Lang. 2007 Jan; 100(1): 79–94. doi: 10.1016/j.bandl.2006.06.012.
Argyriou, P., Byfield, S. and Kita, S. Semantics is crucial for the right-hemisphere involvement in metaphor processing: evidence from mouth asymmetry during speaking. Laterality 2015; 20: 191–210. doi: 10.1080/1357650X.2014.951654.
Arnold, G. and Auvray, M. The graphaesthesia paradigm. Drawing letters on the body to investigate the embodied nature of perspective-taking. Iperception 2017; 8: 2 041669517690163. doi: .
Arzy, S., Overney, L.S., Landis, T. and Blanke, O. Neural mechanisms of embodiment: asomatognosia due to premotor cortex damage. Arch Neurol. 2006; 63: 1022–1025. 10.1001/archneur.63.7.1022.
Aspell, J.E., Heydrich, L., Marillier, G., Lavanchy, T., Herbelin, B. and Blanke, O. Turning body and self inside out: visualized heartbeats alter bodily self-consciousness and tactile perception. Psychol Sci. 2013 Dec; 24(12): 2445–53. doi: 10.1177/0956797613498395.
Atherton, K.E., Nobre, A.C., Lazar, A.S., Wulff, K., Whittaker, R.G., Dhawan, V. et al. Slow wave sleep and accelerated forgetting. Cortex 2016; 84: 80–89. doi: 10.1016/j.cortex.2016.08.013.
Babinski, J. Contribution à l'étude des troubles mentaux dans l'hémiplégie organique cérébrale (anosognosie). Revue Neurologique 1914; 27: 845–848.
Baddeley, A., Emslie, H., Kolodny, J. and Duncan, J. Random generation and the executive control of working memory. Q J Exp Psychol A. 1998 Nov; 51(4): 819–52. doi: 10.1080/713755788.
Baden, T. and Nilsson, D.E. Is our retina really upside down? Curr Biol. 2022 Apr 11; 32(7): R300-R303. doi: 10.1016/j.cub.2022.02.065.
Bairey Merz, C.N. The Yentl syndrome is alive and well. Eur Heart J. 2011; 32: 1313–5. doi: 10.1093/eurheartj/ehr083. Epub 2011 Mar 10.
Balint, R. Seelenlähmung des Schauens, optische Ataxie, räumliche Störung der Aufmerksamkeit. Monatsschrift für Psychiatrie und Neurologie 1909; 25: 51–66. https: //doi.org/10.1159/000210464.
Balota, D.A., Cortese, M.J., Duchek, J.M., Adams, D., Roediger, III H.L. and McDermott, K.B. Veridical and false memories in healthy older adults and in dementia of the Alzheimer's type. Cog. Neuropsychol 1999; 16: 361–384.
Baluška, F. and Levin, M. On Having No Head: Cognition throughout biological systems. Front Psychol. 2016; 7: 902. doi: 10.3389/fpsyg.2016.00902.
Bancaud, J., Brunet-Bourgin, F., Chauvel, P. and Halgren, E. Anatomical origin of déjà vu and vivid 'memories' in human temporal lobe epilepsy. Brain 1994; 117: 71–90. doi: 10.1093/brain/117.1.71.
Banks, W.P. and Hill, D.K. The apparent magnitude of number scaled by random production. Journal of Experimental Psychology 1974; 102: 353–376.
Bardi, L., Kanai, R., Mapelli, D. and Walsh, V. Direct current stimulation (tDCS) reveals parietal asymmetry in local/global and salience-based selection. Cortex 2013; 49: 850–60. doi: 10.1016/j.cortex.2012.04.016.

Barker, B. and Rice, C. Folk illusions: An unrecognized genre of folklore. The Journal of American Folklore 2012; 125: 444–473. Doi: 10.5406/jamerfolk.125.498.0444.

Barker, K.B. and Rice, C. Folk Illusions. Children, Folklore, and Sciences of Perception. Indiana University Press: Bloomington, IN, 2019.

Barkley, R.A. Executive Functions. What They Are, How They Work, and Why They Evolved. Taylor and Francis: New York, 2020.

Barnby, J.M., Park, S., Baxter, T., Rosen, C., Brugger, P. and Alderson-Day, B. The felt-presence experience: from cognition to the clinic. Lancet Psychiatry 2023; 10: 352–362. doi: 10.1016/S2215–0366(23)00034–2.

Barnett, K.J. and Corballis, M.C. Ambidexterity and magical ideation. Laterality. 2002; 7(1): 75–84. doi: 10.1080/13576500143000131.

Baron-Cohen, S., Ashwin, E., Ashwin, C., Tavassoli, T. and Chakrabarti, B. Talent in autism: hyper-systemizing, hyper-attention to detail and sensory hypersensitivity. Philos Trans R Soc Lond B Biol Sci. 2009 May 27; 364(1522): 1377–83. doi: 10.1098/rstb.2008.0337.

Bartolomeo, P., de Vito, S. and Seidel Malkinson, T. Space-related confabulations after right hemisphere damage. Cortex. 2017 Feb; 87: 166–173. doi: 10.1016/j.cortex.2016.07.007.

Barton, J.J., Rubino, C., Albonico, A., Jackson, M. and Davies-Thompson J. Right hemi-alexia. Cortex 2022; 157: 288–303. doi: 10.1016/j.cortex.2022.09.015.

Bartsch, T. and Butler, C. Transient amnesic syndromes. Nat Rev Neurol. 2013 Feb; 9(2): 86–97. doi: 10.1038/nrneurol.2012.264.

Bastian, H.C. Some problems in connection with aphasia and other speech defects. British Medical Journal 1897; April 3: 833–903.

Bates, E. On language savants and the structure of the mind. A review of Neil Smith and Ianthi-Maria Tsimpli, The mind of a savant: language learning and modularity. Int J Bilingualism 1997; 1: 163–179.

Batterink, L.J., Paller, K.A. and Reber, P.J. Understanding the neural bases of implicit and statistical learning. Top Cogn Sci. 2019 Jul; 11(3): 482–503. doi: 10.1111/tops.12420.

Bauby, J.D. The Diving Bell and the Butterfly. Éditions Robert Laffont: Paris, 1997.

Bauer, R.M. and Verfaellie, M. Electrodermal recognition of familiar but not unfamiliar faces in prosopagnosia. Brain and Cognition 1988; 8: 240–252.

Baxendale, S. The Wada test. Curr Opin Neurol 2009; 22: 185–189. doi: 10.1097/wco.0b013e328328f32e

Beale, I.L. and Webster, D.M. The relevance of leg-movement cues to turn alternation in woodlice (Porcellio scaber). Animal Behavior 1971; 19: 353–356. Doi: 10.1016/S0003–3472(71)80016–7.

Bear, D.M. and Fedio, P. Quantitative analysis of interictal behavior in temporal lobe epilepsy. Arch Neurol. 1977; 34: 454–67. doi: 10.1001/archneur.1977.00500200014003.

Beaumont, J.G. (ed.). Divided Visual Field Studies of Cerebral Organization. Academic Press: London, 1982.

Bechara, A., Tranel, D., Damasio, H., Adolphs, R., Rockland, C. and Damasio, A.R. Double dissociation of conditioning and declarative knowledge relative to the amygdala and hippocampus in humans. Science 1995; 269: 1115–8. doi: 10.1126/science.7652558.

Beeman, M., Friedman, R.B., Grafman, J., Perez, E., Diamond, S. and Lindsay, M.B. Summation priming and coarse semantic coding in the right hemisphere. J Cogn Neurosci. 1994 Winter; 6(1): 26–45. doi: 10.1162/jocn.1994.6.1.26

Belin, P., Zilbovicius, M., Crozier, S., Thivard, L., Fontaine, A., Masure, M.C. et al. Lateralization of speech and auditory temporal processing. J Cogn Neurosci. 1998; 10: 536–40. doi: 10.1162/089892998562834.

Bellelli, G., Morandi, A., Davis, D.H., Mazzola, P., Turco, R., Gentile, S. et al. Validation of the 4AT, a new instrument for rapid delirium screening: a study in 234 hospitalised older people. Age Ageing 2014; 43: 496–502. doi: 10.1093/ageing/afu021.

Belli, F. and Fischer, M.H. Respiratory phases modulate visuo-spatial attention: Evidence from line bisection and landmark tasks. Laterality. 2025; 16: 1–28. doi: 10.1080/1357650X.2025.2491319.

Bellone, J.A. and van Patten, R. Becoming a Neuropsychologist. Advice and Guidance for Students and Trainees. Springer: New York, 2021.

Bennett, A. Clayhanger. Methuen: London. 1910.

Benrimoh, D., Pomerleau, V.J., Demoustier, A., Poulin, S., Maltais, J.R., Brouillette, J. et al. Why we still use "organic causes": results from a survey of psychiatrists and residents. J Neuropsychiatry Clin Neurosci. 2019; 31: 57–64. doi: 10.1176/appi.neuropsych.18050099.

Benschop, R. What is a tachistoscope? Historical explorations of an instrument. Science in Context 1998; 11: 23–50.

Benton, A. Jacques Loeb and the method of double stimulation. Journal of the History of Medicine and Allied Sciences 1956; 47–53.

Bergin, P.S., Thompson, P.J., Baxendale, S.A., Fish, D.R. and Shorvon, S.D. Remote memory in epilepsy. Epilepsia. 2000; 41: 231–9. doi: 10.1111/j.1528-1157.2000.tb00145.x.

Bergouignan, L., Nyberg, L. and Ehrsson, H.H. Out-of-body-induced hippocampal amnesia. Proc Natl Acad Sci U S A. 2014; 111: 4421–4426. doi: 10.1073/pnas.1318801111.

Berti, A., Bottini, G., Gandola M, Pia L, Smania N, Stracciari A et al. Shared cortical anatomy for motor awareness and motor control. Science 2005; 309: 488–491. 10.1126/science.1110625.

Berti, A . and Frassinetti, F. When far becomes near: remapping of space by tool use. Journal of Cognitive Neuroscience 2000; 12: 415–420.

Berti, A. Unconscious processing in neglect. In Karnath H-O, Milner AD, Vallar G (eds.). The Cognitive and Neural Bases of Spatial Neglect. Oxford University Press: Oxford, 2002, 312–326.

Best, F. Hemianopsie und Seelenblindheit bei Hirnverletzungen. Graefes Archiv für Ophthalmologie 1917; 93: 49–150.

Beyer, E. Über Verlagerungen im Gesichtsfeld bei Flimmerskotom, Neurologisches Centralblatt 1895; 14: 10–15.

Bhaskaran, R., Kumar, A. and Nayar, P.C. Autoscopy in hemianopic field. J Neurol Neurosurg Psychiatry 1990; 53: 1016–7. doi: 10.1136/jnnp.53.11.1016.

Bickerton, W.L., Samson, D., Williamson, J. and Humphreys, G.W. Separating forms of neglect using the Apples Test: validation and functional prediction in chronic and acute stroke. Neuropsychology. 2011; 25: 567–80. doi: 10.1037/a0023501.

Biklen, D. Communication unbound: autism and praxis. Harvard Educational Review 1990; 60: 291–314. DOI: 10.17763/haer.60.3.013h5022862vu732.

Bilder, R.M. Executive control: balancing stability and flexibility via the duality of evolutionary neuroanatomical trends. Dialogues Clin Neurosci. 2012 Mar; 14(1): 39–47. doi: 10.31887/DCNS.2012.14.1/rbilder.

Binet, A. Reverse illusions of orientation. Psychological Review 1894; 1: 337–350.

Bisaz, R., Travaglia, A. and Alberini, C.M. The neurobiological bases of memory formation: from physiological conditions to psychopathology. Psychopathology. 2014; 47: 347–356. doi: 10.1159/000363702.

Bishop, D.V. Putting language genes in perspective. Trends Genet. 2002; 18: 57–9. doi: 10.1016/s0168–9525(02)02596–9.

Bisiach, E., Geminiani, G., Berti, A. and Rusconi, M.L. Perceptual and premotor factors of unilateral neglect. Neurology 1990; 40: 1278–81. doi: 10.1212/wnl.40.8.1278.

Bisiach, E. and Luzzatti, C. Unilateral neglect of representational space. Cortex 1978; 14: 129–133.

Blackiston, D.J., Silva Casey, E. and Weiss, M.R. Retention of memory through metamorphosis: can a moth remember what it learned as a caterpillar? PLoS One. 2008 Mar 5; 3(3): e1736. doi: 10.1371/journal.pone.0001736.

Blakemore, S.J., Wolpert, D.M. and Frith, C.D. Central cancellation of self-produced tickle sensation. Nat Neurosci. 1998 Nov; 1(7): 635–40. doi: 10.1038/2870.

Blakemore, S.J., Smith, J., Steel, R., Johnstone, C.E. and Frith, C.D. The perception of self-produced sensory stimuli in patients with auditory hallucinations and passivity experiences: evidence for a breakdown in self-monitoring. Psychol Med. 2000 Sep; 30(5): 1131–9. doi: 10.1017/s0033291799002676.

Blanke, O., Ortigue, S., Landis, T. and Seeck, M. Stimulating illusory own-body perceptions. Nature 2002; 419: 269–70. doi: 10.1038/419269a.

Blanke, O., Landis, T., Spinelli, L. and Seeck, M. Out-of-body experience and autoscopy of neurological origin. Brain 2004 Feb; 127(Pt 2): 243–58. doi: 10.1093/brain/awh040.

Blanke, O. and Metzinger, T. Full-body illusions and minimal phenomenal selfhood. Trends Cogn Sci. 2009; 13: 7–13. doi: 10.1016/j.tics.2008.10.003.

Blanke, O. Multisensory brain mechanisms of bodily self-consciousness. Nat Rev Neurosci. 2012 Jul 18; 13(8): 556–71. doi: 10.1038/nrn3292.

Blanken, G. The production of nominal compounds in aphasia. Brain Lang. 2000 Aug; 74(1): 84–102. doi: 10.1006/brln.2000.2338.

Bleecker, M.L. Carbon monoxide intoxication. Handb Clin Neurol. 2015; 131: 191–203. doi: 10.1016/B978–0–444–62627–1.00024–X.

Bless, J.J., Westerhausen, R., von Koss Torkildsen, J., Gudmundsen, M., Kompus, K. and Hugdahl, K. Laterality across languages: Results from a global dichotic listening study using a smartphone application. Laterality 2015; 20: 434–52. doi: 10.1080/1357650X.2014.997245.

Bleuler, E. Extrakampine Halluzinationen. Psychiatrisch-Neurologische Wochenschrift 1903; 25: 261–264.

Block, C. (ed.). The Neuropsychologist's Roadmap. A Training and Career Guide. American Psychological Association: New York, 2021.

Block, N.J. Why do mirrors reverse right/left but not up/down? The Journal of Philosophy 1974; 71: 259–277. DOI: 10.2307/2024963.

Blom, J.D., Nanuashvili, N. and Waters, F. Time Distortions: A systematic review of cases characteristic of Alice in Wonderland Syndrome. Front Psychiatry 2021; 12: 668633. doi: 10.3389/fpsyt.2021.668633.

Blom, J.D., Ter Meulen, B.C., Dool, J. and Ffytche, D.H. A century of prosopometamorphopsia studies. Cortex. 2021; 139: 298–308. doi: 10.1016/j.cortex.2021.03.001.

Blom, J.D. A Dictionary of Hallucinations (2nd ed.). Springer: New York, 2023.

Blom, R.M., Hennekam, R.C. and Denys, D. Body integrity identity disorder. PLoS One. 2012; 7(4): e34702. doi: 10.1371/journal.pone.0034702.

Blom, R.M., van der Wal, S.J., Vulink, N.C. and Denys, D. Role of sexuality in Body Integrity Identity Disorder (BIID): A cross-sectional Internet-based survey study. J Sex Med. 2017 Aug; 14(8): 1028–1035. doi: 10.1016/j.jsxm.2017.06.004.

Blondiaux, E., Heydrich, L. and Blanke, O. Common and distinct brain networks of autoscopic phenomena. Neuroimage Clin. 2021; 30: 102612. doi: 10.1016/j.nicl.2021.102612.

Bobholz, J.A. and Gremley, S. Multiple sclerosis and other demyelinating disorders. In Schoenberg MR and Scott JG. 2011 Chapter 20, 647–661.

Bodamer, J. Die prosop-Agnosie. Archiv für Psychiatrie und Nervenkrankheiten 1947; 179: 6–53.

Bogen, J.E. and Bogen, G.M. The other side of the brain. Ill: The corpus callosum and creativity. Bulletin of the Los Angeles Neurological Society 1969; 34: 191–220.

Bogen, J.E. and Bogen, G.M. Creativity and the corpus callosum. Psychiatric Clinics of North America 1988; 11: 293–301.

Boles, D.B. What bilateral displays do. Brain Cogn 1990; 12: 205–28. doi: 10.1016/0278–2626(90)90016–h.

Bonato, M., Zorzi, M. and Umilta, C. When time is space: Evidence for a mental time line. Neurosci Biobehv Reviews 2012; 36: 2257–2273. doi: 10.1016/j.neubiorev.2012.08.007.

Bonato, M. and Cutini, S. Increased attentional load moves the left to the right. J Clin Exp Neuropsychol 2016; 38: 158–70. doi: 10.1080/13803395.2015.1091065.

Bonnier P. L'aschématie. Rev Neurol (Paris) 1905;13:605–609 (English translation in Epilepsy Behav. 2009 Nov; 16(3): 401-3. doi: 10.1016/j.yebeh.2009.09.020.)

Bordenstein, S.R. The disciplinary matrix of holobiont biology. Uniting life's seen and unseen realms guides a conceptual advance in research. Science 2024; 386: 731–732.

Botvinick, M. and Cohen, J. Rubber hands 'feel' touch that eyes see. Nature 1998; 391: 756. doi: 10.1038/35784.

Boulton, A.A., Baker, G.B. and Hiscock, M. Neuromethods Vol. 17: Neuropsychology. Humana Press: Totowa, 2013 (1st edition 1990).

Bourne, V.J. The divided visual field paradigm: methodological considerations. Laterality 2006 Jul; 11(4): 373–93. doi: 10.1080/13576500600633982.

Bower, J.E. Cancer-related fatigue: Mechanisms, risk factors, and treatments. Nat Rev Clin Oncol. 2014; 11: 597–609. doi: 10.1038/nrclinonc.2014.127

Bracha, H.S. Is there a right hemi-hyper-dopaminergic psychosis? Schizophr Res. 1989 Jul-Oct; 2(4–5): 317–24. doi: 10.1016/0920–9964(89)90022–4.

Bracha, H.S., Livingston, R.L., Clothier, J., Linington, B.B. and Karson, C.N. Correlation of severity of psychiatric patients' delusions with right hemispatial inattention (left-turning behavior). Am J Psychiatry 1993; 150: 330–2. doi: 10.1176/ajp.150.2.330.

Braun, C.M., Dumont, M., Duval, J. and Hamel-Hébert, I. Speech rate as a sticky switch: a multiple lesion case analysis of mutism and hyperlalia. Brain Lang. 2004; 89: 243–52. doi: 10.1016/S0093–934X(03)00402–4.

Brennan, J.R. Language and the Brain. A Slim Guide to Neurolinguistics. Oxford University Press: Oxford, 2022.

Brion, S. and Jedinak, CP. Troubles du transfer interhémisphérique ("callosal disconnection"). A propos de trois observations de tumeurs du corp calleux. Le signe de la main étrangère. Revue Neurologique 1972; 126: 257–266.

Brodmann, K. Vergleichende Lokalisationslehre der Grobhirnrinde in ihren Prinzipien dargestellt auf Grund des Zellenbaues. J.A. Barth: Leipzig, 1909.

Brooks, A., van der Zwan, R., Mercier, M. and Blanke, O. (2009). Visual neuropsychology. *In*: Binder, M.D., Hirokawa, N. and Windhorst, U. (eds.). Encyclopedia of Neuroscience. Springer: Berlin, 2009, 4318–4322garasi8. doi: 10.1007/978–3–540–29678–2_6359.

Brown, A.S. A review of the tip-of-the-tongue experience. Psychol Bull. 1991; 109: 204–23. doi: 10.1037/0033–2909.109.2.204.

Brown, G. S. Probability and psychical research. Nature 1953; 172: 154–156.

Brown, G. S. Probability and Scientific Inference. Longmans, Green & Co.: London, 1957.

Brown, M.B., Abramowicz, A.E., Panzica, P.J. and Weber, G. Anesthetic considerations of organ procurement after brain and cardiac death: A narrative review. Cureus 2023; 15: e40629. doi: 10.7759/cureus.40629.

Brugger, P., Regard, M., Landis, T., Cook, N., Krebs, D. and Niederberger, J. "Meaningful" patterns in visual noise: effects of lateral stimulation and the observer's belief in ESP. Psychopathology. 1993; 26: 261–265. doi: 10.1159/000284831.

Brugger, P., Agosti, R., Regard, M., Wieser, H.G. and Landis, T. Heautoscopy, epilepsy, and suicide. J Neurol Neurosurg Psychiatry 1994a; 57: 838–839. doi: 10.1136/jnnp.57.7.838.

Brugger, P., Dowdey, M.A. and Graves, R.E. From superstitious behavior to delusional thinking: the role of the hippocampus in misattributions of causality. Med Hypotheses 1994b; 43: 397–402. doi: 10.1016/0306–9877(94)90015–9.

Brugger, P. and Gardner M. Perseveration in healthy subjects: an impressive classroom demonstration for educational purposes. Percept Mot Skills 1994 Jun; 78(3 Pt 1): 777–8. doi: 10.1177/003151259407800318.

Brugger, P., Monsch A.U., Salmon D.P. and Butters N. Random number generation in dementia of the Alzheimer type: a test of frontal executive functions. Neuropsychologia. 1996a Feb;34(2): 97–103. doi: 10.1016/0028-3932(95)00066-6.

Brugger, P., Regard, M. and Landis, T. Unilaterally felt "presences": the neuropsychiatry of one's invisible doppelgänger. Neuropsychiatry Neuropsychol Behav Neurol. 1996b; 9: 114–122.

Brugger, P. Pavlov on neuroimaging. J Neurol Neurosurg Psychiatry. 1997 Jun; 62(6): 636. doi: 10.1136/jnnp.62.6.636.

Brugger, P. Variables that influence the generation of random sequences: an update. Percept Mot Skills. 1997; 84: 627–61. doi: 10.2466/pms.1997.84.2.627.

Brugger, P., Regard, M. and Landis T. Illusory reduplication of one's own body: Phenomenology and classification of autoscopic phenomena. Cogn Neuropsychiatry 1997 Feb 1;2(1):19-38. doi: 10.1080/135468097396397.

Brugger, P. Oto-neuropsychiatry: listening to the inner ear. Neuropsychiatry, Neuropsychol Behav Neurol. 1999; 12: 199–200.

Brugger, P., Regard, M., Landis, T. and Oelz, O. Hallucinatory experiences in extreme-altitude climbers. Neuropsychiatry Neuropsychol Behav Neurol. 1999; 12: 67–71.

Brugger, P., Kollias, S.S., Müri, R.M., Crelier, G., Hepp-Reymond, M.C. and Regard, M. Beyond remembering: phantom sensations of congenitally absent limbs. Proc Natl Acad Sci U S A. 2000 May 23; 97(11): 6167–6172. doi: 10.1073/pnas.100510697.

Brugger, P. From haunted brain to haunted science: a cognitive neuroscience view of paranormal and pseudoscientific thought. In Houran J, Lange R (eds): Hauntings and Poltergeists: Multidisciplinary Perspectives. McFarland: Jefferson (NC), 2001, 195–213.

Brugger, P. Reflective mirrors: perspective-taking in autoscopic phenomena. Cogn Neuropsychiatry 2002; 7: 179–94. doi: 10.1080/13546800244000076.

Brugger, P., Macas, E. and Ihlemann, J. Do sperm cells remember? Behav Brain Res 2002 Oct 17; 136(1): 325–8. doi: 10.1016/s0166–4328(02)00127–4.

Brugger, P. and Taylor, K. I. ESP: Extrasensory Perception or Effect of Subjective Probability? Journal of Consciousness Studies 2003; 10(6-7): 221–246.

Brugger, P. From phantom limb to phantom body. Varieties of extracorporeal awareness. *In*: Knoblich, G., Thornton, I., Grosjean, M. and Shiffrar, M. (eds.). Perception of the Human Body from the Inside Out. Oxford University Press: Oxford, 2006, 171–209.

Brugger, P., Blanke, O., Regard, M., Bradford, D.T. and Landis, T. Polyopic heautoscopy: Case report and review of the literature. Cortex. 2006 Jul; 42(5): 666–74. doi: 10.1016/s0010–9452(08)70403–9.

Brugger, P. Hostile interactions between body and self. Dialogues Clin Neurosci. 2007; 9: 210–213.

Brugger, P. and Funk, M. Out on a limb: neglect and confabulation in the study of "aplasic phantoms". *In*: della Sala S. (ed.). Tall tales about the mind and brain. Separating fact from fiction. Oxford: Oxford University Press, 2007, 348–368.

Brugger, P. and Mohr, C. The paranormal mind: how the study of anomalous experiences and beliefs may inform cognitive neuroscience. Cortex. 2008 Nov-Dec; 44(10): 1291–1298. doi: 10.1016/j.cortex.2008.05.008.

Brugger, P. and Weiss, PH. Dermo-optical perception: the non-synesthetic "palpability of colors" a comment on Larner (2006). J Hist Neurosci. 2008; 17: 253–5. doi: 10.1080/09647040601013325.

Brugger, P. SNARC, SCARC, SMARC und SPARC: Gibt es raumfreie Grössen? [Are there spaceless magnitudes?]. Zeitschrift für Neuropsychologie 2008a; 19: 271–274.

Brugger, P. The phantom limb in dreams. Conscious Cogn. 2008b Dec; 17(4): 1272–8. doi: 10.1016/j.concog.2008.01.005.

Brugger, P. and Mohr, C. Out of the body, but not out of mind. Cortex 2009; 45: 137–40. doi: 10.1016/j.cortex.2008.11.004.

Brugger, P., Schubiger, M. and Loetscher, T. Leftward bias in number space is modulated by magical ideation. Cogn Behav Neurol. 2010 Jun; 23(2): 119–23. doi: 10.1097/WNN.0b013e3181d74901.

Brugger, P. Tabula Rama (book review). Cognitive Neuropsychiatry 2012; 17(4): 351–358. doi: 10.1080/13546805.2012.685295.

Brugger, P., Lenggenhager, B. and Giummarra, M.J. Xenomelia: a social neuroscience view of altered bodily self-consciousness. Front Psychol. 2013 Apr 24; 4: 204. doi: 10.3389/fpsyg.2013.00204.

Brugger, P. and Lenggenhager, B. The bodily self and its disorders: neurological, psychological and social aspects. Curr Opin Neurol. 2014 Dec; 27(6): 644–52. doi: 10.1097/WCO.0000000000000151.

Brugger, P. Animal behavior. Chicks with a number sense. Science 2015; 347: 477–478. doi: 10.1126/science.aaa4854.

Brugger, P. and Meier, R. A new illusion at your elbow. Perception 2015; 44: 219–21. doi: 10.1068/p7910.

Brugger, P., Christen, M., Jellestad, L. and Hänggi, J. Limb amputation and other disability desires as a medical condition. Lancet Psychiatry 2016 Dec; 3(12): 1176–1186. doi: 10.1016/S2215-0366(16)30265-6.

Brugger, P., Kurthen, I., Rashidi-Ranjbar, N. and Lenggenhager, B. Grey matter or social matters? Causal attributions in the era of biological psychiatry. Eur Psychiatry 2018; 52: 45–46. doi: 10.1016/j.eurpsy.2018.03.005.

Brugger, P. Are pisces prone to believe in astrology? Academia Letters 2021; article 3891.

Brugger P. Commentary on Sheldrake & Smart: Directional Scopaesthesia. Journal of Scientific Exploration 2024; 38(1): 96–97. https://doi.org/10.31275/20243137.

Brun, R. Ein Fall von Hirntumor bei der Ameise. Schweizer Archiv für Neurologie und Psychiatrie 1925; 16: 86–99.

Bryden, M.P. Correlates of the dichotic right-ear effect. Cortex 1988 Jun; 24(2): 313–9. doi: 10.1016/s0010–9452(88)80039–x.

Bub, D.N., Arguin, M. and Lecours, A.R. Jules Dejerine and his interpretation of pure alexia. Brain Lang. 1993; 45: 531–59. doi: 10.1006/brln.1993.1059.

Budisavljevic, S. and Ramnani, N. Cognitive deficits from a cerebellar tumour: a historical case report from Luria's Laboratory. Cortex. 2012; 48: 26–35. doi: 10.1016/j.cortex.2011.07.001.

Budson, A.E. and Solomon, P.R. Memory Loss, Alzheimer's Disease, and Dementia. A Practical Guide for Clinicians. Elsevier: New York, 2021.

Bueti, D. and Walsh, V. The parietal cortex and the representation of time, space, number and other magnitudes. Philos Trans R Soc Lond B Biol Sci. 2009 Jul 12; 364(1525): 1831–40. doi: 10.1098/rstb.2009.0028.

Bullock, D.N., Hayday, E.A., Grier, M.D., Tang, W., Pestilli, F. and Heilbronner, S.R. A taxonomy of the brain's white matter: twenty-one major tracts for the 21st century. Cereb Cortex 2022; 32: 4524–4548. doi: 10.1093/cercor/bhab500.

Burlakoti, A., Kumaratilake, J., Taylor, J. and Henneberg, M. Asymmetries of total arterial supply of cerebral hemispheres do not exist. Heliyon 2019; 5: e01086. doi: 10.1016/j.heliyon.2018.e01086.

Burrack, A. and Brugger, P. Individual differences in susceptibility to experimentally induced phantom sensations. Body Image 2005 Sep; 2(3): 307–13. doi: 10.1016/j.bodyim.2005.04.002.

Burrack, A., Knoch, D. and Brugger, P. Mitempfindung in synaesthetes: co-incidence or meaningful association? Cortex 2006; 42: 151–4. doi: 10.1016/s0010–9452(08)70339–3.

Cacioppo, S., Bianchi-Demicheli, F., Frum, C., Pfaus, J.G. and Lewis, J. The common neural bases between sexual desire and love: A multilevel Kernel density fMRI Analysis. J Sexual Medicine 2012; 9: 1048–1054. : .

Camenzind, M., Göbel, N., Eberhard-Moscicka, A.K., Knobel, S.E.J., Hegi, H., Single, M. et al. The phenomenology of pareidolia in healthy subjects and patients with left- or right-hemispheric stroke. Heliyon 2024; 10: e27414. doi: 10.1016/j.heliyon.2024.e27414.

Camicioli, R., Howieson, D., Lehman, S. and Kaye, J. Talking while walking. The effect of a dual task in aging and Alzheimer's disease. Neurology 1997; 48: 955–958. doi: 10.1212/WNL.48.4.955.

Cappa, S.F., Sterzi, R., Vallar, G. and Bisiach, E. Remission of hemineglect and anosognosia during vestibular stimulation. Neuropsychologia 1987; 25: 775–782. DOI 10.1016/0028–3932(87)90115–1.

Carlesimo, G.A. Memory disorders in patients with cerebral tumors. J Neurooncol 2012; 108: 253–256. DOI 10.1007/s11060–012–0825–4.

Carlesimo, G.A. The temporal lobes and memory. Handb Clin Neurol 2022: 187: 319–337. doi: 10.1016/B978–0–12–823493–8.00025–0.

Carpenter, R.H.S. Does scopesthesia imply extramission? Journal of Consciousness Studies 2005; 12: 76–78.

Carreiras, M., Lopez, J., Rivero, F. and Corina, D. Linguistic perception: neural processing of a whistled language. Nature 2005; 433: 31–2. doi: 10.1038/433031a.

Carrieu, M. De la fatigue et de son influence pathogénique. Baillière: Paris, 1878.

Carroll, L. Alice's adventures in wonderland and through the looking-glass. New York: Random House, 1862/1984.

Carter, R. The Human Brain Book: An Illustrated Guide to its Structure, Function, and Disorders (3rd ed). Dorling Kindersley Publishing: London, 2019.

Casasanto, D. Embodiment of abstract concepts: Good and bad in right-and left-handers. Journal of Experimental Psychology: General 2009; 138: 351–367. doi: 10.1037/a0015854.

Catani, M. and ffytche, D.H. The rises and falls of disconnection syndromes. Brain 2005; 128: 2224–39. doi: 10.1093/brain/awh622.

Catani, M., Thiebaut de Schotten, M., Slater, D. and Dell'Acqua, F. Connectomic approaches before the connectome. Neuroimage 2013; 80: 2–13. doi: 10.1016/j.neuroimage.2013.05.109

Catsman-Berrevoets, C. and Patay, Z. Cerebellar mutism syndrome. Handb Clin Neurol. 2018; 155: 273–288. doi: 10.1016/B978–0–444–64189–2.00018–4.

Caucheteux, C., Gramfort, A . and King, J.R. Evidence of a predictive coding hierarchy in the human brain listening to speech. Nat Hum Behav. 2023; 7: 430–441. doi: 10.1038/s41562–022–01516–2.

Chae, K.S., Kim, S.C., Kwon, H.J. and Kim, Y. Human magnetic sense is mediated by a light and magnetic field resonance-dependent mechanism. Sci Rep. 2022; 12: 8997. doi: 10.1038/s41598–022–12460–6.

Charles James, J., Schulze, H., Siems, N., Prehn, C., Quast, DR., Trampe, N. et al. Neurological post-COVID syndrome is associated with substantial impairment of verbal short-term and working memory. Sci Rep. 2025 Jan 11; 15(1): 1695. doi: 10.1038/s41598–025–85919–x

Chatterjee, A., Ricci, R. and Calhoun, J. Weighing the evidence for cross over in neglect. Neuropsychologia. 2000; 38(10): 1390–7. doi: 10.1016/s0028–3932(00)00042–7.

Chen, P., Hreha, K., MacPhee, C., Salter, A. and Eskes, G.A. Does prism adaptation treatment reduce spatial neglect and improve function? Front Rehabil Sci. 2025; 6: 1539887. doi: 10.3389/fresc.2025.1539887.

Chen, Y., Jungsuwadee, P., Vore, M., Butterfield, D.A. and St. Clair, D.K. Collateral damage in cancer chemotherapy: Oxidative stress in nontargeted tissues. Molecular Interventions 2007; 7: 147–156. https: //doi.org/10.1124/mi.7.3.6.

Cherbuin, N. and Brinkman, C. Cognition is cool: Can hemispheric activation be assessed by tympanic membrane thermometry? Brain Cogn 2004; 54: 228–231. doi: 10.1016/j.bandc.2004.02.014

Chevreul, M.E. De la baguette divinatoire, du pendule dit explorateur et des tables tournantes, au point de vue de l'histoire de la critique et de la méthode expérimentale. Mallet-Bachelier: Paris, 1854.

Chi, J.G., Dooling, E.C. and Gilles, F.H. Left-right asymmetries of the temporal speech areas of the human fetus. Arch Neurol 1977; 34: 346–348. doi: 10.1001/archneur.1977.00500180040008.

Chiarello, C. Hemisphere dynamics in lexical access: automatic and controlled priming. Brain Lang 1985; 26: 146–172. doi: 10.1016/0093–934x(85)90034–3.

Child, G. The other presence. Backpacker 1989; 17: 68–72.

Chittka, L. and Niven, J. Are bigger brains better? Current Biology 2009; 19: R995– R1008. Doi 10.1016/j.cub.2009.08.023.

Chokron, S., Bartolomeo, P., Perenin, M.T., Helft, G. and Imbert, M. Scanning direction and line bisection: a study of normal subjects and unilateral neglect patients with opposite reading habits. Brain Res Cogn Brain Res 1998; 7: 173–8. doi: 10.1016/s0926–6410(98)00022–6.

Christen, M. and Brugger, P. From control to cooperation - Plea for a new role of IRBs. Cortex 2015; 71: 415–6. doi: 10.1016/j.cortex.2015.05.004.

Christodoulou, G.N., Margariti, M., Kontaxakis, V.P. and Christodoulou, N.G. The delusional misidentification syndromes: strange, fascinating, and insrtuctive. Curr Psychiatry Rep. 2009 Jun 11(3): 185–9. doi: 10.1007/s11920–009–0029–6.

Chronicle, E.P. and Glover, J. A ticklish question: does magnetic stimulation of the primary motor cortex give rise to an 'efference copy'? Cortex 2003; 39: 105–10. doi: 10.1016/s0010–9452(08)70078–9.

Clark, D.G., Guitar, B. and Hoffman, P.R. Reliability of verbal-manual interference. Brain Cogn 1985 Oct; 4(4): 486–91. doi: 10.1016/0278–2626(85)90036–3.

Clarke, S., Regli, L., Janzer, R.C., Assal, G. and de Tribolet, N. Phantom face: conscious correlate of neural reorganization after removal of primary sensory neurones. Neuroreport 1996 Nov 25; 7(18): 2853–2857.

Clifton, P.G., Vickers, S.P. and Somerville, E.M. Little and often: ingestive behavior patterns following hippocampal lesions in rats. Behav Neurosci 1998; 112: 502–11. doi: 10.1037//0735–7044.112.3.502.

Code, C., Lincoln, M. and Dredge, R. Asymmetries in mouth opening during word generation in male stuttering and non-stuttering participants. Laterality 2005; 10: 471–486. doi: 10.1080/13576500442000238.

Cohen, N.J. and Squire, L.R. Preserved learning and retention of pattern analyzing skill in amnesia: dissociation of knowing how and knowing that. Science 1980, 210: 207–210. doi: 10.1126/science.7414331.

Cohen, L. and Dehaene, S. Competition between past and present. Assessment and interpretation of verbal perseverations. Brain. 1998 Sep; 121 (Pt 9): 1641–59. doi: 10.1093/brain/121.9.1641.

Cohen, R.A. The neuropsychology of attention. 2nd edition. Springer: Berlin, 2013.

Cohen, R.L. Memory for action events: The power of enactment. Educ Psychol Rev 1989; 1: 57–80. doi.org/10.1007/BF01326550.

Cole, M., Levitin, K. and Luria, A.R. The autobiography of Alexander Luria. A Dialogue with The Making of Mind. New York: Taylor and Francis, 2005.

Colier, W.N., Quaresima, V., Wenzel, R., van der Sluijs, M.C., Oeseburg, B., Ferrari, M. et al. Simultaneous near-infrared spectroscopy monitoring of left and right occipital areas reveals contra-lateral hemodynamic changes upon hemi-field paradigm. Vision Res. 2001; 41: 97–102. doi: 10.1016/s0042–6989(00)00219–4.

Coltheart, M. Deep dyslexia is right-hemisphere reading. Brain Lang. 2000 Feb, 1; 71(2): 299–309. doi: 10.1006/brln.1999.2183.

Cona, G., Scarpazza, C., Sartori, G., Moscovitch, M. and Bisiacchi, P.S. Neural bases of prospective memory: a meta-analysis and the "Attention to Delayed Intention" (AtoDI) model. Neurosci Biobehav Rev 2015; 52: 21–37. doi: 10.1016/j.neubiorev.2015.02.007.

Conchiglia, G., Della Rocca, G. and Grossi, D. On a peculiar environmental dependency syndrome in a case with frontal-temporal damage: Zelig-like syndrome. Neurocase 2007; 13: 1–5, DOI: 10.1080/13554790601160558

Conway, C.M. and Pisoni, D.B. Neurocognitive basis of implicit learning of sequential structure and its relation to language processing. Ann N Y Acad Sci 2008; 1145: 113–131. doi: 10.1196/annals.1416.009.

Cooper, J.M., Gadian, D.G., Jentschke, S., Goldman, A., Munoz, M., Pitts, G. et al. Neonatal hypoxia, hippocampal atrophy, and memory impairment: evidence of a causal sequence. Cereb Cortex 2015; 25: 1469–76. doi: 10.1093/cercor/bht332.

Corballis, M.C. Much ado about mirrors. Psychon Bull Rev 2000; 7: 163–169. doi: 10.3758/bf03210736.

Corballis, M.C. From Hand to Mouth: The Origins of Language. Princeton University Press: Princeton, 2002.

Corballis, M.C. From mouth to hand. Gesture, speech, and the evolution of right-handedness. Behav Brain Sci. 2003 Apr; 26(2): 199–208; discussion 208–60. doi: 10.1017/s0140525x03000062.

Corballis, M.C. The dual-brain myth. *In*: Della Sala, S. (ed.). Tall Tales about the Mind and Brain: Separating fact from fiction. Oxford University Press: Oxford, 2007, 291–313.

Corcoran, D.W. The phenomena of the disembodied eye or is it a matter of personal geography? Perception 1977; 6: 247–53. doi: 10.1068/p060247.

Coren, S. and Porac, C. Fifty centuries of rlghthandedness: the historical record. Science 1977; 198: 631–632. 10.1126/science.335510.

Coren, S. The Lateral Preference Inventory for measurement of handedness, footedness, eyedness, and earedness: Norms for young adults. Bulletin of the Psychonomic Society 1993; 31: 1–3.

Corkin, S. Acquisition of motor skill after bilateral medial temporal-lobe excision. Neuropsychologia 1968; 6: 225–265.

Cowan, N. What are the differences between long-term, short-term, and working memory? Prog Brain Res. 2008; 169: 323–338. doi: 10.1016/S0079–6123(07)00020–9.

Cox, J.H., Seri, S. and Cavanna, A.E. Sensory aspects of Tourette syndrome. Neurosci Biobehav Rev 2018; 88: 170–176. doi: 10.1016/j.neubiorev.2018.03.016.

Craig, A.D. How do you feel? Interoception: the sense of the physiological condition of the body. Nat Rev Neurosci. 2002; 3: 655–66. doi: 10.1038/nrn894.

Craig, A.D. How do you feel - now? The anterior insula and human awareness. Nat Rev Neurosci. 2009a; 10: 59–70. doi: 10.1038/nrn2555.

Craig, A.D. Emotional moments across time: a possible neural basis for time perception in the anterior insula. Philos Trans R Soc Lond B Biol Sci 2009b; 364: 1933–1942.

Critchley, M. Metamorphopsia of central origin. Transactions of the Ophthalmologic Society of the UK 1949; 69: 111–121.

Critchley, M. The Divine Banquet of the Brain and Other Essays. Raven Press: New York, 1979.

Critchley, M. The Parietal Lobes. Edward Arnold: London, 1953 (Chapter 10: Disorders of spatial thought, 326–355).

Critchley, M. Personification of paralysed limbs in hemiplegics. Br Med J. 1955 Jul 30; 2(4934): 284-6. doi: 10.1136/bmj.2.4934.284.

Croisile, B., Laurent, B., Michel, D. and Trillet, M. Pure agraphia after deep left hemisphere haematoma. J Neurol Neurosurg Psychiatry 1990; 53: 263–265. doi: 10.1136/jnnp.53.3.263.

Crowe, S.F. and Kuttner, M. Differences between schizophrenia and the schizophrenia-like psychosis of temporal lobe epilepsy. Neuropsychiatry, Neuropsychol Behav Neurol 1991; 4: 127–135.

Crowhurst, T., Koszyca, B., Holmes, M. and Holmes-Liew, C.L. Progressive multifocal leukoencephalopathy in a lung transplant recipient presenting with memory impairment: Case report and literature review. Transpl Infect Dis. 2020 Jun; 22(3): e13293. doi: 10.1111/tid.13293.

Cruse, H. and Hübner, D. Selforganizing memory: active learning of landmarks used for navigation. Biol Cybern. 2008; 99: 219–236. doi: 10.1007/s00422–008–0256–7.

Cummings, J.L and Mega, M.S. Neuropsychiatry and Behavioral Neuroscience. Oxford University Press: New York, 2003.

Cunningham, M.G., Goldstein, M., Katz, D., O'Neil, S.Q., Joseph, A. and Price, B. Coalescence of psychiatry, neurology, and neuropsychology: from theory to practice. Harv Rev Psychiatry 2006; 14: 127–40. doi: 10.1080/10673220600748536.

D'esposito, M., Alexander, M.P., Fischer, R., Mcglinchey-Berroth, R. and O'connor, M. Recovery of memory and executive function following anterior communicating artery aneurysm rupture. J. Int. Neuropsychol. Soc. 1996; 2: 565–570. doi: 10.1017/s1355617700001740.

Damiani, D., Nascimento, A.M. and Kühl-Pereira, L. Cortical brain functions – The Brodmann legacy in the 21st century. Neurosurgery 2020; 39: 261–270. DOI: 10.1055/s-0036–1597573.

Danckert, J., Striemer, C. and Rossetti, Y. Blindsight. Handb Clin Neurol. 2021; 178: 297–310. doi: 10.1016/B978–0–12–821377–3.00016–7.

Dary, Z., Lenggenhager, B., Lagarde, S., Medina Villalon, S., Bartolomei, F. and Lopez, C. Neural bases of the bodily self as revealed by electrical brain stimulation: A systematic review. Hum Brain Mapp. 2023; 44: 2936–2959. doi: 10.1002/hbm.26253.

Das, J.M. and Naqvi, I.A. Anton Syndrome. *In*: StatPearls [Internet]. Treasure Island (FL): StatPearls Publishing, 2024.

David, A. Frontal lobology – psychiatry's new pseudoscience. Br J Psychiatry 1992; 161: 244–248. doi: 10.1192/bjp.161.2.244.

Davidoff, J. and Landis, T. Recognition of unfamiliar faces in prosopagnosia. Neuropsychologia. 1990; 28(11): 1143–61. doi: 10.1016/0028–3932(90)90051–o.

de Haan, E.H. and Cowey, A. On the usefulness of "what" and "where" pathways in vision. Trends Cogn Sci. 2011 Oct; 15(10): 460–6. doi: 10.1016/j.tics.2011.08.005.

de Haan, B., Karnath, H.-O. and Driver, J. Mechanisms and anatomy of unilateral extinction after brain injury. Neuropsychologia 2012; 50: 1045–1053. doi: 10.1016/j.neuropsychologia.2012.02.015.

de Haan, E.H.F. and Dijkerman, H.C. Somatosensation in the brain: A theoretical re-evaluation and a new model. Trends Cogn Sci. 2020 Jul; 24(7): 529–541. doi: 10.1016/j.tics.2020.04.003.

de Havas, J., Gomi, H. and Haggard, P. Experimental investigations of control principles of involuntary movement: a comprehensive review of the Kohnstamm phenomenon. Exp Brain Res. 2017 Jul; 235(7): 1953–1997. doi: 10.1007/s00221–017–4950–3.

de Hevia, M.D., Izard, V., Coubart, A., Spelke, E.S. and Streri, A. Representations of space, time, and number in neonates. Proc Natl Acad Sci USA 2014; 111: 4809–13. doi: 10.1073/pnas.1323628111.

de Marco, M. and Venneri, A. Serial recall order of category fluency words: Exploring its neural underpinnings. Front Psychol. 2022; 12: 777838. doi: 10.3389/fpsyg.2021.777838.

de Preester, H. Merleau-Ponty's sexual schema and the sexual component of body integrity identity disorder. Med Health Care Philos. 2013 May; 16(2): 171–84. doi: 10.1007/s11019–011–9367–3.

De Vignemont, F. Mind the Body: An Exploration of Bodily Self-Awareness. Oxford University Press: Oxford, UK, 2020.

Deaner, R. O., Isler, K., Burkart, J. and van Schaik, C. (2007). Overall brain size, and not encephalization quotient, best predicts cognitive ability across non-human primates. Brain Behav. Evol. 70: 115–124.

Deffieux, T., Demené, C. and Tanter, M. Functional Ultrasound Imaging: A new imaging modality for neuroscience. Neuroscience 2021; 15: 110–121. doi: 10.1016/j.neuroscience.2021.03.005.

Dehaene, S., Bossini, S. and Giraux, P. The mental representation of parity and number magnitude. Journal of Experimental Psychology: General 1993; 122: 371–396. https: //doi.org/10.1037/0096–3445.122.3.371.

Dehaene, S. and Changeux, J.P. Development of elementary numerical abilities: A neuronal model. J Cog Neurosci. 1993; 5: 390–407. doi: 10.1162/jocn.1993.5.4.390.

Della Sala, S., Laiacona, M., Spinnler, H. and Trivelli, C. Poppelreuter-Ghent's Overlapping Figures Test: its sensitivity to age, and its clinical use. Arch Clin Neuropsychol 1995 Nov; 10(6): 511–34. doi: 10.1016/0887–6177(94)00049–v.

Dember, W. N. and Richman, C. L. (1989). Spontaneous Alternation Behavior. New York: Springer.

Demené, C., Robin, J., Dizeux, A., Heiles, B., Pernot, M., Tanter, M. et al. Transcranial ultrafast ultrasound localization microscopy of brain vasculature in patients. Nat Biomed Eng 2021; 5: 219–228. doi: 10.1038/s41551–021–00697–x.

Dening, T.R. and Berrios, G.E. Autoscopic phenomena. Br J Psychiatry 1994; 165: 808–817. doi: 10.1192/bjp.165.6.808.

Devinsky, O., Feldmann, E., Burrowes, K. and Bromfield, E. Autoscopic phenomena with seizures. Arch Neurol. 1989; 46: 1080–1088. doi: 10.1001/archneur.1989.00520460060015.

Devinsky, O., Davachi, L., Santchi, C., Quinn, B.T., Staresina, B.P. and Thesen, T. Hyperfamiliarity for faces. Neurology 2010; 74: 970–4. doi: 10.1212/WNL.0b013e3181d5dc22.

Dewdney, D. A specific distortion of the human facial percept in childhood schizophrenia. Psychiatr Q. 1973; 47(1): 82–94. doi: 10.1007/BF01562899.

Di Noto, P.M., Newman, L., Wall, S. and Einstein, G. The hermunculus: what is known about the representation of the female body in the brain? Cereb Cortex 2013 May; 23(5): 1005–13. doi: 10.1093/cercor/bhs005.

Dieguez, S. Doubles everywhere: literary contributions to the study of the bodily self. Front Neurol Neurosci. 2013; 31: 77–115. doi: 10.1159/000345912.

Dieguez, S. and Lopez, C. The bodily self: Insights from clinical and experimental research. Annals of Physical and Rehabilitation Medicine 2017; 60: 198–207. 10.1016/j.rehab.2016.04.007.

Dijkerman, H.C. and de Haan, E.H. Somatosensory processes subserving perception and action. Behav Brain Sci. 2007 Apr; 30(2): 189–201; discussion 201–39. doi: 10.1017/S0140525X07001392.

Doricchi, F., Guariglia, P., Gasparini, M. and Tomaiuolo, F. Dissociation between physical and mental number line bisection in right hemisphere brain damage. Nat Neurosci. 2005; 8: 1663–1665. doi: 10.1038/nn1563.

Doricchi, F., Iaria, G., Silvetti, M., Figliozzi, F. and Siegler I. The "ways" we look at dreams: evidence from unilateral spatial neglect (with an evolutionary account of dream bizarreness). Exp Brain Res. 2007 Apr; 178(4): 450–61. doi: 10.1007/s00221–006–0750–x.

Doss, R.C., Zhang, W., Risse, G.L. and Dickens, G.L. Lateralizing language with magnetic source imaging: validation based on the Wada test. Epilepsia. 2009; 50: 2242–2248. doi: 10.1111/j.1528–1167.2009.02242.x

Downing, P.E., Jiang, Y., Shuman, M. and Kanwisher, N. A cortical area selective for visual processing of the human body. Science 2001; 293: 2470–2473. doi: 10.1126/science.1063414.

Drescher, J. A history of homosexuality and organized psychoanalysis. J Am Acad Psychoanal Dyn Psychiatry. 2008; 36: 443–60. doi: 10.1521/jaap.2008.36.3.443.

Driver, J. and Vuilleumier, P. Unconscious processing in neglect and extinction. *In*: De Gelder, B., De Haan, E.H. and Heywood, C.A. (eds.). Out of Mind: Varieties of unconscious processes. Oxford University Press: Oxford, 2001, 107–139.

Dror, S., Miklósi, Á., Sommese, A. and Fugazza, C. A citizen science model turns anecdotes into evidence by revealing similar characteristics among Gifted Word Learner dogs. Sci Rep 2023; 13: 21747. doi: 10.1038/s41598–023–47864–5.

Du, J., Rolls, E.T., Cheng, W., Li, Y., Gong, W., Qiu, J. et al. Functional connectivity of the orbitofrontal cortex, anterior cingulate cortex, and inferior frontal gyrus in humans. Cortex 2020; 123: 185–199. https: //doi.org/10.1016/j.cortex.2019.10.012.

Dubois, B., Slachevsky, A., Litvan, I. and Pillon, B. The FAB: A Frontal Assessment Battery at bedside. Neurology. 2000 Dec 12; 55(11): 1621–6. doi: 10.1212/wnl.55.11.1621.

Duchêne, A., Graves, R.E. and Brugger, P. Schizotypal thinking and associative processing: a response commonality analysis of verbal fluency. J Psychiatry Neurosci 1998 Jan; 23(1): 56–60.

Dudai, Y. and Eisenberg, M. Rites of passage of the engram: reconsolidation and the lingering consolidation hypothesis. Neuron. 2004. Sep 30; 44(1): 93–100. doi: 10.1016/j.neuron.2004.09.003.

Duecker, F., Formisano, E. and Sack, A.T. Hemispheric differences in the voluntary control of spatial attention: direct evidence for a right-hemispheric dominance within frontal cortex. J Cogn Neurosci. 2013; 25: 1332–1342. doi: 10.1162/jocn_a_00402.

Duecker, F. and Sack, A.T. The hybrid model of attentional control: New insights into hemispheric asymmetries inferred from TMS research. Neuropsychologia 2015; 74: 21–29. doi: 10.1016/j.neuropsychologia.2014.11.023.

Duque, J., Murase, N., Celnik, P., Hummel, F., Harris-Love, M., Mazzocchio, R. et al. Intermanual Differences in movement-related interhemispheric inhibition. J Cogn Neurosci 2007, 19. 204–213. doi: 10.1162/jocn.2007.19.2.204.

Duss, S.B., Reber, T.P., Hänggi, J., Schwab, S., Wiest, R., Müri, R.M. et al. Unconscious relational encoding depends on hippocampus. Brain 2014; 137: 3355–70. doi: 10.1093/brain/awu270.

Dutton, D.L. The cold reading technique. Experientia 1988; 44: 326–32. doi: 10.1007/BF01961271.

Ebisch, S.J., Ferri, F. and Gallese, V. Touching moments: desire modulates the neural anticipation of active romantic caress. Front Behav Neurosci 2014; 8: article 60: 1–12. doi: 10.3389/fnbeh.2014.00060. eCollection 2014.

Eckblad, M. and Chapman, L.J. Magical ideation as an indicator of schizotypy. J Consult Clin Psychol 1983 Apr; 51(2): 215–25. doi: 10.1037//0022–006x.51.2.215.

Editorial. When once is enough. Nat Neurosci 2004; 7: 93. Doi: 10.1038/nn0204–93.

Ehrsson, H.H. The experimental induction of out-of-body experiences. Science 2007 Aug 24; 317(5841): 1048. doi: 10.1126/science.1142175

Eichele, T., Nordby, H., Rimol, L.M. and Hugdahl, K. Asymmetry of evoked potential latency to speech sounds predicts the ear advantage in dichotic listening. Brain Res Cogn Brain Res 2005; 24: 405–412. doi: 10.1016/j.cogbrainres.2005.02.017

Eichenbaum, H. Using olfaction to study memory. Ann N Y Acad Sci 1998 Nov 30; 855: 657–69. doi: 10.1111/j.1749–6632.1998.tb10642.x.

Eide, L.S., Rike, P.O., Reme, S.E., Snekkevik, H., Rossner, S., Rosen, G. et al. Using hypnotic suggestion in the rehabilitation of working memory capacity after acquired brain injury: study protocol for a randomized controlled trial. Trials 2024 Jan 2; 25(1): 11. doi: 10.1186/s13063–023–07867–z.

Eisenberg, L. Mindlessness and brainlessness in psychiatry Br J Psychiatry 1986; 148: 497–508. doi: 10.1192/bjp.148.5.497.

Eisenberger, N.I. Social pain and the brain: controversies, questions, and where to go from here. Annu Rev Psychol. 2015 Jan 3; 66: 601–29. doi: 10.1146/annurev-psych-010213–115146.

Eliot, L., Ahmed, A., Khan, H. and Patel, J. Dump the "dimorphism": Comprehensive synthesis of human brain studies reveals few male-female differences beyond size. Neuroscience & Biobehavioral Reviews 2021; 125: 667–697. doi 10.1016/j.neubiorev.2021.02.026

Endo, K.E., Miyasaka, M., Makishita, H., Yanagisawa, N. and Sugishita, M. Tactile agnosia and tactile aphasia: symptomatological and anatomical differences. Cortex 1992; 28: 445–469.

England, M. and Nicoladis, E. Functional fixedness and body-part-as-object production in pantomime. Acta Psychol (Amst). 2018 Oct; 190: 174–187. doi: 10.1016/j.actpsy.2018.07.010.

Erdelyi, M.H. The Interpretation of Dreams and of Jokes. Routledge: New York, 2023.

Ericsson, K.A., Cheng, X., Pan, Y., Ku, Y. and Hu, Y. Memory skills mediating superior memory in a world-class memorist. Memory 2017; 25: 1294–1302. doi: 10.1080/09658211.2017.1296164.

Faraday, M. Experimental investigation of table-moving. J Franklin Institute 1853; 56: 328–333.

Farah, M. Why does the somatosensory homunculus have hands next to face and feet next to genitals? A hypothesis. Neural Comput. 1998; 10: 1983–5. doi: 10.1162/089976698300016936.

Farnsworth ,D. The Farnsworth-Munsell 100 hue and Dichotomous tests for color vision. J Opt Soc Am 1943; 33: 568–578.

Faurie, C. and Raymond, M. Handedness frequency over more than 10,000 years. Proceedings of the Royal Society of London, Series B 2004; 271: S43–S45.

Fechner, G.T. Elemente der Psychophysik, Vol. 2. Breitkopf & Härtel, Leipzig, 1860.

Fernandez, Y.M. and Frucht, S.J. Primary progressive apraxia: an unusual ideomotor syndrome. J Clin Mov Disord. 2017; 4: 17. doi: 10.1186/s40734–017–0064–0.

Fernandez-Ruiz, A., Sirota, A., Lopes-Dos-Santos, V. and Dupret, D. Over and above frequency: Gamma oscillations as units of neural circuit operations. Neuron. 2023 Apr 5; 111(7): 936–953. doi: 10.1016/j.neuron.2023.02.026.

Ferrè ER, Haggard P. Vestibular cognition: State-of-the-art and future directions. Cogn Neuropsychol. 2020; 37: 413–420. doi: 10.1080/02643294.2020.1736018.

Feuillet, L., Dufour, H. and Pelletier, J. Brain of a white-collar worker. Lancet. 2007 Jul 21; 370(9583): 262. doi: 10.1016/S0140-6736(07)61127-1.

Ffytche, D.H. The hodology of hallucinations. Cortex. 2008; 44: 1067–83. doi: 10.1016/j.cortex.2008.04.005

Finger, S. and Hustwit, M.P. Five early accounts of phantom limb in context: Paré, Descartes, Lemos, Bell and Mitchell. Neurosurgery. 2003 Mar; 52(3): 675–86; discussion 685–6. doi: 10.1227/01.neu.0000048478.42020.97.

Finger, S. and Roe, D. Gustave Dax and the early history of cerebral dominance. Arch Neurol. 1996 Aug; 53(8): 806–13. doi: 10.1001/archneur.1996.00550080132021.

First, M.B. Desire for amputation of a limb: Paraphilia, psychosis, or a new type of identity disorder. Psychol Med 2005 Jun; 35(6): 919–28. doi: 10.1017/s0033291704003320.

Fischer, M.H. Cognition in the bisection task. Trends Cogn Sci. 2001 Nov 1; 5(11): 460–462. doi: 10.1016/s1364–6613(00)01790.3.

Fischer, M.H., Mills, R.A. and Shaki, S. How to cook a SNARC: number placement in text rapidly changes spatial-numerical associations. Brain Cogn 2010; 72: 333–336. doi: 10.1016/j.bandc.2009.10.010.

Fischer, M.H. and Brugger, P. When digits help digits: spatial-numerical associations point to finger counting as prime example of embodied cognition. Front Psychol. 2011 Oct 17; 2: 260. doi: 10.3389/fpsyg.2011.00260.

Fisher, C.A. and Larner, A.J. Jean Langlais (1907–91): An historical case of a blind organist with stroke-induced aphasia and Braille alexia but without amusia. J Med Biogr. 2008; 16: 232–4. doi: 10.1258/jmb.2007.007043.

Fix, J.D. Neuroanatomy (3rd edition). Lippincott, Williams & Wilkins: Philadelphia, 2002.

Fleck, L. Genesis and Development of a Scientific Fact. University of Chicago Press: Chicago, 1979 (originally published 1935 in Polish).

Flegr, J., Ullmann, J. and Toman, J. Parasitic manipulation or side effects? The effects of past *Toxoplasma gondii* and *Borrelia* spp. infections on human personality and cognitive performance are not mediated by impaired health. Folia Parasitol (Praha). 2023; 70: 2023.020. doi: 10.14411/fp.2023.020.

Fleminger J.J. and Bunce L. Investigation of cerebral dominance in 'left-handers' and 'right-handers' using unilateral electroconvulsive therapy. J Neurol Neurosurg Psychiatry. 1975 Jun; 38(6): 541–5. doi: 10.1136/jnnp.38.6.541.

Fleminger, J.J., de Horne, D.J. and Nott, P.N. Unilateral electroconvulsive therapy and cerebral dominance: effect of right- and left-sided electrode placement on verbal memory. J Neurol Neurosurg Psychiatry 1970; 33: 408–411. DOI: 10.1136/jnnp.33.3.408.

Flor, H., Elbert, T., Knecht, S., Wienbruch, C., Pantev, C., Birbaumers, N. et al. Phantom-limb pain as a perceptual correlate of cortical reorganization following arm amputation. Nature 1995; 375: 482–484. https: //doi. org/10.1038/375482a0.

Folstein, M.F., Folstein, S.E. and McHugh, P.R. "Mini-mental state". A practical method for grading the cognitive state of patients for the clinician. J Psychiatr Res. 1975 Nov; 12(3): 189–98. doi: 10.1016/0022–3956(75)90026–6.

Forrester, G.S., Quaresmini, C., Leavens, D.A., Spiezio, C. and Vallortigara, G. Target animacy influences chimpanzee handedness. Anim Cogn. 2012 Nov; 15(6): 1121–7. doi: 10.1007/s10071-012-0536-4.

Fortis, P., Ronchi, R., Velardo, V., Calzolari, E., Banco, E., Algeri, L. et al. A home-based prism adaptation training for neglect patients. Cortex. 2020; 122: 61–80. doi: 10.1016/j.cortex.2018.09.001.

Fox, R.J., Kasner, S.E., Chatterjee, A. and Chalela, J.A. Aphemia: an isolated disorder of articulation. Clin Neurol Neurosurg. 2001 Jul; 103(2): 123–6. doi: 10.1016/s0303–8467(01)00126–3.

Frankland, P.W. and Josselyn, S.A. Neuroscience: Memory and the single molecule. Nature 2013 Jan 17; 493(7432): 312–3. doi: 10.1038/nature11850.

Freud, S. Zur Auffassung der Aphasien – eine kritische Studie. Deiticke: Wien, 1891.

Fridriksson, J., den Ouden, D.B., Hillis, A.E., Hickok, G., Rorden, C., Basilakos, A. et al. Anatomy of aphasia revisited. Brain. 2018 Mar 1; 141(3): 848–862. doi: 10.1093/brain/awx363.

Friedrich, T.E. and Elias, L.J. Behavioural asymmetries on the greyscales task: The influence of native reading direction. Cult Brain 2014; 2: 161–172. doi: 10.1007/s40167–014–0019–3.

Frisby, J. Seeing: Illusion, Brain, and Mind. Oxford University Press: Oxford, 1979.

Fujinaga, N., Muramatsu, T., Ogano, M. and Kato, M. A 3–year follow-up study of "orientation agnosia". Neuropsychologia 2005; 43: 1222–1226.

Funk, M., Shiffrar, M. and Brugger, P. Hand movement observation by individuals born without hands: phantom limb experience constrains visual limb perception. Exp Brain Res. 2005 Jul; 164(3): 341–6. doi: 10.1007/s00221–005–2255–4.

Furnham, A. Personality and handedness. Personality and Individual Differences 1983; 4: 715–716. .

Gagliano, M., Vyazovskiy, V.V., Borbély, A.A., Grimonprez, M. and Depczynski, M. Learning by Association in Plants. Sci Rep. 2016 Dec 2; 6: 38427. doi: 10.1038/srep38427.

Gainotti, G. Lower- and higher-level models of right hemisphere language. A selective survey. Funct Neurol. 2016 Apr-Jun; 31(2): 67–73. doi: 10.11138/fneur/2016.31.2.067.

Gainotti, G. and Trojano, L. Constructional apraxia. Handb Clin Neurol. 2018; 151: 331–348. doi: 10.1016/B978-0-444-63622-5.00016-4.

Gallagher, S., Butterworth, G.E., Lew, A. and Cole, J. Hand-mouth coordination, congenital absence of limb, and evidence for innate body schemas. Brain Cogn. 1998 Oct; 38(1): 53–65. doi: 10.1006/brcg.1998.1020.

Galt, A.H. The evil eye as synthetic image and its meanings on the Island of Pantelleria, Italy. American Ethnologist 1982; 9: 664–681.

Galton, F. Statistics of mental imagery. Oxford Journals 1880; 19: 301–318.

Gaudet, L.A., Rybka, L., Mandonnet, E., Volle, E., Barberis, M., Jonkers, R. et al. Leveraging relatedness-based measures in people with language disorders: A scoping review. J Neuropsychol. 2024 Dec 16. doi: 10.1111/jnp.12405.

Gawda, B. The structure of the concepts related to love spectrum: Emotional verbal fluency technique application, initial psychometrics, and its validation. J Psycholinguist Res. 2019; 48: 1339–1361. doi: 10.1007/s10936–019–09661–y.

Geffen, G., Traub, E. and Stierman, I. Language laterality assessed by unilateral ECT and dichotic monitoring. J Neurol Neurosurg Psychiatry 1978; 41: 354–360. 10.1136/jnnp.41.4.354.

Geiser, N., Kaufmann, B.C., Knobel, S.E.J., Cazzoli, D., Nef, T. and Nyffeler, T. Comparison of uni- and multimodal motion stimulation on visual neglect: A proof-of-concept study. Cortex 2024; 171: 194–203. doi: 10.1016/j.cortex.2023.10.018.

Genner, T. Das Sehen des eigenen Spiegelbildes als epileptisches Äquivalent. Wiener Klinische Wochenschrift 1947; 59: 656–658.

Genova, L. Remember: The Science of Memory and the Art of Forgetting. Harmony: Chatsworth, CA, 2021.

Gerlach, C., Kühn, C.D., Mathiassen, A.B., Kristensen, C.L. and Starrfelt, R. The face inversion effect or the face upright effect? Cognition 2023; 232: 105335. doi: 10.1016/j.cognition.2022.105335

Gerrans, P. Pain Asymbolia as depersonalization for pain experience. An interoceptive active inference account. Front Psychol. 2020; 11: 523710. doi: 10.3389/fpsyg.2020.523710.

Gerstmann, J. Syndrome of finger agnosia, disorientation for right and left, agraphia and acalculia. Arch Neurology Psychiatry 1930; 44: 398–408.

Gerstmann, J. Syndrome of finger agnosia, disorientation for right and left, agraphia and acalculia. Arch Neurol Psychiatry 1940; 44: 398–407.

Gerstmann J. Problem of imperception of disease and of impaired body territories with organic lesions. Arch Neurol Psychiatry 1942; 48: 890–913.

Ghadri, J.R., Sarcon, A., Diekmann, J., Bataiosu, D.R., Camman, V.L., Jurisic, S. et al. Happy heart syndrome: role of positive emotional stress in takotsubo syndrome. Eur Heart J 2016; 37: 2823–2829. doi: 10.1093/eurheartj/ehv757.

Ghika-Schmid, F., van Melle, G., Guex, P. and Bogousslavsky, J. Subjective experience and behavior in acute stroke: the Lausanne Emotion in Acute Stroke Study. Neurology 1999; 52: 22–8. doi: 10.1212/wnl.52.1.22.

Gholipour, T., Koubeissi, M.Z. and Shields, DC. Stereotactic electroencephalography. Clin Neurol Neurosurg. 2020 Feb; 189: 105640. doi: 10.1016/j.clineuro.2019.105640.

Ghose, S., Parmar, T., Dada, T., Vanathi, M. and Sharma, S. A new computer-based Farnsworth Munsell 100–hue test for evaluation of color vision. Int Ophthalmol. 2014; 34: 747–51. doi: 10.1007/s10792–013–9865–9.

Giora, E. and Büttemeyer, W. Roberto Ardigò as a forerunner of George M. Stratton's experiments on inverted vision. Hist Psychol 2020; 23: 26–39. doi: 10.1037/hop0000126. Epub 2019 Apr 11.

Giummarra, M.J., Bradshaw, J.L., Hilti, L.M., Nicholls, M.E. and Brugger, P. Paralyzed by desire: a new type of body integrity identity disorder. Cogn Behav Neurol. 2012 Mar; 25(1): 34–41. doi: 10.1097/WNN.0b013e318249865a.

Gloning, I., Gloning, K., Weingarten, K. and Berner, P. Über einen Fall mit Alexie der BRAILLE-Schrift. Wiener Zeitschrift für Nervenheilkunde 1955; 10: 260–273.

Gloning, I., Jellinger, K., Sluga, W. and Weingarten, K. Über Uhrzeitagnosie. Archiv für Psychiatrie und Zeitschrift für die gesamte Neurologie 1958; 198: 85–95. https://doi.org/10.1007/BF00353165.

Godden, D.R. and Baddeley, A.D. Context-dependent memory in two natural environments: On land and underwater. Br. J. Psychol. 1975; 66: 325–331. DOI 10.1111/j.2044-8295.1975.tb01468.x)

Goebel, S., Fischer, R., Ferstl, R. and Mehdorn, H.M. Normative data and psychometric properties for qualitative and quantitative scoring criteria of the Five-point Test. Clin Neuropsychol. 2009 May; 23(4): 675–90. doi: 10.1080/13854040802389185.

Goel, V., Shuren, J., Sheesley, L. and Grafman, J. Asymmetrical involvment of frontal lobes in social reasoning. Brain 2004; 127: 783–790.

Gogolla, N. The insular cortex. Curr Biol. 2017 Jun 19;2 (12):R580-R586. doi: 10.1016/j.cub.2017.05.010.

Goldberg, E., Harner, R., Lovell, M., Podell, K. and Riggio, S. Cognitive bias, functional cortical geometry, and the frontal lobes: laterality, sex, and handedness. J Cogn Neurosci. 1994; 6: 276–96. doi: 10.1162/jocn.1994.6.3.276.

Goldberg, E. The Executive Brain: Frontal Lobes and the Civilized Mind. Oxford University Press: Oxford, 2001.

Goldberg, E.B. and Hillis, A.E. Sign language aphasia. *In*: Hillis, A.E. and Fridriksson (eds.). Handbook of Clinical Neurology 2022; 185: 297–315. doi: 10.1016/B978–0–12–823384–9.00019–0.

Goldenberg, G. Why should we read the classics? Cortex 2001; 37: 293–294. DOI: 10.1016/S0010–9452(08)70575–6.

Goldenberg, G. Apraxia and beyond: life and work of Hugo Liepmann. Cortex. 2003 Jun; 39(3): 509–24. doi: 10.1016/s0010–9452(08)70261–2.

Goldenberg, G. Apraxia and the parietal lobes. Neuropsychologia. 2009 May; 47(6): 1449–59. doi: 10.1016/j.neuropsychologia.2008.07.014.

Goldenberg, G. Apraxia. The cognitive side of motor control. Oxford Academic: Oxford, 2013. (also as a target article with the same title in a discussion forum in Cortex 2014; 57: 270–274 with comments by more than a dozen experts).

Goldmann Gross, R. and Grossman, M. Update on apraxia. Curr Neurol Neurosci Rep. 2008 Nov; 8(6): 490–6. doi: 10.1007/s11910–008–0078–y.

Goldstein, K. and Gelb, A. Psychologische Analysen hirnpathologischer Fälle auf Grund von Untersuchungen Hirnverletzter. I. Abhandlung: Zur Psychologie des optischen Wahrnehmungs- und Erkennungsvorganges. Zeitschrift für die gesamte Neurologie und Psychiatrie 1918; 41: 1–142.

Gollin, E.S. Developmental studies of visual recognition of incomplete objects. Percept Mot Skills 1960; 11: 289–298. doi 10.2466/pms.1960.11.3.2.

Gonzales Rothi, L.J. and Heilman, K M. (eds.). Apraxia. The neuropsychology of action. Imprint Psychology Press: London, 1997 (eBook 2014, DOI: https://doi.org/10.4324/9781315804545).

Goodale, M.A. and Milner, A.D. Separate visual pathways for perception and action. Trends in Neurosciences 1992; 15: 20–25. Doi: 10.1016/0166–2236(92)90344–8.

Goodwin, D. Homonymous hemianopia: challenges and solutions. 2014; 8: 1919–1927. doi: 10.2147/OPTH.S59452.

Gosselin, F. and Schyns, P.G. Superstitious perceptions reveal properties of internal representations. Psychol Sci. 2003 Sep; 14(5): 505–9. doi: 10.1111/1467–9280.03452.

Grasland, M., Turmel, N., Chesnel, C., Haddad, R., Le Breton, F., Amarenco, G. et al. Complications neurologiques du coït: revue de la littérature [Neurological complications of coitus: Review of literature]. Prog Urol. 2021 Jun; 31(7): 392–405. French. doi: 10.1016/j.purol.2021.01.005.

Graves, R., Goodglass, H. and Landis, T. Mouth asymmetry during spontaneous speech. Neuropsychologia 1982; 20: 371–381. doi: 10.1016/0028–3932(82)90037–9.

Graves, R., Landis, T. and Simpson, C. On the interpretation of mouth asymmetry. Neuropsychologia 1985; 23: 121–122. doi: 10.1016/0028–3932(85)90051–x.

Graybiel, A. and Kellogg, R.S. Inversion illusion in parabolic flight: its probable dependence on otolith function. Aerospace Medicine 1967; 38: 1099–1103.

Green, B.G. Localization of thermal sensation: An illusion and synthetic heat. Perception & Psychophysics 1977; 22: 331–337. https: //doi.org/10.3758/BF03199698.

Greer, D.M., Lewis, A. and Kirschen, M.P. New developments in guidelines for brain death/death by neurological criteria. Nat Rev Neurol. 2024; 20: 151–161. doi: 10.1038/s41582–024–00929–z.

Grossberg, S. Acetylcholine neuromodulation in normal and abnormal learning and memory: Vigilance control in waking, sleep, autism, amnesia and alzheimer's disease. Front Neural Circuits 2017, 11: 82. doi: 10.3389/fncir.2017.00082.

Grosslight, J.H. and Ticknor, W. Variability and reactive inhibition in the mealworm as a function of determined turning sequences. Journal of Comparative and Physiological Psychology 1953; 46(1): 35e38.

Grotstein, J.S. Autoscopy: the experience of oneself as a double. Hillside J Clin Psychiatry 1983; 5: 259–304.

Grüsser, O.-J. and Hagner, M. On the history of deformation phosphenes and the idea of internal light generated in the eye for the purpose of vision. Documenta Ophthalmologica 1990; 74: 57–85.

Grüsser, O.-J. and Landis, T. Visual agnosias and other disturbance of visual perception and cognition. Vol. 12 of Cronly-Dillon, J. (ed.). Vision and Visual Dysfunction. MacMillan: London, 1991.
Guida, A., Abrahamse, E. and van Dijck, J.P. About the interplay between internal and external spatial codes in the mind: implications for serial order. Ann N Y Acad Sci 2020 Oct; 1477(1): 20–33. doi: 10.1111/nyas.14341.
Güntürkün, O., Ströckens, F. and Ocklenburg, S. Brain lateralization: A comparative perspective. Physiol Rev 2020; 100: 1019–1063. doi: 10.1152/physrev.00006.2019.
Guterstam, A., Kean, H.H., Webb, Y.W., Kean, F.S. and Graziano, M.S.A. Implicit model of other people's visual attention as an invisible, force-carrying beam projecting from the eyes. Proceedings of the National Academy of Sciences 2019; 116: 328–333. doi: 10.1073/pnas.1816581115.
Haan, J. and Ferrari, M.D. Picasso's migraine: Illusory cubist splitting or illusion? Cephalalgia 2011; 31: 1057–60. doi: 10.1177/0333102411406752.
Hadi, Z., Mahmud, M., Pondeca, Y., Calzolari, E., Chepisheva, M., Smith, R.M. et al. The human brain networks mediating the vestibular sensation of self-motion. J Neurol Sci. 2022; 443: 120458. doi: 10.1016/j.jns.2022.120458.
Hahamy, A., Macdonald, S.N., van den Heiligenberg, F., Kieliba, P., Emir, U., Malach, R. et al. Representation of multiple body parts in the missing-hand territory of congenital one-handers. Current Biology 2017; 27: 1350–1355. 10.1016/j.cub.2017.03.053.
Hailman, J.P. and Sustare, B.D. What a stuffed toy tells a stuffed shirt. BioScience 1973; 23: 644–651. https: //doi.org/10.2307/1296777.
Halligan, P.W. and Marshall, J.C. Left neglect for near but not far space in man. Nature 1991; 350: 498–500.
Halligan, P.W., Marshall, J.C. and Wade, D.T. Left on the right: allochiria in a case of left visuo-spatial neglect. J Neurol Neurosurg Psychiatry 1992; 55: 717–719.
Halligan, P.W., Marshall, J.C. and Wade, D.T. Unilateral somatoparaphrenia after right hemisphere stroke: a case description. Cortex. 1995 Mar; 31(1): 173–82. doi: 10.1016/s0010–9452(13)80115–3.
Hampshire, A., Azor, A., Atchison, C., Trender, W., Hellyer, P.J., Giunchiglia, V. et al. Cognition and memory after Covid-19 in a large community sample. N Engl J Med 2024 Feb 29; 390(9): 806–818. doi: 10.1056/NEJMoa2311330.
Han, S. and Northoff, G. Culture-sensitive neural substrates of human cognition: a transcultural neuroimaging approach. Nat Rev Neurosci 2008; 9: 646–54. doi: 10.1038/nrn2456.
Hancock, R., Richlan, F. and Hoeft, F. Possible roles for fronto-striatal circuits in reading disorder. Neurosci Biobehav Rev 2017; 72: 243–260. doi: 10.1016/j.neubiorev.2016.10.025.
Hardyck, C. and Petrinovich, L.F. Left-handedness. Psychological Bulletin 1977; 84: 385–404.
Harish, E. and Osherov, N. Fungal priming: Prepare or perish. J Fungi (Basel) 2022; 8: 448. doi: 10.3390/jof8050448.
Harrington, A. Nineteenth-century ideas on hemisphere differences and duality of mind. Behav Brain Sci 1985; 8: 617–34.
Harrington, A. Medicine, mind and the double brain: a study in nineteenth-century thought. Princeton University Press: Princeton, NJ, 1987.
Harrington A. Unfinished business: models of laterality in the nineteenth century. In Davidson RJ, Hugdahl K. Brain Asymmetry, MIT Press: Cambridge, MA, 1995, 3–27.
Harrington, A. How to house a mind inside a brain. Lessons from history. EMBO Rep 2007; 8(Suppl 1): S12–5. doi: 10.1038/sj.embor.7400974.
Harris, L.J. Footedness in parrots: three centuries of research, theory, and mere surmise. Can J Psychol. 1989; 43: 369–396. doi: 10.1037/h0084228.
Harris, L.J. The discovery of cerebral specialization. Front Neurol Neuroscience. 2019: 44: 1–14. doi: 10.1159/000494938.
Harrison, N.A., Johnston, K., Corno, F., Casey, S.J., Friedner, K., Humphreys, K. et al. Psychogenic amnesia: syndromes, outcome, and patterns of retrograde amnesia. Brain 2017; 140: 2498–2510. doi: 10.1093/brain/awx186.
Hartley, T., Lever, C., Burgess, N. and O'Keefe, J. Space in the brain: how the hippocampal formation supports spatial cognition. Philosophical Transactions of the Royal Society B 2014; 369: 201–205 doi: 10.1098/rstb.2012.0510.

Hartwigsen, G. and Siebner, H.R. Novel methods to study aphasia recovery after stroke. Front Neurol Neurosci 2013; 32: 101–11. doi: 10.1159/000346431.

Hatta, T. Recognition of Japanese kanji in the left and right visual fields. Neuropsychologia 1977; 15: 685–688. doi: 10.1016/0028–3932(77)90073–2.

Hauk, O., Johnsrude, I. and Pulvermüller, F. Somatotopic representation of action words in human motor and premotor cortex. Neuron. 2004 Jan 22; 41(2): 301–7. doi: 10.1016/s0896-6273(03)00838-9.

Hausinger, T. and Pletzer, B. Sex hormones modulate sex differences and relate to hemispheric asymmetries in a divided visual field Navon task. Behav Brain Res 2021: 408: 113281. doi: 10.1016/j.bbr.2021.113281.

Hausmann, M. Why sex hormones matter for neuroscience: A very short review on sex, sex hormones, and functional brain asymmetries. J Neurosci Res 2017; 95: 40–49. doi: 10.1002/jnr.23857.

Hayward, V. A brief taxonomy of tactile illusions and demonstrations that can be done in a hardware store. Brain Res Bull. 2008; 75: 742–52. doi: 10.1016/j.brainresbull.2008.01.008.

Hayward, W., Buch, E.R., Norato, G., Iwane, F., Dash, D., Salamanca-Girón, R.F. et al. Procedural motor memory deficits in patients with long-COVID. Neurology 2024; 102: e208073. doi: 10.1212/WNL.0000000000208073.

Healy, B. The Yentl syndrome. The New England Journal of Medicine 1991; 325: 274–276. doi: 10.1056/NEJM199107253250408.

Heilman, K.M. Neglect and related disorders. *In*: Heilman, K.M. and Valenstein, E. (eds.). Clinical Neuropsychology. Oxford University Press: New York, 1979, 268–307.

Heilman, K.M. and Valenstein, E. (eds.). Clinical Neuropsychology (5th edition). Oxford University Press: Oxford, 2011 (1st edition 1979).

Heilman, K.M., Howell, G., Valenstein, E. and Rothi, L. Mirror-reading and writing in association with right-left spatial disorientation. J Neurol Neurosurg Psychiatry 1980; 43: 774–780. DOI: 10.1136/jnnp.43.9.774.

Heilman, K.M., Valenstein, E. and Watson, R.T. Neglect and related disorders. Semin Neurol 2000; 20(4): 463–70. doi: 10.1055/s-2000–13179.

Heilman, K.H. Intentional neglect. Frontiers in Bioscience 2004; 9: 694–705. doi: 10.2741/1261.

Heim, S., Eickhoff, S.B. and Amunts, K. Specialisation in Broca's region for semantic, phonological, and syntactic fluency? Neuroimage 2008 Apr. 15; 40(3): 1362–8. doi: 10.1016/j.neuroimage.2008.01.009.

Hellige, J.B. and Longstreth, L.E. Effects of concurrent hemisphere-specific activity on unimanual tapping rate. Neuropsychologia 1981; 19(3): 395–405. doi: 10.1016/0028–3932(81)90069–5.

Hellige, J.B., Jonsson, J.E. and Michimata, C. Processing from LVF, RVF and bilateral presentations: Examinations of metacontrol and interhemispheric interaction. Brain and Cognition 1988; 7: 39–53.

Hellige, J.B. and Adamson, M.M. Hemispheric differences in processing handwritten cursive. Brain Lang. 2007 Sep; 102(33): 215–27. doi: 10.1016/j.bandl.2006.11.003.

Henderson, S.K., Peterson, K.A., Patterson, K., Lambon Ralph, M.A. and Rowe, J.B. Verbal fluency tests assess global cognitive status but have limited diagnostic differentiation: evidence from a large-scale examination of six neurodegenerative diseases. Brain Commun. 2023; 5: fcad042. doi: 10.1093/braincomms/fcad042.

Henderson, W.R. and Smyth, G.E. Phantom limbs. J Neurol Neurosurg Psychiatry 1948; 11: 88–112. doi: 10.1136/jnnp.11.2.88.

Henke, K. A model for memory systems based on processing modes rather than consciousness. Nat Rev Neurosci 2010; 11: 523–32. doi: 10.1038/nrn2850.

Henley, C. Foundations of Neuroscience. Michigan State University Libraries: East Lansing, 2021.

Henrich, J., Heine, S.J. and Norenzayan, A. The weirdest people in the world? Behav Brain Sci 2010; 33: 61–83; discussion part 83–135. doi: 10.1017/S0140525X0999152X.

Henry, M., Thomas, K.G. and Ross, I.L. Episodic memory impairment in Addison's disease: results from a telephonic cognitive assessment. Metab Brain Dis 2014; 29: 421–430. doi: 10.1007/s11011–014–9511–x.

Hepper, P.G., Shahidullah, S. and White, R. Origins of fetal handedness. Nature 1990; 347: 431. doi: 10.1038/347431b0.

Hepper, P.G., McCartney, G.R. and Shannon, W.A. Lateralised behaviour in first trimester human foetuses. Neuropsychologia 1998; 36: 531–534. doi: 10.1016/s0028–3932(97)00156–5.

Hepper, P.G., Wells, D.L. and Lynch, C. Prenatal thumb sucking is related to postnatal handedness. Neuropsychologia 2005; 43: 313–315. doi: 10.1016/j.neuropsychologia.2004.08.009.

Herculano-Houzel, S., Collins, C.E., Wong, P. and Kaas, J.H. Cellular scaling rules for primate brains. Proc Natl Acad Sci USA 2007; 104: 3562–3567.

Herculano-Houzel, S. The human brain in numbers: a linearly scaled-up primate brain. Front in Hum Neurosci 2009 Nov 9; 3: 31. doi: 10.3389/neuro.09.031.2009.

Hermann, D., Siccoli, M., Brugger, P., Wachter, K., Mathis, J., Achermann, P. et al. Evolution of neurological, neuropsychological and sleep-wake disturbances after paramedian thalamic stroke. Stroke 2008; 39: 62–68. doi: 10.1161/STROKEAHA.107.494955.

Hermsdörfer, J., Mai, N., Spatt, J., Marquardt, C., Veltkamp, R. and Goldenberg, G. Kinematic analysis of movement imitation in apraxia. Brain 1996; 119: 1575–86. doi: 10.1093/brain/119.5.1575.

Hershkowitz, M. and Samuel, D. The retention of learning during metamorphosis of the crested newt (*Triturus cristatus*). Anim Behav. 1973 Feb; 21(1): 83–5. doi: 10.1016/s0003–3472(73)80043–0.

Hertz, R. La prééminence de la main droite. Etude sur la polarité religieuse. Revue Philosophique de la France et de l'étranger 1909; 68: 553–580. English translation: The pre-eminence of the right hand. A study in religious polarity. HAU: Journal of Ethnographic Theory 2013; 3: 335–357. DOI: 10.14318/hau3.2.024.

Heuts, B.A. and Brunt, T. Behavioural left- right asymmetry extends to arthropods. Behavioural and Brain Sciences 2005; 28(4), 601–602. DOI: 10.1017/S0140525X05350102.

Heux, L., Rathbone, C., Gensburger, S., Clifford, R. and Souchay, C. Collective memory and autobiographical memory: Perspectives from the humanities and cognitive sciences. Wiley Interdiscip Rev Cogn Sci 2023 May-Jun; 14(3): e1635. doi: 10.1002/wcs.1635.

Heydrich, L. and Blanke, O. Distinct illusory own-body perceptions caused by damage to posterior insula and extrastriate cortex. Brain 2013; 136: 790–803. doi: 10.1093/brain/aws364.

Hickok, G. and Poeppel, D. The cortical organization of speech processing. Nat Rev Neurosci. 2007; 8: 393–402. doi: 10.1038/nrn2113.

Hickok, G. The dual stream model of speech and language processing. Handb Clin Neurol 2022; 185: 57–69. doi: 10.1016/B978–0–12–823384–9.00003–7.

Hicks, R.A. and Pellegrini, R.J. Handedness and anxiety. Cortex 1978 Mar; 14(1): 119–21. doi: 10.1016/s0010-9452(78)80014-8.

Hillis, A.E. and Caramazza, A. A framework for interpreting distinct patterns of hemispatial neglect, Neurocase 1995; 1: 189–207. DOI: 10.1080/13554799508402364.

Hilti, L.M. and Brugger, P. Incarnation and animation: physical versus representational deficits of body integrity. Exp Brain Res 2010; 204: 315–26. doi: 10.1007/s00221–009–2043–7.

Hilti, L.M., Hänggi, J., Vitacco, D.A., Kraemer, B., Palla, A., Luechinger, R., et al. The desire for healthy limb amputation: structural brain correlates and clinical features of xenomelia. Brain. 2013 Jan; 136(Pt 1): 318–29. doi: 10.1093/brain/aws316.

Hirnstein, M. and Hausmann, M. Sex/gender differences in the brain are not trivial—A commentary on Eliot et al. (2021). Neuroscience and Biobehavioral Reviews 2021; 130: 408–409.

Hiscock, M. and Kinsbourne, M. Attention and the right-ear advantage: what is the connection? Brain Cogn 2011; 76: 263–75. doi: 10.1016/j.bandc.2011.03.016.

Hoffman, D.D. Visual Intelligence. How we create what we see. WW Norton & Co.: New York, 1998. Introduction to Sensation and Perception Copyright © 2022 by Students of PSY 3031 and Edited by Dr. Cheryl Olman is licensed under a Creative Commons Attribution 4.0 International License, except where otherwise noted.

Hogeveen, J. and Grafman, J. Alexithymia. Handb Clin Neurol. 2021: 183: 47–62. doi: 10.1016/B978–0–12–822290–4.00004–9.

Holden, C. Sex and the suffering brain. Science 2005; 308: 1574–1577. doi: 10.1126/science.308.5728.1574

Holé, J., Reilly, K.T., Nash, S. and Rode, G. Caloric vestibular stimulation reduces the directional bias in representational neglect. Brain Sci. 2020; 10: 323. doi: 10.3390/brainsci10060323.

Holmes, G. Disturbances of visual orientation. British Journal of Opthalmology 1918; 2: 449–468, 506–516.

Hook-Costigan, M.A. and Rogers, L.J. Lateralized use of the mouth in production of vocalizations by marmosets. Neuropsychologia 1998; 36: 1265–1273. doi: 10.1016/s0028–3932(98)00037–2.

Hopfinger, J.B. and Slotnick, S.C. (eds.). The cognitive neuroscience of attention. Current debates and research. Routledge: New York, 2022.
Hori, M. and Takahashi, S. (eds.). Lateral Asymmetry in Animals. Predator-Prey Interactions, Dynamics, and Evolution. Springer Nature: Singapore, 2022.
Houben, M., Chettouf, S., Van Der Werf, Y.D. and Stins, J. Theta-burst transcranial magnetic stimulation for the treatment of unilateral neglect in stroke patients: A systematic review and best evidence synthesis. Restor Neurol Neurosci 2021; 39(6): 447–465. doi: 10.3233/RNN-211228.
Hoyt, A.T. and Smith, K.A. Selective amygdalohippocampectomy. Neurosurg Clin N Am 2016; 27: 1–17. doi: 10.1016/j.nec.2015.08.009.
Hubel, D.H. and Wiesel, T.N. Receptive fields of single neurons in the cat's striate cortex. J Physiol. 1959; 148: 574–591.
Hugdahl, K. "Hearing voices": auditory hallucinations as failure of top-down control of bottom-up perceptual processes. Scand J Psychol 2009; 50: 553–60. doi: 10.1111/j.1467–9450.2009.00775.x.
Humphreys, G.W. and Forde, E.M.E. Disordered action schema and action disorganization syndrome. Cog Neuropsychology 1998; 15: 771–811.
Hunter, Z.R. and Brysbaert, M. Visual half-field experiments are a good measure of cerebral language dominance if used properly: evidence from fMRI. Neuropsychologia 2008; 46: 316–25. doi: 10.1016/j.neuropsychologia.2007.07.007.
Hunziker, E. and Mazzola, G. Ansichten eines Hirns: Aktuelle Perspektiven der Hirnforschung. Birkhäuser: Basel, 1990.
Husain, M., Mattingley, J.B., Rorden, C., Kennard, C. and Driver, J. Distinguishing sensory and motor biases in parietal and frontal neglect. Brain 2000; 123: 1643–1659. doi: 10.1093/brain/123.8.1643.
Igarashi, K.M. Entorhinal cortex dysfunction in Alzheimer's disease. Trends Neurosci 2023; 46: 124–136. doi: 10.1016/j.tins.2022.11.006.
Infeld, B., Davis, S.M., Lichtenstein, M., Mitchell, P.J. and Hopper, J.L. Crossed cerebellar diaschisis and brain recovery after stroke. Stroke 1995; 26: 90–95. doi: 10.1161/01.str.26.1.90.
Inoue, S. and Matsuzawa, T. Working memory of numerals in chimpanzees. Curr Biol 2007; 17: R1004–5. doi: 10.1016/j.cub.2007.10.027.
Ionta, S. and Blanke, O. Differential influence of hands posture on mental rotation of hands and feet in left and right handers. Exp Brain Res 2009 May; 195(2): 207–17. doi: 10.1007/s00221–009–1770–0.
Iorio, L., Falanga, A., Fragassi, N.A. and Grossi, D. Visual associative agnosia and optic aphasia. A single case study and a review of the syndromes. Cortex 1992; 28: 23–37. doi: 10.1016/s0010–9452(13)80163–3.
Iosa, M., Morone, G. and Paolucci, S. Phi in physiology, psychology and biomechanics: The golden ratio between myth and science. Biosystems 2018; 165: 31–39. doi: 10.1016/j.biosystems.2018.01.001.
Isaacson, R.L. The limbic system. 2nd edition. Springer: New York, 2013.
Ishida, M., Imanishi, J., Yazawa, Y., Sunakawa, Y., Torigoe, T. and Onishi, H. "Phantom akathisia" in an amputated leg of a sarcoma patient: a case report. Biopsychosoc Med 2020; 14: 4. doi: 10.1186/s13030–020–00178–8.
Ishihara, S. Tests for Color-Blindness (6th edition) Kanehara & Co.: Tokyo, 1932.
Ishihawa, T., Saito, M., Fujimoto, S. and Imahashi, H. Ictal increased writing preceded by dysphasic seizures. Brain Dev 2000; 22: 398–402. doi: 10.1016/s0387–7604(00)00163–7.
Jackson, J.H. On the nature of the duality of the brain (first section). Medical Press and Circular 1874; 17: 19–21.
Jaffe, S.L. Commentary on Price et al. Neurology 54, 2000. Neurology 2000; 55: 603–604.
Price, B.H., Adams, R.D. and Coyle, J.T. Neurology and psychiatry: Closing the great divide. Neurology 2000; 54: 8–8. doi: 10.1212/wnl.54.1.8.
Jain, H. and Ghosh, S. Imprinting reversible deformations on a compressed soft rod network. Soft Matter 2024; 20: 5053–5059. doi: 10.1039/d4sm00099d.
James, J.C., Schulze, H., Siems, N., Prehn, C., Quast, D.R., Trampe, N. et al. Neurological post-COVID syndrome is associated with substantial impairment of verbal short-term and working memory. Sci Rep. 202515: 1695. doi: 10.1038/s41598–025–85919–x.
James, W. The principles of psychology. Holt: New York, 1890.
Janet, P. Le renversement de l'orientation ou l'allochirie des représentations. Journal de Psychologie 1908; 5: 89–97.

Javadi, A.H. and Wee, N. Cross-category adaptation: Objects produce gender adaptation in the perception of faces. PLoS ONE 2012; 7: e46079. doi: 10.1371/journal.pone.0046079.

Jawinski, P., Markett, S., Sander, C., Huang, J., Ulke, C., Hegerl, U. et al. The big five personality traits and brain arousal in the resting state. Brain Sci. 2021; 11: 1272. doi: 10.3390/brainsci11101272.

Jenkinson, P.M., Papadaki, C., Besharati, S., Moro, V., Gobbetto, V., Crucianelli, L. et al. Welcoming back my arm: affective touch increases body ownership following right-hemisphere stroke. Brain Commun. 2020; 2: fcaa034. doi: 10.1093/braincomms/fcaa034.

Jerison, H. Evolution of the Brain and Intelligence. Academic Press: New York, 1973.

Jewell, G. and McCourt, M.E. Pseudoneglect: a review and meta-analysis of performance factors in line bisection tasks. Neuropsychologia 2000; 38: 93–110. doi: 10.1016/S0028 -3932(99)00045 -7.

Johannsen, L., Ackermann, H. and Karnath, H.O. Lasting amelioration of spatial neglect by treatment with neck muscle vibration even without concurrent training. J Rehabil Med 2003; 35: 249–253. doi: 10.1080/16501970310009972.

Johnson, J.K. and Graziano, A.B. Some early cases of aphasia and the capacity to sing. Prog Brain Res 2015; 216: 73–89. doi: 10.1016/bs.pbr.2014.11.004.

Johnson, T.W. and Irwig, M.S. The hidden world of self-castration and testicular self-injury. Nat Rev Urol 2014; 11: 297–300. doi: 10.1038/nrurol.2014.84.

Johnstone, L.T., Karlsson, E.M. and Carey, D.P. Left-handers are less lateralized than right-handers for both left and right hemispheric functions. Cereb Cortex 2021; 31: 3780–3787. doi: 10.1093/cercor/bhab048.

Jones, B.J., Chen, M.E., Simoncini, L. and Spencer, R.M.C. Sleep enhances reconsolidation-based strengthening of visuospatial memories. Sci Rep 2022; 12: 7307. doi: 10.1038/s41598–022–11135–6.

Jones, L.A. Motor illusions: what do they reveal about proprioception? Psychol Bull. 1988 Jan; 103(1): 72–86. doi: 10.1037/0033–2909.103.1.72.

Jordan, T.R., Patching, G.R. and Milner, A.D. Central fixations are inadequately controlled by instructions alone: implications for studying cerebral asymmetry. Q J Exp Psychol A 1998; 51: 371–391.

Jung-Beeman, M. Bilateral brain processes for comprehending natural language. Trends in Cognitive Sciences 2005; 9: 512–518. DOI: 10.1016/j.tics.2005.09.009.

Kammers, M.P., van der Ham, I.J. and Dijkerman, H.C. Dissociating body representations in healthy individuals: differential effects of a kinaesthetic illusion on perception and action. Neuropsychologia 2006; 44: 2430–2436. doi: 10.1016/j.neuropsychologia.2006.04.009.

Kandell, E.R. In Search of Memory. The Emergence of a New Science of Mind. WW Norton & Co.: New York, 2006.

Kanizsa, G. Organization in Vision. Essays on Gestalt Perception. Praeger: New York, 1979.

Kanwisher, N., McDermott, J. and Chun, M.M. The fusiform face area: a module in human extrastriate cortex specialized for face perception. J Neurosci. 1997; 17: 4302–4311. doi: 10.1523/JNEUROSCI.17–11–04302.1997.

Kapasi, A., Leurgans, S.E., James, B.D., Boyle, P.A., Arvanitakis, Z., Nag, S. et al. Watershed microinfarct pathology and cognition in older persons. Neurobiol Aging. 2018; 70: 10–17. doi: 10.1016/j.neurobiolaging.2018.05.027.

Kapur, N. Focal retrograde amnesia and the attribution of causality: An exceptionally benign commentary. Cogn Neuropsychol. 2000 Oct 1; 17(7): 623–37. doi: 10.1080/026432900750002181.

Karavatos, A., Kaprinis, G. and Tzavaras, A. Hemispheric specialization for language in the congenitally blind: The influence of the Braille system. Neuropsychologia 1984; 22: 521–525. DOI: 10.1016/0028–3932(84)90048–4.

Karmakar, S., Kesh, A. and Muniyandi, M. Thermal illusions for thermal displays: a review. Front Hum Neurosci. 2023; 17: 1278894. doi: 10.3389/fnhum.2023.1278894.

Karnath, H.O. Pusher syndrome - a frequent but little-known disturbance of body orientation perception. J Neurol 2007; 254: 415–24. doi: 10.1007/s00415–006–0341–6.

Karnath, H.-O., Milner, A. D. and Vallar, G. (eds.). The Cognitive and Neural Bases of Spatial Neglect. Oxford University Press: Oxford, 2002.

Karnath, H.O. and Baier, B. Right insula for our sense of limb ownership and self-awareness of actions. Brain Struct Funct. 2010; 214: 411–7. doi: 10.1007/s00429–010–0250–4.

Kasprian, G., Langs, G., Brugger, P.C., Bittner, M., Weber, M., Arantes, M. et al. The prenatal origin of hemispheric asymmetry: an in utero neuroimaging study. Cereb Cortex 2011; 21: 1076–83. doi: 10.1093/cercor/bhq179.

Kawamura, M. and Miller, M.W. History of Amusia. Front Neurol Neurosci 2019; 44: 83–88. doi: 10.1159/000494955.

Kawano, T., Ushifusa, Y., Mancuso, S., Baluška, F., Sylvain-Bonfanti, L., Arbelet-Bonnin, D. et al. Plants have two minds as we do. Plant Signal Behav 2025; 20: 2474895. doi: 10.1080/15592324.2025.2474895.

Kelly, M.P., Kriznik, N.M., Kinmonth, A.L. and Fletcher, P.C. The brain, self and society: a social-neuroscience model of predictive processing. Soc Neurosci 2019; 14: 266–276. doi: 10.1080/17470919.2018.1471003.

Kemp, S., Young, A.W., Szulecka, K. and de Pauw, K.W. A case of paraprosopia and its treatment. Cogn Neuropsychiatry 2003; 8: 43–56. doi: 10.1080/713752239.

Keogh, R., Pearson, J. and Zeman, A. Aphantasia: The science of visual imagery extremes. Handb Clin Neurol. 2021; 178: 277–296. doi: 10.1016/B978–0–12–821377–3.00012–X.

Kennedy, D.N., O'Craven, K.M., Ticho, B.S., Goldstein, A.M., Makris, N. and Henson, J.W. Structural and functional brain asymmetries in human situs inversus totalis. Neurology. 1999 Oct 12; 53(6): 1260–5. doi: 10.1212/wnl.53.6.1260

Kerkhoff, G. and Schindler, I. Hemineglekt vs. Hemianopsie. Fortschritte der Neurologie und Psychiatrie 1997; 65: 278–289.

Kerkhoff, G. and Bucher, L. Line bisection as an early method to assess homonymous hemianopia. Cortex 2008; 44: 200–2005.

Kesler, S.R., Janelsins, M.C., Koovakkattu, D., Palesh, O., Mustian, K., Morrow, G. et al. Reduced hippocampal volume and verbal memory performance associated with interleukin-6 and tumor necrosis factor-alpha levels in chemotherapy-treated breast cancer survivors. Brain Behavior Immunity 2013; 30: S109–S116. doi.org/10.1016/j.bbi.2012.05.017.

Kety, S.S. A biologist examines the mind and behavior. Science 1960; 132: 1861–1870. doi: 10.1126/science.132.3443.1861.

Khalsa, S.S., Adolphs, R., Cameron, O.G., Critchley, H.D., Davenport, P.W., Feinstein, J.S. et al. Interoception and mental health: A roadmap. Biol Psychiatry Cogn Neurosci Neuroimaging 2018; 3: 501–513. doi: 10.1016/j.bpsc.2017.12.004.

Kieseler, M.L. and Duchaine, B. Persistent prosopagnosia following COVID-19. Cortex 2023; 162: 56–64. doi: 10.1016/j.cortex.2023.01.012.

Kim, B., Zilpelwar, S., Sie, E.J., Marsili, F., Zimmermann, B., Boas, D.A. et al. Measuring human cerebral blood flow and brain function with fiber-based speckle contrast optical spectroscopy system. Commun Biol 2003a; 6: 844.

Kim, Y.S., Kim, Y.E., Choung, Y.H., Kim, H., Kim, H.J., Jung, N.Y. et al. Pearls and oysters: Familial verbal auditory agnosia due to C9orf72 repeat expansion. Neurology 2003b; 101: e2046–e2050. doi: 10.1212/WNL.0000000000207832.

Kimura, D. Cerebral dominance and the perception of verbal stimuli. Canadian Journal of Psychology 1961; 15: 166–171.

Kimura, K., Tachibana, N., Aso, T., Kimura, J. and Shibasaki, H. Subclinical REM sleep behavior disorder in a patient with corticobasal degeneration. Sleep 1997; 20: 891–4. doi: 10.1093/sleep/20.10.891.

Kinsbourne, M. and Cook, J. Generalized and lateralized effects of concurrent verbalization on a unimanual skill. The Quarterly Journal of Experimental Psychology 1971; 23: 341–345.

Kinsbourne, M. and Hicks, R.E. Functional cerebral space: a model for overflow, transfer, and interference effects in human performance. Attention and Performance 1978; 7; 45–362.

Kinsbourne, M. Unity and diversity in the human brain: evidence from injury. Daedalus 1998; 127: 233–256.

Kircher, T.T., Brammer, M.J., Levelt, W., Bartels, M. and McGuire, P.K. Pausing for thought: engagement of left temporal cortex during pauses in speech. Neuroimage 2004; 21: 84–90. doi: 10.1016/j.neuroimage.2003.09.041.

Kircher, T.T., Brammer, M.J., Williams, S.C. and McGuire, P.K. Lexical retrieval during fluent speech production: an fMRI study. Neuroreport 2000; 11: 4093–4096. doi: 10.1097/00001756–200012180–00036.

Knecht, S., Dräger, B., Deppe, M., Bobe, L., Lohmann, H., Flöel, A. et al. Handedness and hemispheric language dominance in healthy humans. Brain 2000; 123: 2512–2518. doi: 10.1093/brain/123.12.2512.

Knoch, D., Gianotti, L.R., Mohr, C. and Brugger, P. Synesthesia: When colors count. Brain Res Cogn Brain Res 2005 Sep; 25(1): 372–4. doi: 10.1016/j.cogbrainres.2005.05.005.

Kohnstamm, O. Demonstration einer katatonieartigen Erscheinung beim Gesunden (Katatonusversuch). Neurologisches Centralblatt 1915; 34: 290–291.

Kolb, B. and Whishaw, I.Q. Fundamentals of Human Neuropsychology, 6th edition. Worth Publishers: New York, 2008.

Kong, F. Space-valence associations depend on handedness: evidence from a bimanual output task. Psychol Res 2013; 77: 773–779. doi: 10.1007/s00426–012–0471–7.

Kopelman, M.D. Focal retrograde amnesia and the attribution of causality: An exceptionally critical view. Cogn Neuropsychol. 2000; 17: 585–621. doi: 10.1080/026432900750002172.

Kopelman, M.D., Bright, P., Fulker, H., Hinton, N., Morrison, A. and Verfaellie, M. Remote semantic memory in patients with Korsakoff's syndrome and herpes encephalitis. Neuropsychology 2009 Mar; 23(2): 144-57. doi: 10.1037/a0014447.

Kortte, K.B., McWhorter, J.W., Pawlak, M.A., Slentz, J., Sur, S. and Hillis, A.E. Anosognosia for hemiplegia: The contributory role of right inferior frontal gyrus. Neuropsychology 2015; 29: 421–32. doi: 10.1037/neu0000135.

Kosaka, B., Hiscock, M., Strauss, E., Wada, J.A. and Purves, S. Dual task performance by patients with left or right speech dominance as determined by carotid amytal tests. Neuropsychologia 1993; 31: 127–136. doi: 10.1016/0028–3932(93)90041–w.

Kosslyn, S.M. and Miller, G.W. Top Brain, Bottom Brain. Simon & Schuster: New York, 2013.

Kramer, P. and Bressan, P. Humans as superorganisms: How microbes, viruses, imprinted genes, and other selfish entities shape our behavior. Perspect Psychol Sci. 2015 Jul; 10(4): 464-81. doi: 10.1177/1745691615583131.

Kramer, P. Icono: a universal language that shows what it says. Front Psychol 2023; 14: 1149381. doi: 10.3389/fpsyg.2023.1149381.

Krummenacher, P., Candia, V., Folkers, G., Schedlowski, M. and Schönbächler, G. Prefrontal cortex modulates placebo analgesia. Pain 2010a; 148: 368–374. doi: 10.1016/j.pain.2009. 09.033. Epub 2009 Oct 28.

Krummenacher, P., Mohr, C., Haker, H. and Brugger, P. Dopamine, paranormal belief, and the detection of meaningful stimuli. J Cogn Neurosci 2010b; 22: 1670–81. doi: 10.1162/jocn.2009.21313.

Kumar, M., Singh, S., Modi, S., Rana, P., D'souza, M., Sekhri, T. et al. Cortical gray matter thickness and volume changes and their association with memory functions in hyperthyroid patients. Neuroendocrinology 2024; 114(4): 348–355. doi: 10.1159/000536027

Lackner, J.R. Some proprioceptive influences on the perceptual representation of body shape and orientation. Brain 1988; 111: 281–97. doi: 10.1093/brain/111.2.281.

Làdavas, E., Tosatto, L. and Bertini, C. Behavioural and functional changes in neglect after multisensory stimulation. Neuropsychol Rehabil 2022; 32: 662–689. doi: 10.1080/09602011.2020.1786411.

Làdavas, E., Zeloni, G. and Farnè, A. Visual peripersonal space centred on the face in humans. Brain 1998 Dec; 121 (Pt 12): 2317–26. doi: 10.1093/brain/121.12.2317.

Lai, C.S., Fisher, S.E., Hurst, J.A., Levy, E.R., Hodgson, S., Fox, M. et al. The SPCH1 region on human 7q31: genomic characterization of the critical interval and localization of translocations associated with speech and language disorder. Am J Hum Genet 2000; 67: 357–68. doi: 10.1086/303011.

Lalonde, R. The neurobiological basis of spontaneous alternation. Neuroscience and Biobehavioral Reviews 2002; 26: 91e104. https: //doi.org/10.1016/S0149–7634(01)00041–0.

Landis, T., Cummings, J.L., Benson, D.F. and Palmer, E.P. Loss of topographic familiarity. An environmental agnosia. Arch Neurol. 1986 Feb; 43(2): 132–6. doi: 10.1001/archneur.1986.00520020026011.

Landis, T., Graves, R., Benson, F. and Hebben, N. Visual recognition through kinaesthetic mediation. Psychological Medicine 1982; 12: 515–531.

Landis, T., Regard, M., Bliestle, A. and Kleihues, P. Prosopagnosia and agnosia for noncanonical views. An autopsied case. Brain 1988; 111: 1287–1297. doi: 10.1093/brain/111.6.1287.

Landis, T., Regard, M., Graves, R. and Goodglass, H. Semantic paralexia: a release of right hemispheric function from left hemispheric control. Neuropsychologia 1983; 21: 359–264. doi: 10.1016/0028–3932(83)90022–2.

Lane, R.D., Weihs, K.L., Herring, A., Hishaw, A. and Smith, R. Affective agnosia: Expansion of the alexithymia construct and a new opportunity to integrate and extend Freud's legacy. Neurosci Biobehav Rev 2015; 55: 594–611. doi: 10.1016/j.neubiorev.2015.06.007.

Langer, N., Weber, M., Hebling Vieira, B., Strzelczyk, D., Wolf, L., Pedroni, A. et al. A deep learning approach for automated scoring of the Rey-Osterrieth complex figure. Elife 2024; 13: RP96017. doi: 10.7554/eLife.96017.

Lecourtier, L., De Vasconcelos, A.P., Leroux, E., Cosquer, B., Geiger, K., Lithfous, S. et al. Septohippocampal pathways contribute to system consolidation of a spatial memory: sequential implication of GABAergic and cholinergic neurons. Hippocampus 2011; 21: 1277–1289. doi: 10.1002/hipo.20837.

Lee, H.J., Lee, J., Kim, C.J., Kim, G.J., Kim, E.S. and Whang, M. Brain process for perception of the "out of the body" tactile illusion for virtual object interaction. Sensors (Basel) 2015; 15: 7913–7932. doi: 10.3390/s150407913.

Lefeuvre, M., Gouat, P., Mulot, B., Cornette, R. and Pouydebat, E. Analogous laterality in trunk movements in captive African elephants: A pilot study. Laterality 2022; 27: 101–126. DOI: 10.1080/1357650X.2021.1999253.

Lehrer, D.S. and Lorenz, J. Anosognosia in schizophrenia: hidden in plain sight. Innov Clin Neurosci 2014; 11: 10–17.

Leiguarda, R.C. and Marsden, C.D. Limb apraxias: higher-order disorders of sensorimotor integration. Brain 2000; 123: 860–79. doi: 10.1093/brain/123.5.860.

Lem, S. Summa Technologiae. University of Minnesota Press: Minneapolis, 2013. (Polish original 1964).

Lenggenhager, B., Busch, C. and Brugger, P. Minding gaps on the skin: Opposite bisection biases on forehead and back of one's head. Conscious Cog 2016; 42: 9–14. doi: 10.1016/j.concog.2016.03.001.

Lenggenhager, B., Tadi, T., Metzinger, T. and Blanke, O. Video ergo sum: manipulating bodily self-consciousness. Science 2007; 317: 1096–1099. doi: 10.1126/science.1143439.

León Méndez, M., Padrón, I., Fumero, A. and Marrero, R.J. Effects of internet and smartphone addiction on cognitive control in adolescents and young adults: A systematic review of fMRI studies. Neurosci Biobehav Rev 2024; 159: 105572. doi: 10.1016/j.neubiorev.2024.105572.

Leonhard, D. and Brugger, P. Creative, paranormal, and delusional thought: a consequence of right hemisphere semantic activation? Neuropsychiatry Neuropsychol Behav Neurol 1998; 11: 177–83.

Leuenberger, K., Hofmann, R., Brugger, P. and Gassert, R. Measurement of human rotation behavior for psychological and neuropsychological investigations. Behav Res Methods 2015; 47: 1425–1435. doi: 10.3758/s13428–014–0554–z.

Leuthard, J., Bächtold, D. and Brugger, P. Is "left" always where the thumb is right?: Stimulus-response compatibilities as a function of posture and location of the responding hand. Cogn Behav Neurol 2005 Sep; 18(3): 173–8.

Levin, M. Life, death, and self: Fundamental questions of primitive cognition viewed through the lens of body plasticity and synthetic organisms. 2021; 564: 114–133.

Levine, B. Autobiographical memory and the self in time: brain lesion effects, functional neuroanatomy, and lifespan development. Brain Cogn 2004; 55(1): 54–68. doi: 10.1016/S0278–2626(03)00280–X.

Levin, B. and Craik, F. (eds.). Mind and the Frontal Lobes. Cognition, Behavior, and Brain Imaging. Oxford University Press: Oxford, 2011.

Levine, D.A., Whitney, R.T., Ye, W., Briceño, E.M., Gross, A.L., Giordani, B.J. et al. Associations between stroke type, ischemic stroke subtypes, and poststroke cognitive trajectories. Stroke 2025; 56: 898–907. doi: 10.1161/STROKEAHA.124.047640.

Levy, D.L., Holzman, P.S. and Proctor, L.R. Vestibular dysfunction and psychopathology. Schizophr Bull. 1983; 9(3): 383–438. doi: 10.1093/schbul/9.3.383

Levy, Y. Modularity of language reconsidered. Brain Lang. 1996; 55: 240–263. doi: 10.1006/brln.1996.0103.

Lewin, R. Is your brain really necessary? Science 1980; 210: 1232–1234. DOI: 10.1126/science.7434023.

Lezak, M.D., Howieson, D.B. and Loring, D.W. Neuropsychological Assessment (4th edition). Oxford University Press: New York, 2004.

Lhermitte, F. Human autonomy and the frontal lobes. Part II: Patient behavior in complex and social situations: the "environmental dependency syndrome". Ann Neurol 1986; 19: 335–43. doi: 10.1002/ana.410190405.

Lhermitte, J. Les visions d'un poète. Revue d'oto-neuro-ophtalmologie 1939; 17: 81–96.
Lhermitte, J. Syndrome de la calotte du "pédoncule cérébral. Les troubles psycho-sensoriels dans les lésions du mésocéphale. Rev Neurol 1922; 38: 1359–1365. https: //ci.nii.ac.jp/naid/10012678114.
Liederman, J., Kohn, S., Wolf, M. and Goodglass, H. Lexical creativity during instances of word-finding difficulty: Broca's vs. Wernicke's aphasia. Brain Lang 1983; 20: 21–32. doi: 10.1016/0093–934x(83)90029–9.
Liepmann, H. and Maas, O. Fall von linksseitiger Agraphie und Apraxie bei rechtsseitiger Lähmung. J Psychol Neurol 1907; 10: 214–227.
Lindell, A.K. In your right mind: right hemisphere contributions to language processing and production. Neuropsychol Rev 2006 Sep; 16(3): 131–48. doi: 10.1007/s11065–006–9011–9.
Lisotto, C., Mainardi, F., Maggioni, F. and Zanchin, G. Did Picasso and De Chirico really suffer from migraine auras? J Headache Pain 2015; 16(Suppl 1): A192. doi: 10.1186/1129–2377–163S1–A192.
Lissauer, H. Ein Fall von Seelenblindheit nebst einem Beitrag zur Theorie derselben. Arch Psychiatrie Nervenkr 1890; 21: 222–270.
Litch, J.A. and Bishop, R.A. High-altitude global amnesia. Wilderness Environ Med 2000; 11: 25–8. doi: 10.1580/1080–6032(2000)011[0025: haga]2.3.co; 2.
Liu, Y.F., Rapp, B. and Bedny, M. Reading Braille by touch recruits posterior parietal cortex. J Cogn Neurosci 2023; 35: 1593–1616. doi: 10.1162/jocn_a_02041.
Livio, M. The Golden Ratio. The Story of Phi, the World's Most Atonishing Number. Broadway Books: New York, 2002.
Llaurens, V., Raymond, M. and Faurie, C. Why are some people left-handed? An evolutionary perspective. Philos Trans R Soc Lond B Biol Sci. 2009 Apr 12; 364(1519): 881–94. doi: 10.1098/rstb.2008.0235.
Lobmaier, J.S. and Mast, F.W. The Thatcher illusion: rotating the viewer instead of the picture. Perception 2007; 36: 537–46. doi: 10.1068/p5508.
Loetscher, T., Regard, M. and Brugger, P. Misoplegia: a review of the literature and a case without hemiplegia. J Neurol Neurosurg Psychiatry 2006 Sep; 77(9): 1099–100. doi: 10.1136/jnnp.2005.087163.
Loetscher, T. and Brugger, P. Exploring number space by random digit generation. Experimental Brain Research 2007a; 180: 655–665.
Loetscher, T. and Brugger, P. A disengagement deficit in representational space. Neuropsychologia 2007b; 45: 1299–1304.
Loetscher, T., Bockisch, C.J. and Brugger, P. Looking for the answer: the mind's eye in number space. Neuroscience 2008; 151: 725–9. doi: 10.1016/j.neuroscience.2007.07.068.
Loetscher, T., Tamagni, C., Baumann, H., Nicholls, M.E. and Brugger, P. Impaired memory for early relative to late events in spatial neglect. Cognitive Processing 2012; 13: 62.
Loffing, F. Left-handedness and time pressure in elite interactive ball games. Biol Lett. 2017 Nov; 13(11): 20170446. doi: 10.1098/rsbl.2017.0446.
Loftus, E.F. Eyewitness Testimony. Harvard University Press: Cambridge, MA, 1979.
Logothetis, N.K. and Pauls, J. Psychophysical and physiological evidence for viewer-centered object representations in the primate. Cerebral Cortex 1995; 5: 270–288.
Logue, S.F. and Gould, T.J. The neural and genetic basis of executive function: attention, cognitive flexibility, and response inhibition. Pharmacol Biochem Behav 2014 Aug; 123: 45–54. doi: 10.1016/j.pbb.2013.08.007
Lokhorst, G.J.C. Classics in neurology. The oldest printed text on hemispheric specialization. Neurology 1982; 32: 762. doi: 10.1212/wnl.32.7.762.
Lokhorst, G.J.C. The first theory about hemispheric specialization: fresh light on an old codex. J Hist Medicine and Allied Sciences 1996; 51: 293–312. doi: 10.1093/jhmas/51.3.293.
Longcamp, M., Hlushchuk, Y. and Hari, R. What differs in visual recognition of handwritten vs. printed letters? An fMRI study. Hum Brain Mapp 2011; 32: 1250–1259. doi: 10.1002/hbm.21105.
Longo, M.R., Azañón, E. and Haggard, P. More than skin deep: body representation beyond primary somatosensory cortex. Neuropsychologia 2010; 48: 655–68. doi: 10.1016/j.neuropsychologia.2009.08.022.
Longo, M.R., Pernigo, S. and Haggard, P. Vision of the body modulates processing in primary somatosensory cortex. Neurosci Lett 2011; 489: 159–63. doi: 10.1016/j.neulet.2010.12.007.

Longo, M.R. Distortion of mental body representations. Trends Cogn. Sci., 2022 Mar; 26(3): 241–254. doi: 10.1016/j.tics.2021.11.005.
Looi, J.C.L. and Anderson, K. COVID-19 isolated-academic logorrhoea: an emergent debilitating disorder afflicting medical academics. Australasian Psychiatry 2021; 29: 104–105. doi: 10.1177/1039856220968397.
Lopez, C. The vestibular system: balancing more than just the body. Curr Opin Neurol. 2016; 29: 74–83. doi: 10.1097/WCO.0000000000000286.
Loriga, L. Body Integrity Dysphoria and the Ethical Dilemma of On-Demand Amputation. Redefining Wholeness: Identity, Autonomy, and the Moral Boundaries of the Human Body. Springer: New York, 2025. https://doi.org/10.1007/978-3-031-82443-2.
Luauté, J., Villeneuve, L., Roux, A., Nash, S., Bar, J.Y., Chabanat, E. et al. Adding methylphenidate to prism-adaptation improves outcome in neglect patients. A randomized clinical trial. Cortex. 2018; 106: 288–298. doi: 10.1016/j.cortex.2018.03.028.
Lubben, N., Ensink, E., Coetzee, G.A. and Labrie, V. The enigma and implications of brain hemispheric asymmetry in neurodegenerative diseases. Brain Commun 2021; 3: fcab211.
Luhrmann, T.M., Chen, X.A., Baumeister, D. and Peters, E. When spirit calls: A phenomenological approach to healthy voice-hearers. Schizophr Bull Open 2023; 4: sgad025. doi: 10.1093/schizbullopen/sgad025.
Luo, Q., Zhang, J.X., Huang, S., Hu, Y.H., Wang, H. and Chen, X. Effects of long-term exposure to high altitude on brain structure in healthy people: an MRI-based systematic review and meta-analysis. Front Psychiatry 2023; 14: 1196113. doi: 10.3389/fpsyt.2023.1196113.
Luria, A.R. The Mind of a Mnemonist. Basic Books: New York, 1968 (Russian original 1968).
Luria, A.R. The Man with a Shattered World. Basic Books: New York, 1972 (Russian original 1971).
Luria, A.R. Higher Cortical Functions in Man. Basic Books: New York, 1980 (Russian original 1962).
Luria, A.R. The Mind of a Mnemonist. A little book about a vast memory. Harvard University Press, 1987 (originally published in 1965).
Maass, A. and Russo, A. Directional bias in the mental representation of spatial events: nature or culture? Psychol Sci 2003; 14: 296–301. doi: 10.1111/1467–9280.14421.
Macknik, S.L., King, M., Randi, J., Robbins, A., Teller, Thompson, J. et al. Attention and awareness in stage magic: turning tricks into research. Nat Rev Neurosci 2008; 9: 871–879. doi: 10.1038/nrn2473.
Macmillan, M. Nineteenth-century inhibitory theories of thinking: Bain, Ferrier, Freud (and Phineas Gage). Hist Psychol 2000; 3: 187–217. doi: 10.1037/1093–4510.3.3.187.
MacPherson, S.E. and Della Sala, S. (eds.). Handbook of Frontal Lobe Assessment. Oxford University Press: Oxford, 2015.
Magrí, R. and Mocchetti, E. Asomatoscopia (autoscopia negativa) in una epilettica. Arch Psicol Neurol Psichiatr 1967; 28: 572–85.
Maher, C. Plant Minds. A Philosophical Defense. Routledge: London, 2017.
Mailend, M.L. and Maas, E. To lump or to split? possible subtypes of apraxia of speech. Aphasiology 2020; 35(4): 592–613. doi: 10.1080/02687038.2020.1836319.
Makin, T.R., Scholz, J., Filippini, N., Henderson Slater, D., Tracey, I. and Johansen-Berg, H. Phantom pain is associated with preserved structure and function in the former hand area. Nature Communications 2013; 4: 1570.
Makin, T.R. and Flor, H. Brain (re)organisation following amputation: Implications for phantom limb pain. Neuroimage 2020 Sep; 218: 116943. doi: 10.1016/j.neuroimage.2020.116943.
Makin, T.R. and Krakauer, J.W. Against cortical reorganization. eLife 2023; 12: e84716. DOI: https: //doi.org/10.7554/eLife.84716
Makin, T.R., Scholz, J., Henderson Slater, D., Johansen-Berg, H. and Tracey, I. Reassessing cortical reorganization in the primary sensorimotor cortex following arm amputation. Brain 2015; 138: 2140–2146. Doi: 10.1093/brain/awv161.
Maldonado, I.L., Parente de Matos, V., Castro Cuesta, T.A., Herbet, G. and Destrieux, C. The human cingulum: From the limbic tract to the connectionist paradigm. Neuropsychologia 2020; 144: 107487. doi: 10.1016/j.neuropsychologia.2020.107487
Mamiya, Y., Nishio, Y., Watanabe, H., Yokoi, K., Uchiyama, M., Baba, T. et al. The pareidolia test: A simple neuropsychological test measuring visual hallucination-like illusions. PLoS One 2016; 11: e0154713. doi: 10.1371/journal.pone.0154713.

Manfredi, M., Bini, G., Cruccu, G., Accornero, N., Berardelli, A. and Medolago, L. Congenital absence of pain. Archives of Neurology 1981; 38: 507–511. doi: 10.1001/archneur.1981.00510080069010.

Marchetti, C. and Della Sala, S. Disentangling the alien and anarchic hand. Cognitive Neuropsychiatry 2010; 3: 191–207. Doi: 10.1080/135468098396143.

Mark, V.W., Kooistra, C.A. and Heilman, K.M. Hemispatial neglect affected by non-neglected stimuli. Neurology 1988; 38: 1207–1211.

Markowitsch, H.J., Kalbe, E., Kessler, J., von Stockhausen, H.M., Ghaemi, M. and Heiss, W.D. Short-term memory deficit after focal parietal damage. J Clin Exp Neuropsychol 1999; 21: 784–97. doi: 10.1076/jcen.21.6.784.853.

Markowitsch, H.J. Transient psychogenic amnesia. Ital J Neurol Sci 1988; Suppl 9: 49–51.

Marks, C.E. Commissurotomy, Consciousness, and Unity of Mind. Bradford Books: Montgomery (Vermont), 1980.

Marshall, J., Atkinson, J., Smulovitch, E., Thacker, A. and Woll, B. Aphasia in a user of British sign language: Dissociation between sign and gesture. Cogn Neuropsychol 2004; 21: 537–54. doi: 10.1080/02643290342000249.

Marshall, J.C. and Halligan, P.W. Blindsight and insight in visuo-spatial neglect. Nature 1988; 336: 766–7. doi: 10.1038/336766a0.

Martin, A., Wiggs, C.L., Ungerleider, L.G. and Haxby, J.V. Neural correlates of category-specific knowledge. Nature 1996; 379: 649–52. doi: 10.1038/379649a0.

Martin, M. and Jones, G.V. Motor imagery theory of a contralateral handedness effect in recognition memory: Toward a chiral psychology of cognition. Journal of Experimental Psychology: General 1999a; 128: 265–282.

Marshall, J. Aphasia. The Basics. Taylor & Francis: New York, 2024.

Martin, M. and Jones, G.V. Hale-Bopp and handedness: Individual differences in memory for orientation. Psychological Sci 1999b; 10: 267–270. DOI: 10.1111/1467–9280.00148.

Mateen, F.J., Josephs, K.A., Trenerry, M.R., Felmlee-Devine, M.D., Weaver, A.L., Carone, M. et al. Long-term cognitive outcomes following out-of-hospital cardiac arrest: a population-based study. Neurology 2011; 77: 1438–45. doi: 10.1212/WNL.0b013e318232ab33.

Mathews, M.S., Linskey, M.E., Binder, D.K. and William, P. Van Wagenen and the first corpus callosotomies for epilepsy. J Neurosurg 2008; 108: 608–613. doi: 10.3171/JNS/2008/108/3/0608.

Matias-Guiu, J.A., Pytel, V., Delgado-Álvarez, A., Delgado-Alonso, C., Cortés-Martínez, A., Fernández-Oliveira, A. et al. The Five-Point Test: Normative data for middle-aged and elderly Spaniards. Appl Neuropsychol Adult 2022; 29: 1323–1331. doi: 10.1080/23279095.2021.1873137.

Maunder, L., Schoemaker, D. and Pruessner, J.C. Frequency of penile-vaginal intercourse is associated with verbal recognition performance in adult women. Arch Sex Behav 2017; 46: 441–453. doi: 10.1007/s10508–016–0890–4.

Maximov, K. Epilepsie occipitale avec hallucinations héautoscopiques. Acta Neurol Belg. 1973; 73: 320–323.

McCarthy, R.A. and Beaumont, J.G. Analysis of sequential letter matching tasks utilising lateralised tachistoscopic presentation. Cortex 1983; 19: 79–98. DOI: 10.1016/S0010–9452(83)80052–5

McConnell, J.V., Jacobson, A.L. and Kimble, D.P. The effects of regeneration upon retention of a conditioned response in the planarian. J Comp Physiol Psychol 1959; 52: 1–5. doi: 10.1037/h0048028.

McConnell, J.V. Memory transfer through cannibalism in planaria. J Neuro-Psychiatry 1962; 3: 45.

McGeoch, P.D., Brang, D., Song, T., Lee, R.R., Huang, M. and Ramachandran, V.S. Xenomelia: a new right parietal lobe syndrome. J Neurol Neurosurg Psychiatry 2011; 82: 1314–9. doi: 10.1136/jnnp-2011–300224.

McGeown, W.J., Mazzoni, G., Vannucci, M. and Venneri, A. Structural and functional correlates of hypnotic depth and suggestibility. Psychiatry Res 2015; 231: 151–159. doi: 10.1016/j.pscychresns.2014.11.015.

McGilchrist, I. The Master and his Emissary: The Divided Brain and the Making of the Western World. Yale University Press: New Haven (CT), 2009.

McGilchrist, I. Reciprocal organization of the cerebral hemispheres. Dialogues Clin Neurosci 2010; 12: 503–15. doi: 10.31887/DCNS.2010.12.4/imcgilchrist.

McGonigle, D.J., Hänninen, R., Salenius, S., Hari, R., Frackowiak, R.S. and Frith, C.D. Whose arm is it anyway? An fMRI case study of supernumerary phantom limb. Brain 2002; 125: 1265–1274. doi: 10.1093/brain/awf139.

McGorry, P.D., Campbell, R. and Copolov, D.L. The Zelig phenomenon: a specific form of identity disturbance. Aust N Z J Psychiatry 1987; 21: 532–538. doi: 10.3109/00048678709158921.

McHugh, T.P. and Stout, M. Near-hanging injury. Ann Emerg Med. 1983; 12: 774–776. doi: 10.1016/s0196–0644(83)80256–x.

McKay, R., Tamagni, C., Palla, A., Krummenacher, P., Hegemann, S.C., Straumann, D. et al. Vestibular stimulation attenuates unrealistic optimism. Cortex 2013; 49: 2272–2275. doi: 10.1016/j.cortex.2013.04.005.

McManus, C. Right Hand, Left Hand. The Origins of Asymmetry in Brains, Bodies, Atoms and Culture. Harvard University Press, Cambridge (MA), 2002.

McManus, I.C. Right hand, left hand. The Origins of Asymmetry in Brains, Bodies, Atoms and Cultures. Weidenfeld and Nicolson: London, 2002.

McManus, I.C., Martin, N., Stubbings, G.F., Chung, E.M. and Mitchison, H.M. Handedness and situs inversus in primary ciliary dyskinesia. Proc Royal Soc Biol Sci 2004; 271: 2579–2582. doi: 10.1098/rspb.2004.2881.

McManus, I.C. The history and geography of human handedness. *In*: Sommer, I.C. and Kahn, R.S. (eds.). Language Lateralization and Psychosis. Cambridge University Press: Cambridge, 2009 (37–58). DOI 10.1017/CBO9780511576744.004.

McManus, I.C. The inheritance of left-handedness. *In*: Bock, G.R. and Marsh, J. (eds.). Biological Asymmetry and Handedness. Wiley; Chichester, UK, 1991, 251–281.

Meador, K.J., Allen, M.E., Adams, R.J. and Loring, D.W. Allochiria vs allesthesia. Is there a misperception? Arch Neurol. 1991; 48: 546–549. doi: 10.1001/archneur.1991.00530170110029.

Medina, J., Kannan, V., Pawlak, M.A., Kleinman, J.T., Newhart, M., Davis, C. et al. Neural substrates of visuospatial processing in distinct reference frames: evidence from unilateral spatial neglect. J Cogn Neurosci 2009; 21: 2073–2084. doi: 10.1162/jocn.2008.21160.

Medland, S.E., Geffen, G. and McFarland, K. Lateralization of speech production using verbal/manual dual tasks: meta-analysis of sex differences and practice effects. Neuropsychologia 2002; 40: 1233–1239. doi: 10.1016/s0028–3932(01)00228–7.

Medland, S.E., Duffy, D.L., Wright, M.J., Geffen, G.M., Hay, D.A., Levy, F. et al. Genetic influences on handedness: data from 25,732 Australian and Dutch twin families. Neuropsychologia 2009; 47: 330–337. doi: 10.1016/j.neuropsychologia.2008.09.005.

Mehling, W.E., Price, C., Daubenmier, J.J., Acree, M., Bartmess, E. and Stewart, A. The Multidimensional Assessment of Interoceptive Awareness (MAIA). PLoS One 2012; 7(11): e48230. doi: 10.1371/journal.pone.0048230.

Meichtry, J.R., Cazzoli, D., Chaves, S., von Arx, S., Pflugshaupt, T., Kalla, R. et al. Pure optic ataxia and visual hemiagnosia - extending the dual visual hypothesis. J Neuropsychol 2018; 12: 271–290. doi: 10.1111/jnp.12119.

Meister, I.G., Wilson, S.M., Deblieck, C., Wu, A.D. and Iacoboni, M. The essential role of premotor cortex in speech perception. Curr Biol 2007 Oct 9; 17(19): 1692–6. doi: 10.1016/j.cub.2007.08.064.

Melzack, R. and Bromage, P.R. Experimental phantom limbs. Exp Neurol 1973; 39: 261–9. doi: 10.1016/0014–4886(73)90228–8.

Melzack, R., Israel, R., Lacroix, R. and Schultz, G. Phantom limbs in people with congenital limb deficiency or amputation in early childhood. Brain 1997; 120: 1603–20. doi: 10.1093/brain/120.9.1603.

Menninger-Lerchenthal, E. Das Truggebilder der eigenen Gestalt. Karger: Berlin, 1935.

Menninger-Lerchenthal, E. Der eigene Doppelgänger. Huber: Bern, 1946.

Merzenich, M.M., Nelson, R.J., Stryker, M.P., Cynader, M.S., Schoppmann, A. and Zook, J.M. Somatosensory cortical map changes following digit amputation in adult monkeys. J Comp Neurol 1984; 224: 591–605. doi: 10.1002/cne.902240408.

Mesulam, M. (1981). A cortical network for directed attention and unilateral neglect. Annals of Neurology, 10(4): 309–325. doi: 10.1002/(ISSN)15318249.

Mesulam, M.M. Principles of Behavioural and Cognitive Neurology. Oxford University Press: Oxford, 2000, p. 332.

Mevorach, C., Humphreys, G.W. and Shalev, L. Attending to local form while ignoring global aspects depends on handedness: evidence from TMS. Nat Neurosci 2005; 8: 276–7. doi: 10.1038/nn1400.
Meyer, J., Magnasco, M.O. and Reiss, D. The relevance of human whistled languages for the analysis and decoding of dolphin communication. Frontiers In Psychology 2021; 12: article 689501.
Miedel, C.J., Patton, J.M., Miedel, A.N., Miedel, E.S. and Levenson, J.M. Assessment of spontaneous alternation, novel object recognition and limb clasping in transgenic mouse models of amyloid-β and tau neuropathology. J Vis Exp 2017; 123: 55523. doi: 10.3791/55523.
Miller, G.A. The magical number seven plus or minus two: some limits on our capacity for processing information. Psychol Rev. 1956; 63: 81–97.
Miller, G.A., Galanter, E. and Pribram, K.H. Plans and the Structure of Behavior. Holt, Rinehart and Winston: New York, 1960.
Millonig, A., Bodner, T., Donnemiller, E., Wolf, E. and Unterberger, I. Supernumerary phantom limb as a rare symptom of epileptic seizures - case report and literature review. Epilepsia 2011; 52: e97–e100. doi: 10.1111/j.1528–1167.2011.03156.x.
Milner, B. Some effects of frontal lobectomy in man. *In*: Warren, J.M. and Akert, K. (eds.). The Frontal Granular Cortex and Behavior. McGraw-Hill: New York, 1964.
Mitchell, A.J. A meta-analysis of the accuracy of the mini-mental state examination in the detection of dementia and mild cognitive impairment. J Psychiatr Res 2009; 43: 411–31. doi: 10.1016/j.jpsychires.2008.04.014.
Mitchell, K.J. and Johnson, M.K. Source monitoring 15 years later: what have we learned from fMRI about the neural mechanisms of source memory? Psychol Bull 2009; 135: 638–77. doi: 10.1037/a0015849.
Miyake, A., Friedman, N.P., Emerson, M.J., Witzki, A.H., Howerter, A. and Wager, T.D. The unity and diversity of executive functions and their contributions to complex "Frontal Lobe" tasks: a latent variable analysis. Cogn Psychol 2000 Aug; 41(1): 49–100. doi: 10.1006/cogp.1999.0734.
Miyashita, Y. Perirhinal circuits for memory processing. Nat Rev Neurosci. 2019; 20: 577–592. doi: 10.1038/s41583–019–0213–6.
Mohr, C., Bracha, H.S. and Brugger, P. Magical ideation modulates spatial behavior. J Neuropsychiatry Clin Neurosci 2003a; 15: 168–74. doi: 10.1176/jnp.15.2.168.
Mohr, C., Landis, T., Bracha, H.S. and Brugger, P. Opposite turning behavior in right-handers and non-right-handers suggests a link between handedness and cerebral dopamine asymmetries. Behavioral Neuroscience 2003b; 117: 1448–1452. Doi: 10.1037/0735–7044.117.6.1448.
Mohr, C., Thut, G., Landis, T. and Brugger, P. Hands, arms, and minds: interactions between posture and thought. J Clin Exp Neuropsychology 2003c; 25: 1000–1010. Doi: 10.1076/jcen.25.7.1000.16491.
Mole, J.A., Tunnard, C., Wilson, B.A. and Gainotti, G. Selective hyperfamiliarity for voices. Cortex 2021; 136: 147–149. doi: 10.1016/j.cortex.2020.12.006.
Monsch, A.U., Bondi, M.W., Butters, N., Paulsen, J.S., Salmon, D.P., Brugger, P. et al. A comparison of category and letter fluency in Alzheimer's disease and Huntington's disease. Neuropsychology 1994; 8: 25–30.
Moon, J.H. Endocrine risk factors for cognitive impairment. Endocrinol Metab (Seoul) 2016; 31: 185–192. doi: 10.3803/EnM.2016.31.2.185.
Mooney, C.M. Closure with negative after-images under flickering light. Can J Psychol. 1956; 10: 191–9. doi: 10.1037/h0083671.
Morais, L.H., Schreiber, H.L. and Mazmanian, S.K. The gut microbiota–brain axis in behaviour and brain disorders. Nat Rev Microbiol 2021; 19: 241–255.
Moreno-Verdú, M., McAteer, S.M., Waltzing, B.M., Van Caenegem, E.E. and Hardwick, R.M. Development and validation of an open-source hand laterality judgement task for in-person and online studies. Neuroscience 2025; 572: 93–107. doi: 10.1016/j.neuroscience.2025.02.056.
Morin, C. Stroke, Body Image, and Self-Representation. Psychoanalytic and Neurological Perspectives. Taylor and Francis: New York, 2017.
Morris, J.S., DeGelder, B., Weiskrantz, L. and Dolan, R.J. Differential extrageniculostriate and amygdala responses to presentation of emotional faces in a cortically blind field. Brain 2001; 124: 1241–1252. doi: 10.1093/brain/124.6.1241.
Morrison, I., Löken, L. and Olausson, H. The skin as a social organ. Experimental Brain Research 2010; 204: 305–314. doi: 10.1007/s00221–009–2007–y.

Moscovitch, M. and Behrmann, M. Coding of spatial information in the somatosensory system: Evidence from patients with neglect following parietal lobe damage. J Cogn Neurosci 1994; 6: 151–155. doi: 10.1162/jocn.1994.6.2.151.

Motz, B.A., James, K.H. and Busey, T.A. The Lateralizer: a tool for students to explore the divided brain. Adv Physiol Educ 2012; 36: 220–225. doi: 10.1152/advan.00060.2012.

Moulin, C.J.A., Bell, N., Turunen, M., Baharin, A. and O'Connor, A.R. The induction of jamais vu in the laboratory: word alienation and semantic satiation. Memory 2021; 29: 933–942. doi: 10.1080/09658211.2020.1727519.

Mueller, T. and Kanis-Seyfried, U. On the life and work of Korbinian Brodmann (1868–1918). J Hist Neurosci 2019; 28: 307–318. doi: 10.1080/0964704X.2019.1589689.

Murdock, B.B. The serial position effect of free recall. Journal of Experimental Psychology 1962; 64: 482–488.

Myachykov, A., Chapman, A.J., Beal, J. and Scheepers, C. Random word generation reveals spatial encoding of syllabic word length. Br J Psychol 2020; 111: 357–368. doi: 10.1111/bjop.12399.

Nachev, P. and Husain, M. Comment on "Detecting awareness in the vegetative state". Science 2007; 315: 1221; author reply 1221. doi: 10.1126/science.1135096.

Nakamura, K., Dehaene, S., Jobert, A., Le Bihan, D. and Kouider, S. Subliminal convergence of Kanji and Kana words: further evidence for functional parcellation of the posterior temporal cortex in visual word perception. J Cogn Neurosci 2005; 17: 954–68. doi: 10.1162/0898929054021166.

Naqvi, N.H. and Bechara, A. The hidden island of addiction: the insula. Trends Neurosci 2009; 32: 56–67. doi: 10.1016/j.tins.2008.09.009.

Nasreddine, Z.S., Phillips, N.A., Bédirian, V., Charbonneau, S., Whitehead, V., Collin, I. et al. The Montreal Cognitive Assessment, MoCA: a brief screening tool for mild cognitive impairment. J Am Geriatr Soc. 2005; 53: 695–9. doi: 10.1111/j.1532–5415.2005.53221.x.

Naveh, T. and Arzy, S. The neuroanatomy of age perception. Behav Brain Res 2019; 372: 112052. doi: 10.1016/j.bbr.2019.112052.

Nelson, P.C. Old and new results about single-photon sensitivity in human vision. Phys Biol 2016; 13: 025001.

Neophytou K,, Wiley, R., Litovsky, C., Tsapkini, K. and Rapp, B. The right hemisphere's capacity for language: evidence from primary progressive aphasia. Cereb Cortex 2023; 33: 9971–9985. doi: 10.1093/cercor/bhad258.

Nersesjan, V., Martens, P., Truelsen, T. and Kondziella, D. After stroke, apraxia of eyelid opening is associated with high mortality and right hemispheric infarction. J. Neurol. Sci. 2020 Nov 15; 418: 117145. doi: 10.1016/j.jns.2020.117145

Neumann, L.M. A simple exercise for teaching mirror vision skills. J Dent Educ. 1988; 52: 170–172.

Nichelli, P., Venneri, A., Pentore, R. and Cubelli, R. Horizontal and vertical neglect dyslexia. Brain Lang 1993; 44: 264–83. doi: 10.1006/brln.1993.1018.

Nicholls, M.E.R., Orr, C.A., Okubo, M. and Loftus, A. Satisfaction guaranteed. The effect of spatial biases on responses to Likert scales. Psychological Science 2006; 17: 1027–1028.

Nigro, G. and Neisser, U. Point of view in personal memories. Cognitive Psychology 1983; 15: 467–482.

Nisbett, R.E. and Miyamoto, Y. The influence of culture: holistic versus analytic perception. Trends Cogn Sci 2005; 9: 467–73. doi: 10.1016/j.tics.2005.08.004.

Nisbett, R.E. and Wilson, T.D. Telling more than we can know: Verbal reports on mental processes. Psychological Review 1977; 84: 231–259.

Nix, K.C., Oh, A., Goad, B.S., Wu, W., Lucas, M.V. and Baumer, F.M. Detection of language lateralization using spectral analysis of EEG. J Clin Neurophysiol 2024; 41: 334–343. doi: 10.1097/WNP.0000000000000988.

Noll, S. and Kasten, E. Body Integrity Identity Disorder (BIID): How Satisfied are Successful Wannabes? Psychology and Behavioral Sciences 2014; 3(6): 222–232. DOI: 10.11648/j.pbs.20140306.17

Noorizadeh, N., Rezaie, R., Varner, J.A., Wheless, J.W., Fulton, S.P., Mudigoudar, B.D. et al. Concordance between Wada, transcranial magnetic stimulation, and magnetoencephalography for determining hemispheric dominance for language: A retrospective study. Brain Sci 2024; 14: 336–347. DOI: 10.3390/ brainsci14040336.

Nordmann, G.C., Hochstoeger, T. and Keays, D.A. Magnetoreception - a sense without a receptor. PLoS Biol. 2017 Oct 23; 15(10): e2003234. doi: 10.1371/journal.pbio.2003234.

Nyffeler, T., Vanbellingen, T., Kaufmann, B.C., Pflugshaupt, T., Bauer, D., Frey, J. et al. Theta burst stimulation in neglect after stroke: functional outcome and response variability origins. Brain 2019; 142: 992–1008. doi: 10.1093/brain/awz029.

Ocklenburg, S., Schmitz, J., Moinfar, Z., Moser, D., Klose, R., Lor, S. et al. Epigenetic regulation of lateralized fetal spinal gene expression underlies hemispheric asymmetries. Elife 2017 Feb 1; 6: e22784. doi: 10.7554/eLife.22784.

Ocklenburg, S. Tachistoscopic Viewing and Dichotic Listening. *In*: Rogers, L.J. and Vallortigara, G. (eds.). Lateralized Brain Functions (Neuromethods Vol. 217). Humana: New York, NY, 2025. Doi: 10.1007/978–1–0716–4240–5_1.

Ogden, J. Contralesional neglect of constructed visual images in right and left brain-damaged patients. Neuropsychologia 1985; 23: 273–277. doi: 10.1016/0028–3932(85)90112–5.

Ogden, J.A., Mee, E.W. and Henning, M. A prospective study of impairment of cognition and memory and recovery after subarachnoid hemorrhage. Neurosurgery 1993; 33: 572–86; discussion 586–7. doi: 10.1227/00006123–199310000–00004.

Ogle, W. On dextral pre-eminence. Medico-Chirurgical Transactions 1871; 54: 279–301. doi: 10.1177/095952877105400114.

Olausson, H., Wessberg, J., McGlone, F. and Vallbo, Å. The neurophysiology of unmyelinated tactile afferents. Neuroscience & Biobehavioral Reviews 2010; 34: 185–191.

Oldendorf, W.H. The quest for an image of brain: a brief historical and technical review of brain imaging techniques. Neurology 1978; 28: 517–533. DOI 10.1212/WNL.28.6.517.

Olsen, T.S., Bruhn, P. and Oberg, R.G. Cortical hypoperfusion as a possible cause of "subcortical aphasia". Brain 1986; 109: 393–410. doi: 10.1093/brain/109.3.393.

Oosterveer, D.M., Wermer, M.J.H., Volker, G. and Vlieland, T. Are there differences in long-term functioning and recovery between hemorrhagic and ischemic stroke patients receiving rehabilitation? J Stroke Cerebrovasc Dis 2022; 31: 106294. doi: 10.1016/j.jstrokecerebrovasdis.2021.106294.

Opdam, H. and Radford, S. Understanding the organ donation conversation landscape in the UK. Anaesthesia 2021; 76: 1567–1571. doi: 10.1111/anae.15497.

Oppenheimer, J.M. Studies of brain asymmetry: Historical perspective. Annals of the New York Academy of Sciences 1977; 299: 4–17. DOI: 10.1111/j.1749–6632.1977.tb41892.x.

Ornstein, R. The Right Mind. Making Sense of the Hemispheres. Harcourt Brace and Co.: San Diego, 1997.

Osiurak, F., Reynaud, E., Baumard, J., Rossetti, Y., Bartolo, A. and Lesourd, M. Pantomime of tool use: looking beyond apraxia. Brain Commun 2021; 3: fcab263. doi: 10.1093/braincomms/fcab263.

Osterrieth, P.A. Le test de copie d'une figure complexe. Arch Psychol. 1944; 30: 206–356.

Otaiku, A.I. Religiosity and risk of Parkinson's disease in England and the USA. Journal of Religion and Health 2023; 62: 4192–4208. doi: 10.1007/s10943–022–01603–8.

Owen, A.M., Coleman, M.R., Boly, M., Davis, M.H., Laureys, S. and Pickard, J.D. Detecting awareness in the vegetative state. Science 2006; 313: 1402. doi: 10.1126/science.1130197.

Ozsancak, C., Auzou, P., Dujardin, K., Quinn, N. and Destée, A. Orofacial apraxia in corticobasal degeneration, progressive supranuclear palsy, multiple system atrophy and Parkinson's disease. J Neurol 2004; 251: 1317–23. doi: 10.1007/s00415–004–0530–0.

Pailhès, A. and Kuhn, G. Influencing choices with conversational primes: How a magic trick unconsciously influences card choices. Proc Natl Acad Sci U S A 2020; 117: 17675–17679. doi: 10.1073/pnas.2000682117.

Paller, K.A. and Oudiette, D. Sleep learning gets real. Sci Am. 2018; 319: 26–31. doi: 10.1038/scientificamerican1118–26.

Paller, K.A. Cross-cortical consolidation as the core defect in amnesia: prospects for hypothesis-testing with neuropsychology and neuroimaging. *In*: Squire, L.R. and Schacter, D.L. (eds.) Neuropsychology of Memory (Vol. 3), Guilford Press: NewYork, 2002, 73–87.

Pande, B.S. and Singh, I. One-sided dominance in the upper limbs of human fetuses as evidenced by asymmetry in muscle and bone weight. J Anat 1971; 109: 457–459.

Papez, J.W. A proposed mechanism of emotion. Arch Neurol Psychiatry 1937; 38: 725–743. 10.1001/archneurpsyc.1937.02260220069003.

Parent, A. Comparative Neurobiology of the Basal Ganglia. Wiley: New York, 1986.

Parker, A.J., Woodhead, Z.V.J., Thompson, P.A. and Bishop, D.V.M. Assessing the reliability of an online behavioural laterality battery: a pre-registered study. Laterality 2021; 26: 359–397. doi: 10.1080/1357650X.2020.1859526.

Parks, P.C. (translator) On the determination of functions from their integral values along certain manifolds. (work originally published in German by Johann Radon in 1917). IEEE Transactions on Medical Imaging 1986; MI-5: 170–176.

Parris, B.A. The role of frontal executive functions in hypnosis and hypnotic suggestibility. Psychology of Consciousness: Theory, Research, and Practice 2017; 4: 211–229.

Parsons, M.W. and Braun, M.M. (eds.) (4th edition). Clinical Neuropsychology. A Pocket Handbook for Assessment. American Psychological Association: New York, 2024.

Passaro, A.D., Rezaie, R., Moser, D.C., Li, Z., Dias, N. and Papanicolaou, A.C. Optimizing estimation of hemispheric dominance for language using magnetic source imaging. Brain Res 2011 Oct 6; 1416: 44–50. doi: 10.1016/j.brainres.2011.08.017.

Passeri, A., Capotosto, P. and Di Matteo, R. The right hemisphere contribution to semantic categorization: a TMS study. Cortex 2015; 64: 318–26. doi: 10.1016/j.cortex.2014.11.014.

Pavlova, M.A. and Sokolov A.A. Reading language of the eyes. Neurosci Biobehav Rev. 2022 Sep; 140: 104755. doi: 10.1016/j.neubiorev.2022.104755.

Pedrazzini, E., Schnider, A. and Ptak, R. A neuroanatomical model of space-based and object-centered processing in spatial neglect. Brain Struct Funct 2017; 222: 3605–3613. doi: 10.1007/s00429–017–1420–4.

Pedroli, E., Serino, S., Cipresso, P., Pallavicini, F. and Riva, G. Assessment and rehabilitation of neglect using virtual reality: a systematic review. Front Behav Neurosci 2015; 9: 226. doi: 10.3389/fnbeh.2015.00226.

Peirce, J., Gray, J.R., Simpson, S., MacAskill, M., Höchenberger, R., Sogo, H. et al. PsychoPy2: Experiments in behavior made easy. Behavior Research Methods 2019; 51: 195–203. https: //doi.org/10.3758/s13428–018–01193–y.

Peltier, M.R., Cosgrove, S.J., Ohayagha, K., Crapanzano, K.A. and Jones, G.N. Do they see dead people? Cultural factors and sensitivity in screening for schizophrenia spectrum disorders. Ethn Health 2017 Apr; 22(2): 119–129. doi: 10.1080/13557858.2016.

Penagos-Corzo, J.C., Cosio van-Hasselt, M., Escobar, D., Vázquez-Roque, R.A. and Flores, G. Mirror neurons and empathy-related regions in psychopathy: Systematic review, meta-analysis, and a working model. Soc Neurosci 2022 Oct; 17(5): 462–479. doi: 10.1080/17470919.2022.2128868.

Pendergrass, J.C., Targum, S.D. and Harrison, J.E. Cognitive impairment associated with cancer: A brief review. Innov Clin Neurosci 2018 Feb 1; 15(1–2): 36–44.

Penfield, W. and Boldrey, E. Somatic motor and sensory representation in the cerebral cortex of man as studied by electrical stimulation. Brain 1937; 60: 389–443.

Penfield, W. and Jasper, H. Epilepsy and the Functional Anatomy of the Human Brain. J and A Churchill: London, 1954, 105–106.

Penfield, W. The electrode, the brain and the mind. Z Neurol 1972; 201: 297–307. doi: 10.1007/BF00316235.

Penner, I.K., Raselli, C., Stöcklin, M., Opwis, K., Kappos, L. and Calabrese, P. The Fatigue Scale for motor and cognitive functions (FSMC): validation of a new instrument to assess multiple sclerosis-related fatigue. Mult Scler 2009; 15: 1509–17. doi: 10.1177/1352458509348519.

Pepperberg, I.M. Alex and Me: How a Scientist and a Parrot Discovered a Hidden World of Animal Intelligence - and Formed a Deep Bond in the Process. Harper-Collins: New York, 2008.

Péran, P., Cardebat, D., Cherubini, A., Piras, F., Luccichenti, G., Peppe, A. et al. Object naming and action-verb generation in Parkinson's disease: a fMRI study. Cortex 2009; 45: 960–71. doi: 10.1016/j.cortex.2009.02.019.

Perkin, G.D. Arnold Bennett and medicine: with particular reference to his description of dressing apraxia. Br Med J (Clin Res Ed) 1981 Dec 19–26; 283(6307): 1666–8. doi: 10.1136/bmj.283.6307.1666

Perrett, D.I., Smith, P.A., Potter, D.D., Mistlin, A.J., Head, A.S., Milner, A.D. et al. Visual cells in the temporal cortex sensitive to face view and gaze direction. Proc R Soc Lond B Biol Sci 1985; 223: 293–317. doi: 10.1098/rspb.1985.0003.

Persinger, M.A. Neuropsychological Bases of God Beliefs. Praeger: New York, 1987.

Persinger, M.A. and Makarec, K. Temporal lobe epileptic signs and correlative behaviors displayed by normal populations. J Gen Psychol 1987; 114: 179–95. doi: 10.1080/00221309.1987.9711068.

Petersson, K.M., Silva, C., Castro-Caldas, A., Ingvar, M. and Reis, A. Literacy: a cultural influence on functional left-right differences in the inferior parietal cortex. Eur J Neurosci 2007; 26: 791–799. doi: 10.1111/j.1460–9568.2007.05701.x.

Petri, M., Messinis, L., Patrikelis, P., Nousia, A. and Nasios, G. Illiteracy, neuropsychological assessment, and cognitive rehabilitation: A narrative review. Adv Exp Med Biol 2023; 1425: 477–484. doi: 10.1007/978–3–031–31986–0_46.

Pflugshaupt, T., Gutbrod, K., Wurtz, P., von Wartburg, R., Nyffeler, T., de Haan, B. et al. About the role of visual field defects in pure alexia. Brain 2009; 132: 1907–1917. doi: 10.1093/brain/awp141.

Pflugshaupt, T., Nösberger, M., Gutbrod, K., Weber, K.P., Linnebank, M. and Brugger, P. Bottom-up Visual Integration in the Medial Parietal Lobe. Cerebral Cortex 2016; 26: 943–949. doi: 10.1093/cercor/bhu256.

Pflugshaupt, T., Nyffeler, T., von Wartburg, R., Wurtz, P., Lüthi, M., Hubl, D. et al. When left becomes right and vice versa: mirrored vision after cerebral hypoxia. Neuropsychologia 2007; 45: 2078–91. doi: 10.1016/j.neuropsychologia.2007.01.018.

Pick, A. Über eine besondere Form von Orientierungsstörung und deren Vorkommen bei Geisteskrankheiten. Deutsche Medizinische Wochenschrift 1909; 34: 2014–2017.

Pinker, S. and Jackendoff, R. The faculty of language: What's special about it? Cognition 2005; 95: 201–236. Doi 10.1016/j.cognition.2004.08.004.

Pinker, S. The Language Instinct: How the Mind Creates Language. HarperCollins, London 2007 (originally published 1994).

Pinti, P., Siddiqui, M.F., Levy, A.D., Jones, E.J.H. and Tachtsidis I. An analysis framework for the integration of broadband NIRS and EEG to assess neurovascular and neurometabolic coupling. Sci Rep 2021; 11: 3977. doi: 10.1038/s41598–021–83420–9.

Pinto, Y., de Haan, E.H.F. and Lamme, V.H.F. The split-brain phenomenon revisited: A single conscious agent with split perception. Trends Cogn Sci 2017; 21: 835–851. doi: 10.1016/j.tics.2017.09.003.

Pizzagalli, D., Lehmann, D. and Brugger, P. Lateralized direct and indirect semantic priming effects in subjects with paranormal experiences and beliefs. Psychopathology 2001; 34: 75–80. doi: 10.1159/000049284.

Pobric, G., Mashal, N., Faust, M. and Lavidor, M. The role of the right cerebral hemisphere in processing novel metaphoric expressions: a transcranial magnetic stimulation study. J Cogn Neurosci 2008; 20: 170–81. doi: 10.1162/jocn.2008.20005.

Podoll, K. and Robinson, D. Out-of-body experiences and related phenomena in migraine art. Cephalalgia 1999; 19: 886–96. doi: 10.1046/j.1468–2982.1999.1910886.x.

Poeppel, D. The analysis of speech in different temporal integration windows: cerebral lateralization as 'asymmetric sampling in time'. Speech Communication 2003; 41: 245–255. DOI: 10.1016/S0167–6393(02)00107–3.

Poffenberger, A.T. Reaction time to retinal stimulation with special reference to the time lost in conduction through nerve centers. Archives of Psychology 1912; 23: 1–73.

Poizner, H., Klima, E.S. and Bellugi, U. What the Hands Reveal about the Brain. MIT Press/Bradford Books: Cambridge, MA, 1987.

Poore, G.V. On fatigue. Lancet 1875; 2: 163–164.

Poppelreuter, W. Die psychischen Schädigungen durch Kopfschuss in Kriege 1914/1916. Vol. 1: Die Störungen der niederen und höheren Sehleistungen durch Verletzungen des Okzipitalhirns. Voss: Leipzig, 1917.

Popper, K.R. The Logic of Scientific Discovery. Routledge & Kegan Paul: London, 1980. (Original work published 1959).

Posner, M.I. (ed.). Cognitive neuroscience of attention. 2nd edition. Taylor and Francis: New York, 2011.

Poston, M. Zenith Man: Death, Love, and Redemption in a Georgia Courtroom. Citadel Books, Kensington Publishing: New York, 2024.

Pratt, R.T.C. and Warrington, E.K. The assessment of cerebral dominance with unilateral ECT. Br J Psychiatry 1972; 121: 327–328. DOI: 10.1192/bjp.121.3.327.

Prete, G. and Tommasi, L. The own-race bias and the cerebral hemispheres. Soc Neurosci 2019 Oct; 14(5): 549–558. doi: 10.1080/17470919.2018.1534755.

Price, B.H., Adams, R.D. and Coyle, J.T. Neurology and psychiatry: Closing the great divide. Neurology 2000; 54: 8–14. doi: 10.1212/wnl.54.1.8.

Price, E.H. A critical review of congenital phantom limb cases and a developmental theory for the basis of body image. Conscious Cogn 2006; 15: 310–22. doi: 10.1016/j.concog.2005.07.003.

Priftis, K., Algeri, L., Barachetti, L., Magnani, S., Gobbo, M. and De Pellegrin, S. Acquired neurogenic foreign accent syndrome after right-hemisphere lesion with left cerebellar diaschisis: A longitudinal study. Cortex 2020; 130: 220–230. doi: 10.1016/j.cortex.2020.05.019.

Priftis, K., Rusconi, E., Umiltà, C. and Zorzi, M. Pure agnosia for mirror stimuli after right inferior parietal lesion. Brain 2003; 126: 908–19. doi: 10.1093/brain/awg075.

Priftis, K., Velardo, V., Vascello, M.G.F., Villella, S., Galeri, S., Spada, M.S. et al. Limited evidence for neuropsychological dysfunction in patients initially affected by severe COVID-19. Neurol Sci. 2022; 43: 6661–6663. doi: 10.1007/s10072–022–06373–5.

Pruccoli, J., Spadoni, C., Orsenigo, A. and Parmeggiani, A. Should echolalia be considered a phonic stereotypy? A narrative review. Brain Sciences 2021; 11: 862–870.

Puccetti, R. Brain bisection and personal identity. British Journal for the Philosophy of Science 1973; 24: 339–355.

Pulvermüller, F. The Neuroscience of Language: On Brain Circuits of Words and Serial Order. Cambridge University Press: Cambridge, 2003.

Pulvermüller, F. The Neuroscience of Language. On Brain Circuits of Words and Serial Order. Cambridge University Press: New York, 2007.

Qiang, S., Li, F.Y., Zhou, Y. and Li, Q. Golden Section in Vulva. Plast Reconstr Surg Glob Open 2020 May 22; 8(5): e2844. doi: 10.1097/GOX.0000000000002844.

Quaranta, A., Siniscalchi, M. and Vallortigara, G. Asymmetric tail-wagging responses by dogs to different emotive stimuli. Curr Biol 2007; 17: R199–201. doi: 10.1016/j.cub.2007.02.008.

Quellet, M., Santiago, J., Israeli, Z. and Gabay, S. Is the future the right time? Exp Psychol 2010; 57: 308–14. doi: 10.1027/1618–3169/a000036.

Railo, H. and Hurme, M. Is the primary visual cortex necessary for blindsight-like behavior? Review of transcranial magnetic stimulation studies in neurologically healthy individuals. Neurosci Biobehav Rev 2021; 127: 353–364. doi: 10.1016/j.neubiorev.2021.04.038.

Ramachandran, V.S., Rogers-Ramachandran, D. and Stewart, M. Perceptual correlates of massive cortical reorganization. Science 1992 Nov 13; 258(5085): 1159–60. doi: 10.1126/science.1439826.

Ramachandran, V.S. and Rogers-Ramachandran, D. Synaesthesia in phantom limbs induced with mirrors. Proc Biol Sci 1996; 263: 377–86. doi: 10.1098/rspb.1996.0058.

Ramachandran, V.S. and Blakeslee, S. Phantoms in the Brain: Probing the Mysteries of the Human Mind. Harper Collins: New York, 1999.

Ramachandran, V.S. and Hubbard, E.M. Psychophysical investigations into the neural basis of synaesthesia. Proceedings of the Royal Society 2001; 268: 979–983. doi: 10.1098/rspb.2001.1576.

Ramchandani, M.S., Cannon, C.A. and Marra, C.M. Syphilis: A modern resurgence. Infect Dis Clin North Am 2023; 37: 195–222. doi: 10.1016/j.idc.2023.02.006. Epub 2023 Mar 31. PMID: 37005164.

Ramon, M., Bobak, A.K. and White, D. Super-recognizers: from the lab to the world and back again. Br J Psychol 2019; 110: 461–479. Doi: 10.111/bjop.12368.

Rand, G. and Wapner, S. Postural status as a factor in memory. Journal of Verbal Learning & Verbal Behavior 1967; 6: 268–271.

Rao, K.N. Internal autoscopy : a case report. Indian J Psychiatry 1992; 34: 280–282.

Rapport, L.J., Webster, J.S. and Dutra, R.L. Digit span performance and unilateral neglect. Neuropsychologia 1994; 32: 517–525.

Rasoulzadeh, V., van Dijck, J.P., Khosrowabadi, R. and Fias, W. Reading direction interacts with spatial processes of temporal order verbal working memory: evidence from Iranian right-to-left readers. Language, Cognition and Neuroscience 2024; 39: 1072–1089. https: //doi.org/10.1080/23273798. 2024.2370046.

Ray, N.J., Bradburn, S., Murgatroyd, C., Toseeb, U., Mir, P., Kountouriotis, G.K. et al. *In vivo* cholinergic basal forebrain atrophy predicts cognitive decline in *de novo* Parkinson's disease. Brain 2018; 141: 165–176.

Ray, W.J. Fundamentals of Brain and Behavior: An Introduction to Human Neuroscience. Routledge: New York, 2024.

Ready, R.E., Kapoulea, E., Orlovsky, I., Whaley, M. and Rafiuddin, H.S. Cultural and multicultural topics in Neuropsychology Journal titles from 2010–2020. Clin Neuropsychol 2025 Feb; 39(2): 237–251. doi: 10.1080/13854046.2024.2385506.

Redlich, F.C. and Dorsey, J.E. Denial of blindness by patients with cerebral disease. Archives of Neurology and Psychiatry 1945; 53: 407–417.

Reed, C.L., Caselli, R.J. and Farah, M.J. Tactile agnosia. Underlying impairment and implications for normal tactile object recognition. Brain 1996; 119: 875–888.

Regard, M., Strauss, E. and Knapp, P. Children's production on verbal and non-verbal fluency tasks. Percept Mot Skills 1982; 55: 839–44. doi: 10.2466/pms.1982.55.3.839.

Regard, M., Landis, T. and Graves, R. Dissociated hemispheric superiorities for reading stenography vs print. Neuropsychologia 1985; 23: 431–435. doi: 10.1016/0028–3932(85)90031–4.

Regard, M., Oelz, O., Brugger, P. and Landis, T. Persistent cognitive impairment in climbers after repeated exposure to extreme altitude. Neurology 1989; 39: 210–213. doi: 10.1212/wnl.39.2.210.

Regard, M., Cook, N.D., Wieser, H.G. and Landis, T. The dynamics of cerebral dominance during unilateral limbic seizures. Brain 1994; 117: 91–104. doi: 10.1093/brain/117.1.91.

Regard, M. and Landis, T. The Smiley. A graphical expression of mood in right anterior cerebral lesions. Neuropsychiatry, Neuropsychology & Behavioral Neurology 1994; 7: 303–307.

Rennig, J., Cornelsen, S., Wilhelm, H., Himmelbach, M. and Karnath, H.O. Preserved expert object recognition in a case of visual hemiagnosia. J Cogn Neurosci 2018; 30: 131–143. doi: 10.1162/jocn_a_01193.

Resch, C., Keulers, E., Martens, R., van Breukelen, G., van Heugten, C.M., Hou, W. et al. Young children's performance on a design fluency task: longitudinal data on total number of designs, clustering and switching, and regression-based norms. J Clin Exp Neuropsychol 2024; 46: 449–465. doi: 10.1080/13803395.2024.2359744.

Révész, B. Geschichte des Seelenbegriffes und der Seelenlokalisation. Enke: Stuttgart, 1917.

Rey, A. L'examen psychologique dans les cas encéphalopathc traumatique. Arch. Psychol. 1941; 28: 286–340.

Rey, A. L'examen Clinique en Psychologie. Presses Universitaires de France: Paris, 1964.

Reynolds, E.H. and Kinnier Wilson, J.V. Psychoses of epilepsy in Babylon: the oldest account of the disorder. Epilepsia. 2008 Sep.; 49(9): 1488–90. doi: 10.1111/j.1528-1167.2008.01614.x.

Ribot, T. Diseases of Memory. Kegan Paul, Trench and Co: London, 1885.

Rich, T.J. and Palmer, J. Neglect dyslexia: whole-word and within-word errors with parafoveal and foveal viewing. Exp Brain Res 2023; 241: 2655–2668. doi: 10.1007/s00221–023–06708–4.

Richter, C.P. Mysterious form of referred sensation in man. Proc Natl Acad Sci U S A. 1977 Oct; 74(10): 4702–5. doi: 10.1073/pnas.74.10.4702.

Riddoch, G. Dissociation of visual perceptions due to occipital injuries, with special reference to appreciation of movement. Brain 1917; 40: 15–57.

Riley, E.P., Infante, M.A. and Warren, K.R. Fetal alcohol spectrum disorders: an overview. Neuropsychol Rev 2011; 21: 73–80. doi: 10.1007/s11065–011–9166–x.

Rinaldi, L., Brugger, P., Bockisch, C.J., Bertolini, G. and Girelli, L. Keeping an eye on serial order: ocular movements bind space and time. Cognition 2015; 142: 291–298. doi: 10.1016/j.cognition.2015.05.022.

Rinaldi, L., Bertolini, G., Bockisch, C.J., Maravita, A., Girelli, L. and Brugger, P. More far is more right: manual and ocular line bisections, but not the Judd ilusion, depend on radial space. Brain and Cognition 2018; 22: 34–44.

Rizzolatti, G. and Sinigaglia, C. Mirror neurons and motor intentionality. Funct Neurol 2007; 22: 205–10.

Roberts, G.W., Done, D.J., Bruton, C. and Crow, T.J. A "mock up" of schizophrenia: temporal lobe epilepsy and schizophrenia-like psychosis. Biol Psychiatry 1990; 28: 127–43. doi: 10.1016/0006–3223(90)90630–k.

Robertson, I. Digit span and visual neglect: A puzzling relationship. Neuropsychologia 1990; 28: 217–222. DOI: 10.1016/0028–3932(90)90103–U.

Robertson, I.H. and Marshall, J.C. Unilateral Neglect: Clinical and Experimental Studies. Taylor and Francis: New York, 2013.

Roby-Brami, A., Hermsdörfer, J., Roy, A.C. and Jacobs, S. A neuropsychological perspective on the link between language and praxis in modern humans. Philos Trans R Soc Lond B Biol Sci 2012; 367: 144–60. doi: 10.1098/rstb.2011.0122.

Rodel, M., Cook, N.D., Regard, M. and Landis, T. Hemispheric dissociation in judging semantic relations: complementarity for close and distant associates. Brain Lang 1992; 43: 448–459. doi: 10.1016/0093–934x(92)90111–q.

Roebuck-Spencer, T.M., Glen, T., Puente, A.E., Denney, R.L., Ruff, R.M., Hostetter, G. et al. Cognitive screening tests versus comprehensive neuropsychological test batteries: A National Academy of Neuropsychology education paper. Archives of Clinical Neuropsychology 2017; 32: 491–498, https://doi.org/10.1093/arclin/acx021.

Roediger, H.L. and McDermott, K.B. Creating false memories: Remembering words not presented in lists. Journal of Experimental Psychology: Learning, Memory, and Cognition 1995; 21: 803–814. DOI: https: //doi.org/10.1037/0278–7393.21.4.803.

Rogers, L.J., Zucca, P. and Vallortigara, G. Advantages of having a lateralized brain. Proc Biol Sci 2004; 271, Suppl 6: S420–S422. doi: 10.1098/rsbl.2004.0200.

Rogers, L.J., Vallortigara, G. and Andrew, R.J. Divided brains. The Biology and Behaviour of Brain Asymmetries. Cambridge University Press: New York, 2013.

Rolls, E.T. The cingulate cortex and limbic systems for action, emotion, and memory. Handb Clin Neurol 2019; 166: 23–37. doi: 10.1016/B978–0–444–64196–0.00002–9.

Romanski, L.M., Tian, B., Fritz, J., Mishkin, M., Goldman-Rakic, P.S. and Rauschecker, J.P. Dual streams of auditory afferents target multiple domains in the primate prefrontal cortex. Nat Neurosci 1999; 2: 1131–1136. doi: 10.1038/16056.

Root, J.C., Wong, P.S. and Kinsbourne, M. Left hemisphere specialization for response to positive emotional expressions: A divided output methodology. Emotion 2006; 6: 473–483. DOI: 10.1037/1528–3542.6.3.473.

Rorden, C., Fruhmann Berger, M. and Karnath, H-O. Disturbed line bisection is associated with posterior brain lesions. Brain Res 2006; 1080: 17–25. doi: 10.1016/j.brainres.2004.10.071.

Ross, E.D. Sensory-specific amnesia and hypoemotionality in humans and monkeys: gateway for developing a hodology of memory. Cortex 2008; 44: 1010–22. doi: 10.1016/j.cortex.2008.02.002.

Ross, E.D. Disorders of vocal emotional expression and comprehension: The aprosodias. Handb Clin Neurol 2021; 183: 63–98. doi: 10.1016/B978–0–12–822290–4.00005–0.

Rothacher, Y., Nguyen, A., Lenggenhager, B., Kunz, A. and Brugger, P. Walking through virtual mazes: Spontaneous alternation behaviour in human adults. Cortex 2020; 127: 1–16. doi: 10.1016/j.cortex.2020.01.018.

Rudoler, J.H., Herweg, N.A. and Kahana, M.J. Hippocampal theta and episodic memory. J Neurosci 2023; 43: 613–620. doi: 10.1523/JNEUROSCI.1045–22.2022.

Rugani, R., Vallortigara, G., Priftis, K. and Regolin, L. Number-space mapping in the newborn chick resembles humans' mental number line. Science 2015; 347: 534–536. DOI: 10.1126/science.aaa1379.

Ruiz Yanzi MA, Alessandro L. Acute toxic reversible amnesia. Acta Neurol Belg. 2020; 120: 1185–1187. doi: 10.1007/s13760–019–01234–z.

Rumiati, R.I., Weiss, P.H., Tessari, A., Assmus, A., Zilles, K., Herzog, H. et al. Common and differential neural mechanisms supporting imitation of meaningful and meaningless actions. Journal of Cognitive Neuroscience 2005; 17: 1420–1431. doi: 10.1162/0898929054985374.

Rumiati, R.I., Carmo, J.C. and Corradi-Dell'Acqua, C. Neuropsychological perspectives on the mechanisms of imitation. Philos Trans R Soc Lond B Biol Sci 2009; 364(1528): 2337–2347. doi: 10.1098/rstb.2009.0063.

Rusconi, E., Pinel, P., Dehaene, S. and Kleinschmidt, A. The enigma of Gerstmann's syndrome revisited: a telling tale of the vicissitudes of neuropsychology. Brain 2010; 133: 320–332. doi: 10.1093/brain/awp281.

Rusconi, E., Tamè, L., Furlan, M., Haggard, P., Demarchi, G., Adriani, M. et al. Neural correlates of finger gnosis. J Neurosci 2014; 34: 9012–9023. doi: 10.1523/JNEUROSCI.3119–13.2014.

Sachdeva, A., Chandra, M., Choudhary, M., Dayal, P. and Anand, K.S. Alcohol-related dementia and neurocognitive impairment: A review study. Int J High Risk Behav Addict 2016 Feb 7; 5(3): e27976. doi: 10.5812/jhrba.27976.eCollection 2016 Sep.

Sadato, N., Pascual-Leone, A., Grafman, J., Ibañez, V., Deiber, M.P., Dold, G. et al. Activation of the primary visual cortex by Braille reading in blind subjects. Nature 1996; 380: 526–8. doi: 10.1038/380526a0.

Sadler, R.M. and Rahey, S. Prescience as an aura of temporal lobe epilepsy. Epilepsia 2004; 45: 982–4. doi: 10.1111/j.0013–9580.2004.68603.x.

Saetta, G., Grond, I., Brugger, P., Lenggenhager, B., Tsay, A.J. and Giummarra, M.J. Apparent motion perception in lower limb amputees with phantom sensations: "obstacle shunning" and "obstacle tolerance". Cortex 2018; 104: 220–231. doi: 10.1016/j.cortex.2018.03.003.

Saetta, G., Hänggi, J., Gandola, M., Zapparoli, L., Salvato, G., Berlingeri, M. et al. Neural correlates of body integrity dysphoria. Curr Biol 2020; 30: 2191–2195.e3. doi: 10.1016/j.cub.2020.04.001.

Saetta, G., Michels, L. and Brugger, P. Where in the Brain is "the Other's" Hand? Mapping Dysfunctional Neural Networks in Somatoparaphrenia. Neuroscience 2021 Nov 10; 476: 21–33. doi: 10.1016/j.neuroscience.2021.09.007

Saj, A., Fuhrman, O., Vuilleumier, P. and Boroditsky, L. Patients with left spatial neglect also neglect the "left side" of time. Psychol Sci 2014; 25: 207–214. doi: 10.1177/0956797612475222.

Salmon, A. Nuove osservazioni sui movimenti automatici che si compiono dopo gli sforzi muscolari e del loro valore in neuropatologia. Atti della Accademia Medico-Fisica Fiorentina 1914; [no volume]: 78–91.

Saltafossi, M., Zaccaro, A., Perrucci, M.G., Ferri, F. and Costantini, M. The impact of cardiac phases on multisensory integration. Biol Psychol 2023; 182: 108642. doi: 10.1016/j.biopsycho.2023.108642.

Sandson, J. and Albert, M.L. Varieties of perseveration. Neuropsychologia 1984; 22: 715–32. doi: 10.1016/0028–3932(84)90098–8.

Scaliti, E., Gruppioni, E. and Becchio, C. And yet it moves: What we currently know about phantom arm movements. Neuroscientist 2020; 26: 328–342. doi: 10.1177/1073858420904326.

Schacter, D., Dobbins, I. and Schnyer, D. Specificity of priming: a cognitive neuroscience perspective. Nat Rev Neurosci 2004; 5: 853–862.

Schacter, D.L., Wang, P.L., Tulving, E. and Friedman, M. Functional retrograde amnesia: a quantitative case study. Neuropsychologia 1982; 20: 523–532.

Schacter, D.L., Addis, D.R. and Buckner, R.L. Remembering the past to imagine the future: the prospective brain. Nature Reviews Neuroscience 2007; 8: 657–661.

Schacter, D.L. The Seven Sins of Memory: How the Mind Forgets and Remembers. Updated edition (1st edition 2002). Mariner Books, HarperCollins: Boston, MA, 2021.

Schaeffer, A.A. Spiral movement in man. Journal of Morphology 1928; 45: 293–398. https: //doi.org/10.1002/jmor.1050450110.

Schärli, H., Brugger, P., Regard, M., Mohr, C. and Landis, T. Localisation of "unseen" visual stimuli: Blindsight in normal observers? Swiss Journal of Psychology 2003; 62: 159–165. https: //doi.org/10.1024/1421–0185.62.3.159.

Schechter, E. Self-consciousness and "Split" Brains. The Minds' I. Oxford University Press: New York, 2018.

Schenkenberg, T., Bradford, D.C. and Ajax, E.T. Line bisection and unilateral visual neglect in patients with neurologic impairment. Neurology 1980; 30: 509–517.

Schilder, P. and Stengel, E. Asymbolia for pain. Arch. Neurol. Psychiatry 1931; 25: 598–600. doi: 10.1001/archneurpsyc.1931.02230030156007.

Schindler, I., Kerkhoff, G., Karnath, H.O., Keller, I. and Goldenberg, G. Neck muscle vibration induces lasting recovery in spatial neglect. J Neurol Neurosurg Psychiatry 2002; 73: 412–9. doi: 10.1136/jnnp.73.4.412.

Schlich, T. Cutting the body to cure the mind. Lancet Psychiatry 2015; 2: 390–2. doi: 10.1016/S2215–0366(15)00188–1.

Schmahmann, J. The cerebellar cognitive affective syndrome. Brain 1998; 121: 561–579.

Schnider, A. and Ptak, R. Spontaneous confabulators fail to suppress currently irrelevant memory traces. Nat Neurosci 1999; 2: 677–81. doi: 10.1038/10236.

Schnider, A., von Däniken, C. and Gutbrod, K. Disorientation in amnesia: a confusion of memory traces. Brain 1996; 119: 1627–1632.

Schnider, A. Verhaltensneurologie. Die neurologische Seite der Neuropsychologie. Thieme: Stuttgart, 1997.

Schnider, A. Spontaneous confabulation and the adaptation of thought to ongoing reality. Nat Rev Neurosci 2003; 4: 662–671. doi: 10.1038/nrn1179.

Schnider, A. The Confabulating Mind (2nd edition). Oxford Academic Press: New York, 2017.

Schoenberg, M.R., Marsh, P.J. and Lerner, A.J. Neuroanatomy primer: Structure and function of the human nervous system. *In*: Schoenberg, M.R. and Scott, J.G. (eds.). The Little Black Book of Neuropsychology, Springer: New York, 2011, 59–126.

Schott, G.D. Distant referral of cutaneous sensation (Mitempfindung). Brain 1988; 111: 1187–1198. doi: 10.1093/brain/111.5.1187.

Schott, G.D. Penfield's homunculus: a note on cerebral cartography. J Neurol Neurosurg Psychiatry 1993; 56: 329–33. doi: 10.1136/jnnp.56.4.329.

Schüeli, H., Henn, V. and Brugger, P. Vestibular stimulation affects dichotic lexical decision performance. Neuropsychologia 1999; 37: 653–659. doi: 10.1016/s0028–3932(98)00129–8.

Schwartz, S. and Maquet, P. Sleep imaging and the neuro-psychological assessment of dreams. Trends Cogn Sci 2002; 6: 23–30. doi: 10.1016/s1364–6613(00)01818–0.

Schwartz, S., Baldo, J., Graves, R.E. and Brugger, P. Pervasive influence of semantics in letter and category fluency: a multidimensional approach. Brain Lang 2003; 87: 400–11. doi: 10.1016/s0093–934x(03)00141–x.

Schwarz, U. Neuro-ophthalmology: A short primer. *In*: Baert, A.L., Sartor, K. and Müller-Forell, W.S. (eds.). Imaging of Orbital and Visual Pathway Pathology. Springer: Berlin/New York, 2002.

Schwiedrzik, C.M., Melloni, L. and Schurger, A. Mooney face stimuli for visual perception research. PLoS One 2018; 13: e0200106. doi: 10.1371/journal.pone.0200106.

Scott, J.G. and Schoenberg, M.R. Frontal lobe/executive functioning. *In*: Schoenberg, M.R. and Scott, J.G. (eds.). The Little Black Book of Neuropsychology, Springer: New York, 2011, 219–248.

Scrivener, S. Two for the price of one. Perception 1983; 12: 769. doi: 10.1068/p120769.

Searleman, A. A review of right hemisphere linguistic capabilities. Psychological Bulletin 1977; 84: 503–528.

Sears, R. The Beautiful Poetry of Donald Trump. Canongate Books: Edinburgh, 2017.

Sedda, A, Scarpina F. Dorsal and ventral streams across sensory modalities. Neurosci Bull. 2012; 28: 291–300. doi: 10.1007/s12264–012–1223–9.

Segalowitz, S.J. and Stewart, C. Left and right lateralization for letter matching: strategy and sex differences. Neuropsychologia 1979; 17: 521–525. doi: 10.1016/0028–3932(79)90060–5.

Sekiyama, K. Importance of head axes in perception of cutaneous patterns drawn on vertical body surfaces. Percept Psychophys. 1991; 49: 481–92. doi: 10.3758/bf03212182.

Sergent, J. The cerebral balance of power: confrontation or cooperation? J Exp Psychol Hum Percept Perform 1982; 8: 253–72. doi: 10.1037//0096–1523.8.2.253.

Serino, A., Cecere, R., Dundon, N., Bertini, C., Sanchez-Castaneda, C. and Làdavas, E. When apperceptive agnosia is explained by a deficit of primary visual processing. Cortex 2014; 52: 12–27. doi: 10.1016/j.cortex.2013.12.011.

Shah, P., Gaule, A., Sowden, S., Bird, G. and Cook, R. The 20–item prosopagnosia index (PI20): A self-report instrument for identifying developmental prosopagnosia. Royal Society Open Science, 2015; 6: 140343.

Shallice, T. Cognitive neuropsychology and its vicissitudes: The fate of Caramazza's axioms. Cogn Neuropsychol. 2015; 32: 385–411. doi: 10.1080/02643294.2015.1131677.

Shanmugam, S., Haver, H.L., Knecht, S.M., Rajjoub, R., Ali, O. and Chow, R. Bilateral occipital lobe infarct neglect deficit (BLIND) syndrome. J Community Hosp Intern Med Perspect 2021; 11: 678–681. doi. 10.1080/20009666.2021.1971730.

Shapiro, M. Plasticity, hippocampal place cells, and cognitive maps. Arch Neurol. 2001; 58: 874–81. doi: 10.1001/archneur.58.6.874.

Sharot, T., Korn, C.W. and and Dolan, R.J. How unrealistic optimism is maintained in the face of reality. Nature Neuroscience 2011; 14: 1475–1479. doi 10.1038/nn.2949.

Sharot, T. The optimism bias. Curr Biol 2011; 21: R941–5. doi: 10.1016/j.cub.2011.10.030.

Sheldrake, R. and Smart, P. Directional scopaesthesia and its implications for theories of vision. Journal of Scientific Exploration 2023; 37(3): 312–329. https://doi.org/10.31275/20232897

Shiffrar, M. and Fredy, J.J. Apparent motion of the human body. Psychological Science 1990; 1(4): 257–264. https://doi.org/10.1111/j.1467-9280.1990.tb00210.x.

Shomrat, T. and Levin, M. An automated training paradigm reveals long-term memory in planarians and its persistence through head regeneration. J Exp Biol 2013; 216: 3799–810. doi: 10.1242/jeb.087809.
Shprecher, D. and Mehta, L. The syndrome of delayed post-hypoxic leukoencephalopathy. NeuroRehabilitation 2010; 26(1): 65–72.
Sierra, M. and Berrios, G.E. Flashbulb memories and other repetitive images: a psychiatric perspective. Compr Psychiatry 1999; 40: 115–25. doi: 10.1016/s0010–440x(99)90115–3.
Sillito, A.M., Jones, H.E., Gerstein, G.L. and West, D.C. Feature-linked synchronization of thalamic relay cell firing induced by feedback from the visual cortex. Nature 1994; 369: 479–482.
Simard, S., Rouleau, I., Brosseau, J., Laframboise, M. and Bojanowsky, M. Impact of executive dysfunctions on episodic memory abilities in patients with ruptured aneurysm of the anterior communicating artery. Brain Cogn 2003 Nov; 53(2): 354–8. doi: 10.1016/s0278–2626(03)00142–8.
Simfukwe, C., Youn, Y.C. and Jeong, H.T. A machine-learning algorithm for predicting brain age using Rey-Osterrieth complex figure tests of healthy participants. Appl Neuropsychol Adult 2025; 32: 225–230. doi: 10.1080/23279095.2022.2164198.
Simon, T., Frasnelli, E., Guo, K., Barber, A., Wilkinson, A. and Mills, D.S. Is there an association between paw preference and emotionality in pet dogs? Animals 2022; 12: 1153. doi: 10.3390/ani12091153.
Simunovic, M.P. Metamorphopsia and its quantification. Retina 2015 Jul; 35(7): 1285–1291. doi: 10.1097/IAE.0000000000000581.
Singh, S.P. Magnetoencephalography: Basic principles. Ann Indian Acad Neurol 2014 Mar; 17(Suppl 1): S107–S112. doi: 10.4103/0972–2327.128676
Sirigu, A., Grafman, J., Bressler, K. and Sunderland, T. Multiple representations contribute to body knowledge processing. Evidence from a case of autotopagnosia. Brain 1991; 114: 629–642. doi: 10.1093/brain/114.1.629.
Sitek, E.J., Sołtan, W., Wieczorek, D., Schinwelski, M., Robowski, P., Reilmann, R. et al. Self-awareness of motor dysfunction in patients with Huntington's disease in comparison to Parkinson's disease and cervical dystonia. J Int Neuropsychol Soc 2011; 17: 788–95. doi: 10.1017/S1355617711000725.
Sladky, R., Höflich, A., Küblböck, M., Kraus, C., Baldinger, P., Moser, E. et al. Disrupted effective connectivity between the amygdala and orbitofrontal cortex in social anxiety disorder during emotion discrimination revealed by dynamic causal modeling for FMRI. Cereb Cortex 2015; 25: 895–903. doi: 10.1093/cercor/bht279.
Smetacek, V. and Mechsner, F. Making sense. Nature 2004; 432: 21. doi: 10.1038/432021a.
Smith, G. and CNS. Education and training in clinical neuropsychology: Recent developments and documents from the clinical neuropsychology synarchy. Clin Neuropsychol 2019; 33: 447–465. doi: 10.1080/13854046.2018.1552437.
Smith, N. and Tsimpli. I.-M. The Mind of a Savant: Language Learning and Modularity. Basil Blackwell: Oxford, 1995.
Smith, P.E., Shah, P., Sharpe, J., Todd, A. and Goringe, A.P. Palinopsia. Lancet 2003; 361: 1098. doi: 10.1016/S0140–6736(03)12896–6.
Society for Neuroscience, Brain Facts. A Primer on the Brain and Nervous System. Society for Neuroscience: Washington, DC, 2005.
Sokolov, A.A., Erb, M., Grodd, W. and Pavlova, M.A. Structural loop between the cerebellum and the superior temporal sulcus: evidence from diffusion tensor imaging. Cereb Cortex 2014; 24: 626–32. doi: 10.1093/cercor/bhs346.
Sollier, P. Les Phénomènes d'Autoscopie. Alcan: Paris, 1903.
Sollier, P. Autoscopie interne vérifiée expérimentalement. Journal de Psychologie Normale et Pathologique 1908; 5: 354–358.
Solms, M., Kaplan-Solms, K., Saling, M. and Miller, P. Inverted vision after frontal lobe disease. Cortex 1988; 24: 499–509. doi: 10.1016/s0010–9452(88)80044–3.
Solonen, K.A. The phantom phenomenon in amputated Finnish war veterans. Acta Orthopaedica Scandinavica 1962; 33: sup54: 5–37. doi: 10.3109/ ort.1962.33.suppl-54.01.
Spence, S.A. Prefrontal white matter--the tissue of lies?: Invited commentary on... Prefrontal white matter in pathological liars. Br J Psychiatry. 2005 Oct; 187: 326–7. doi: 10.1192/bjp.187.4.326.
Spencer, D. and Burchiel, K. Selective amygdalohippocampectomy. Epilepsy Res Treat 2012; 2012: 382095. doi: 10.1155/2012/382095.

Sperry, R.W. The great cerebral commissure. Sci Am 1964; 210: 42–52. doi: 10.1038/scientificamerican0164–42.

Springer, S.P. and Deutsch, G. Left Brain, Right Brain: Perspectives from Cognitive Neuroscience (5th edition, originally published 1985). W. H. Freeman and Co.: New York, 1998

Spitzer, M., Braun, U., Hermle, L. and Maier, S: Associative semantic network dysfunction in thought-disordered schizophrenic patients: Direct evidence from indirect semantic priming. Biol Psychiatry 1993; 34: 864–877.

Sporns, O. The human connectome: origins and challenges. NeuroImage 2013; 80: 53–61.

Stammler, B., Flammer, K., Schuster, T., Lambert, M., Neumann, O., Lux, M. et al. Spatial neglect therapy with the augmented reality app "Negami" for active exploration training: A randomized controlled trial on 20 stroke patients with spatial neglect. Arch Phys Med Rehabil 2023; 104: 1987–1994. doi: 10.1016/j.apmr.2023.07.017.

Staniloiu, A. and Markowitsch, H. Dissociative amnesia. Lancet Psychiatry 2014; 1: 226–241. Doi: 10.1016/S2215–0366(14)70279–2.

Stevens, J.A., Fonlupt, P., Shiffrar, M. and Decety, J. New apsects of motion perception: selective neural encoding of apparent human movements. Neuroreport 2000 Jan 17; 11(1): 109–15. doi: 10.1097/00001756–200001170/00022.

Stigler, S.M. Stigler's law of eponymy. Transactions of the New York Academy of Sciences 1980; 39: 147–157.

Stilson, D.W., Haaseth, K., Walsmith, C.R., Schneider, R.J. and Rogers, M. Marbe's law, stimulus expectancy, and response competition. The American Journal of Psychology 1972; 85: 549–567.

Stoléru, S., Fonteille, V., Cornélis, C., Joyal, C. and Moulier, V. Functional neuroimaging studies of sexual arousal and orgasm in healthy men and women: a review and meta-analysis. Neurosci Biobehav Rev 2012; 36: 1481–509. doi: 10.1016/j.neubiorev.2012.03.006.

Stone, S.P., Patel, P. and Greenwood, R.J. Selection of acute stroke patients for treatment of visual neglect. J Neurol Neurosurg Psychiatry 1993; 56: 463–6. doi: 10.1136/jnnp.56.5.463.

Stracke, J. Über die Wahrnehmung von Körper, Raum und Bewegung durch die Haut. Der Nervenarzt 1947; 10: 433–437.

Stratton, G.M. Some preliminary experiments on vision without inversion of the retinal image. Psychological Review 1896; 3: 611–617.

Stratton, G.M. The spatial harmony of touch and sight. Mind 1899; 8: 492–493.

Strauss, E., Wada, J. and Kosaka, B. Visual laterality effects and cerebral speech dominance determined by the carotid amytal test. Neuropsychologia 1985; 23: 567–70. doi: 10.1016/0028–3932(85)90010–7.

Street, R.F. A Gestalt Completion Test. Teachers College, Columbia University: New York, 1931.

Streiff, R.A. Institutional neglect of anosognosia is a critical barrier in the treatment of psychosis-related disorders. J Clin Psychopharmacol 2023; 43: 200–203. doi: 10.1097/JCP.0000000000001695.

Striem-Amit, E. Brain Plasticity: When the feet and mouth replace the hand. Curr Biol 2017; 27: R356–R358. doi: 10.1016/j.cub.2017.03.057.

Ströckens, F., Güntürkün, O. and Ocklenburg, S. Limb preferences in non-human vertebrates. Laterality 2013; 18: 536–575. doi: 10.1080/1357650X.2012.723008.

Stromeyer, C.F. and Psotka, J. The detailed texture of eidetic images. Nature 1970; 24: 346–349. doi: 10.1038/225346a0.

Stroobant, N. and Vingerhoets, G. Transcranial Doppler ultrasonography monitoring of cerebral hemodynamics during performance of cognitive tasks: a review. Neuropsychol Rev 2000; 10: 213–231. doi: 10.1023/a: 1026412811036.

Stuss, D.T. Traumatic brain injury: relation to executive dysfunction and the frontal lobes. Curr Opin Neurol. 2011 Dec; 24(6): 584–9. doi: 10.1097/WCO.0b013e32834c7eb9.

Sun, T. and Walsh, C.A. Molecular approaches to brain asymmetry and handedness. Nat Rev Neurosci 2006; 7: 655–662. doi: 10.1038/nrn1930.

Sung, K., Gordon, B., Yang, S. and Schretlen, D.J. Evidence of semantic clustering in letter-cued word retrieval. J Clin Exp Neuropsychol 2013; 35(10): 1015–23. doi: 10.1080/13803395.2013.845141.

Suri, R. and Ali, A. Coining the Pablo Picasso Syndrome. Headache 2020; 60: 626–629. doi: 10.1111/head.13702.

Swift, A.B. Tympanic thermometry: an index of hemispheric activity. Percept Mot Skills. 1991; 73: 275–93. doi: 10.2466/pms.1991.73.1.275.

Tafreshi, D. Adolphe Quetelet and the legacy of the "average man" in psychology. Hist Psychol 2022; 25: 34–55. doi: 10.1037/hop0000202.
Taïb, S., Yrondi, A., Lemesle, B., Péran, P. and Pariente, J. What are the neural correlates of dissociative amnesia? A systematic review of the functional neuroimaging literature. Front Psychiatry 2023; 14: 1092826. doi: 10.3389/fpsyt.2023.1092826.
Tallon-Baudry, C. Interoception: Probing internal state is inherent to perception and cognition. Neuron 2023; 111: 1854–1857. doi: 10.1016/j.neuron.2023.04.019.
Tamagni, C., Mantei, T. and Brugger, P. Emotion and space: lateralized emotional word detection depends on line bisection bias. Neuroscience 2009 Sep 15; 162(4): 1101–5. doi: 10.1016/j.neuroscience.2009.05.072.
Tamagni, C., Mondadori, C.R., Valko, P.O., Brugger, P., Schuknecht, B. and Linnebank, M. Cerebellum and source memory. European Neurology 2010; 63: 234–236. doi: 10.1159/000282277.
Tamura, I., Hamada, S., Soma, H., Moriwaka, F. and Tashiro, K. A case of pure autotopagnosia following Creutzfeldt-Jakob disease. Cogn Neuropsychol 2016; 33: 398–404. doi: 10.1080/02643294.2016.1234450.
Tanabe, H., Sawada, T., Asai, H., Okuda, J. and Shiraishi J. Lateralization phenomenon of complex auditory hallucinations. Acta Psychiatr Scand 1986 Aug; 74(2): 178–82. doi: 10.1111/j.1600–0447.1986.tb10602.x.
Tanaka, Y., Albert, M.L., Hara, H., Miyashita, T. and Kotani, N. Forced hyperphasia and environmental dependency syndrome. J Neurol Neurosurg Psychiatry 2000 Feb; 68(2): 224–6. doi: 10.1136/jnnp.68.2.224.
Tascedda, F. and Ottaviani, E. Tumors in invertebrates. Invertebrate Survival Journal 2014; 11: 197–203.
Taylor, K.I., Zäch, P. and Brugger, P. Why is magical thinking related to a leftward shift in implicit line bisection? Cortex 2002; 38: 247–252.
Taylor, K.I., Brugger, P. and Schwarz, U. Audiovisual peduncular hallucinations: a release of cross-modal integration sites? Cogn Behav Neurol 2005; 18: 135–136.
Teasdale, G. and Jennett, B. Assessment of coma and impaired consciousness. A practical scale. Lancet 1974; 2: 81–4. doi: 10.1016/s0140–6736(74)91639–0.
Terhune, D.B. and Brugger, P. Doing better by getting worse: posthypnotic amnesia improves random number generation. PLoS One 2011; 6(12): e29206. doi: 10.1371/journal.pone.0029206.
Tetsuka, S., Ogawa, T., Hashimoto, R. and Kato, H. Clinical features, pathogenesis, and management of stroke-like episodes due to MELAS. Metab Brain Dis 2021 Dec; 36(8): 2181–2193. doi: 10.1007/s11011–021–00772–x.
Thiebaut de Schotten, M., Dell'Acqua, F., Ratiu, P., Leslie, A., Howells, H., Cabanis, E. et al. From Phineas Gage and monsieur Leborgne to H.M.: Revisiting disconnection syndromes. Cereb Cortex 2015; 25: 4812–27. doi: 10.1093/cercor/bhv173.
Thiele, K., Quinting, J.M. and Stenneken, P. New ways to analyze word generation performance in brain injury: A systematic review and meta-analysis of additional performance measures. J Clin Exp Neuropsychol 2016 Sep; 38(7): 764–81. doi: 10.1080/13803395.2016.1163327.
Thompson, A.E. and Thompson, P.D. Frontal lobe motor syndromes. Handb Clin Neurol 2023; 196: 443–455. doi: 10.1016/B978–0–323–98817–9.00008–9.
Thompson, R., Emmorey, K. and Gollan, T.H. "Tip of the fingers" experiences by deaf signers: insights into the organization of a sign-based lexicon. Psychol Sci 2005 Nov; 16(11): 856–60. doi: 10.1111/j.1467–9280.2005.01626.x.
Thum, A.S. and Gerber, B. Connectomics and function of a memory network: the mushroom body of larval drosophila. Curr Opin Neurobiol 2019 Feb; 54: 146–154. doi: 10.1016/j.conb.2018.10.007.
Toglia, J. and Chen, P. Spatial exploration strategy training for spatial neglect: A pilot study. Neuropsychol Rehabil 2022 Jun; 32(5): 792–813. doi: 10.1080/09602011.2020.1790394.
Torfs, K., Vancleef, K., Lafosse, C., Wagemans, J. and de-Wit, L. The Leuven Perceptual Organization Screening Test (L-POST), an online test to assess mid-level visual perception. Behav Res Methods 2014 Jun; 46(2): 472–87. doi: 10.3758/s13428-013-0382-6.
Tremblay, P. and Dick, A.S. Broca and Wernicke are dead, or moving past the classic model of language neurobiology. Brain Lang. 2016 Nov; 162: 60–71. doi: 10.1016/j.bandl.2016.08.004.

Treu, S., Barcia, J.A., Torres, C., Bierbrauer, A., Gonzalez-Rosa, J.J., Nombela, C. et al. Deep-brain stimulation of the human nucleus accumbens-medial septum enhances memory formation. Res Sq [Preprint]. 2023 Nov 20: rs.3.rs-3476665. doi: 10.21203/rs.3.rs-3476665/v1.

Tronson, N.C. and Taylor, J.R. Molecular mechanisms of memory reconsolidation. Nat Rev Neurosci 2007; 8: 262–275.

Tsanov, M. Basal forebrain impairment: Understanding the mnemonic function of the septal region translates in therapeutic advances. Front Neural Circuits 2022; 16: 916499. doi:

Tseng, W.S. and Tzeng, N.S. Dromosagnosia, or why some people lose their sense of direction while driving. Med Hypotheses 2013; 81(5): 839–841. doi: 10.1016/j.mehy.2013.09.010.

Tubridy, N., Hutchinson, M. and Murphy, R.P. Transient global amnesia: 'amnesia by the seaside' revisited. J Neurol 1999 Jun; 246(6): 500–1. doi: 10.1007/s004150050391.

Tucha, O., Steup, A., Smely, C. and Lange, K.W. Toe agnosia in Gerstmann syndrome. J Neurol Neurosurg Psychiatry 1997 Sep; 63(3): 399–403. doi: 10.1136/jnnp.63.3.399.

Turbyne, C., Koning, P., Zantvoord, J. and Denys, D. Body integrity identity disorder using augmented reality: a symptom reduction study. BMJ Case Rep 2021; 14: e238554. doi: 10.1136/bcr-2020–238554.

Turgut, M., Yurttas, C. and Tubbs, R.S. (eds.). Island of Reil (Insula) in the Human Brain. Anatomical, Functional, Clinical and Surgical Aspects. Springer, Berlin, 2018.

Turk, E., van den Heuvel, M.I., Benders, M.J., de Heus, R., Franx, A., Manning, J.H. et al. Functional connectome of the fetal brain. J Neurosci 2019; 39: 9716–9724. doi: 10.1523/JNEUROSCI.2891–18.2019.

Turnbull, O.H., Beschin, N. and Della Sala. Agnosia for object orientation: implictions for theories of object recognition. Neuropsychologia 1997; 35: 153–163.

Turnbull, O.H., Fotopoulou, A. and Solms, M. Anosognosia as motivated unawareness: the 'defence' hypothesis revisited. Cortex 2014; 61: 18–29. doi: 10.1016/j.cortex.2014.10.008.

Ungerleider, L.G. and Haxby, J.V. 'What' and 'where' in the human brain. Curr Opin Neurobiol 1994; 4: 157–65. doi: 10.1016/0959–4388(94)90066–3.

Uomini, N. and Ruck, L. Manual laterality and cognition through evolution: An archeological perspective. Prog Brain Res 2018; 238: 295–323. doi: 10.1016/bs.pbr.2018.06.015.

Uysal, H., Şavkliyildiz, A., Özkan, Ö., Çolak, Ö.H., Barçin, E., Şenol, U. et al. Complementary phenomena: phantom hand and phantom face. Cogn Behav Neurol 2021; 34: 150–159. doi: 10.1097/WNN.0000000000000258.

Valdes, C.P., Varma, H.M., Kristoffersen, A.K., Dragojevic, T., Culver, J.P. and Durduran, T. Speckle contrast optical spectroscopy, a non-invasive, diffuse optical method for measuring microvascular blood flow in tissue. Biomed. Opt. Express 2014; 5: 2769–2784. doi: 10.1364/BOE.5.002769.

Valentin, G. Über die subjectiven Gefühle von Personen, welche mit mangelhaften Extremitäten geboren sind. Repertorium für Anatomie und Physiologie 1836; 1: 328–337.

Vallar, G., Guariglia, C., Nico, D. and Bisiach, E. Spatial hemineglect in back space. Brain 1995; 118: 467–472.

Vallar, G. and Ronchi, R. Somatoparaphrenia: a body delusion. A review of the neuropsychological literature. Exp Brain Res 2009; 192: 533–51. doi: 10.1007/s00221–008–1562–y.

Vallebona, A. Stratigrafia selletiva e stratigrafia in rilievo. Informatore Medico: Genoa, 1955. [in Italian].

Vallortigara, G., Cozzutti, C., Tommasi, L. and Rogers, L.J. How birds use their eyes: Opposite left-right specialization for the lateral and frontal visual hemifield in the domestic chick. Current Biology 2001; 11: 29–33. Doi: 10.1016/s0960–9822(00)00027–0.

van Bogaert, L. L'hallucinose pédonculaire. Rev. Neurol. 1927; 43: 608–617.

van der Knaap, L.J., van der Ham, I.J.M. How does the corpus callosum mediate interhemispheric transfer? A review. Behav Brain Res 2011; 223: 211–221. doi: 10.1016/j.bbr.2011.04.018.

van Elk, M., van Schie, H.T. and Bekkering, H. From left to right: Processing acronyms referring to names of political parties activates spatial associations. The Quarterly Journal of Experimental Psychology 2010; 63: 2202–2219. Doi: 10.1080/17470218.2010.495160.

van Elk, M. and Blanke, O. The relation between body semantics and spatial body representations. Acta Psychol (Amst) 2011; 138: 347–58. doi: 10.1016/j.actpsy.2011.09.001.

van Essen, D.C., Ugurbil, K., Auerbach, E., Barch, D., Behrens, T.E., Buchholz, R. et al. The human connectome project: a data acquisition perspective. NeuroImage 2012; 62: 2222–2231.

van Patten, R., Bellone, J.A., Schmitt, T.R., Gaynor, L. and Block, C. Digital methods of delivering education and training in neuropsychology. Arch Clin Neuropsychol 2022; 37: 1103–1117. doi: 10.1093/arclin/acac033.

van Vuuren, A.J., Saling, M., Rogerson, S., Anderson, P., Cheong, J. and Solms, M. Cerebral arterial asymmetries in the neonate: Insight into the pathogenesis of stroke. Symmetry 2022; 14: 456–469. Doi: 10.3390/sym14030456.

Vanbellingen, T., Kersten, B., Van de Winckel, A., Bellion, M., Baronti, F., Müri, R. et al. A new bedside test of gestures in stroke: the apraxia screen of TULIA (AST). J Neurol Neurosurg Psychiatry 2011; 82: 389–92. doi: 10.1136/jnnp.2010.213371.

Vanderah, T.W. and Gould, D.J. Nolte's The Human Brain: An Introduction to Its Functional Anatomy. (9th edition). Elsevier: New York, 2025.

Vannasing, P., Florea, O., González-Frankenberger, B., Tremblay, J., Paquette, N., Safi, D. et al. Distinct hemispheric specializations for native and non-native languages in one-day-old newborns identified by fNIRS. Neuropsychologia 2016; 84: 63–69. doi: 10.1016/j.neuropsychologia.2016.01.038.

Varney, N.R. Colour association and "colour amnesia" in aphasia. J Neurol Neurosurg Psychiatry 1982; 45: 248–252.

Veale, J.F. Edinburgh handedness inventory - short form: A revised version based on confirmatory factor analysis. Laterality 2014; 19: 164–177.

Verrelli, C.M., Caprioli, L. and Iosa, M. Hidden time-patterns in cyclic human movements: a matter of temporal Fibonacci sequence generation and harmonization. Front Hum Neurosci 2025; 19: 1525403. doi: 10.3389/fnhum.2025.1525403.

Viaud-Delmon, I., Brugger, P. and Landis, T. Hemineglect: Take a look at the back space. Ann Neurol 2007 Oct; 62(4): 418–422. doi: 10.1002/ana.21176.

Villringer, A. and Chance, B. Non-invasive optical spectroscopy and imaging of human brain function. Trends Neurosci 1997; 20: 435–42. doi: 10.1016/s0166–2236(97)01132–6.

Vingerhoets, G. and Luppens, E. Cerebral blood flow velocity changes during dichotic listening with directed or divided attention: a transcranial Doppler ultrasonography study. Neuropsychologia. 2001; 39(10): 1105–11. doi: 10.1016/s0028-3932(01)00030-6.

Vingerhoets, G., Verhelst, H., Gerrits, R., Badcock, N., Bishop, D.V.M., Carey, D. et al. Laterality indices consensus initiative (LICI): A Delphi expert survey report on recommendations to record, assess, and report asymmetry in human behavioural and brain research. Laterality 2023; 28: 122–191. doi: 10.1080/1357650X.2023.2199963.

Voigt, K., Liang, E.X., Misic, B., Ward, P.G.D., Egan, G.F. et al. Metabolic and functional connectivity provide unique and complementary insights into cognition-connectome relationships. Cereb Cortex 2023; 33: 1476–1488. doi: 10.1093/cercor/bhac150.

von Stockert, F.G. Lokalisation und klinische Differenzierung des Symptoms der Nichtwahrnehmung einer Körperhälfte. Deutsche Zeitschrift für Nervenheilkunde 1934; 134: 1–13.

Vuilleumier, P., Valenza, N., Mayer, E., Reverdin, A. and Landis, T. Near and far visual space in unilateral neglect. Ann Neurol 1998; 43: 406–410.

Vuilleumier, P., Mohr, C., Valenza, N., Wetzel, C. and Landis, T. Hyperfamiliarity for unknown faces after left lateral temporo-occipital venous infarction: a double dissociation with prosopagnosia. Brain 2003; 126: 889–907. doi: 10.1093/brain/awg086.

Vuilleumier, P. Anosognosia: the neurology of beliefs and uncertainties. Cortex 2004; 40: 9–17. doi: 10.1016/s0010–9452(08)70918–3.

Vygotsky, L.S. Thought and Language. MIT Press: Cambridge, MA, 1962 (originally published in 1943).

Wada, J. and Rasmussen, T. Intracarotid injection of sodium amytal for the lateralization of cerebral speech dominance: experimental and clinical observations. J Neurosurg 1960; 17: 266–282.

Walker, D.R. Memory transfer in planarians: An artifact of the experimental variables. Psychonomic Science 1966; 5: 357–358.

Walker, M.L., Patel, K., Li, T. and Kassir, M. Opioid-associated amnestic syndrome. Cureus 2021; 13: e20056. doi: 10.7759/cureus.20056.

Walker, S.C., Trotter, P.D., Woods, A. and McGlone, F. Vicarious ratings of social touch reflect the anatomical distribution and velocity tuning of C-tactile afferents: A hedonic homunculus? Behavioural Brain Research 2017; 320: 91–96.

Wallas, G. The Art of Thought. Butler and Tanner: London, 1926.

Walsh, V. A theory of magnitude: common cortical metrics of time, space and quantity. Trends Cogn Sci 2003 Nov; 7(11): 483–8. doi: 10.1016/j.tics.2003.09.002.

Warrington, E.K. and James, M. Visual Object and Space Perception Battery (VOSP). Pearson: London, 1991.

Wason, P.C. Reasoning. *In*: Foss, B. (ed.). New Horizons in Psychology. Penguin: Harmondsworth 1966, 135–151.

Waxman, S.G. and Geschwind, N. The interictal behavior syndrome of temporal lobe epilepsy. Arch Gen Psychiatry 1975; 32(12): 1580–1586. 10.1001/archpsyc.1975.01760300118011

Wegner, D.M., Fuller, V.A. and Sparrow, B. Clever hands: uncontrolled intelligence in facilitated communication. J Pers Soc Psychol 2003 Jul; 85(1): 5–19. doi: 10.1037/0022–3514.85.1.5.

Weil, R.S., Schwarzkopf, D.S., Bahrami, B., Fleming, S.M., Jackson, B.M., Goch, T.J.C. et al. Assessing cognitive dysfunction in Parkinson's disease: An online tool to detect visuo-perceptual deficits. Mov Disord. 2018 Apr; 33(4): 544–553. doi: 10.1002/mds.27311.

Weinberg, J., Diller, L., Gerstman, L. and Schulman, P. Digit span in right and left hemiplegics. J Clin Psychol 1972; 28: 361. doi: 10.1002/1097–4679(197207)28: 3+<361: : aid-jclp2270280315>3.0.co; 2–5.

Weiner, M.F., Hynan, L.S., Rossetti, H. and Falkowski, J. Luria's three-step test: what is it and what does it tell us? Int Psychogeriatr 2011 Dec; 23(10): 1602–6. doi: 10.1017/S1041610211000767.

Weinstein, N.D. Unrealistic optimism about future life events. Journal of Personality and Social Psychology 1980; 39(5): 806–820. https://doi.org/10.1037/0022-3514.39.5.806

Weisbrod, M., Maier, S., Harig, S., Himmelsbach, U. and Spitzer, M: Lateralised semantic and indirect semantic priming effects in people with schizophrenia. Br J Psychiatry 1998; 172: 142–146. doi: 10.1192/bjp.172.2.142.

Weiskrantz, L., Warrington, E.K., Sanders, M.D. and Marshall, J. Visual capacity in the hemianopic field following a restricted occipital ablation. Brain 1974; 97: 709–728.

Weng, Y.-L. and Lee, C.-L. Reduced right-hemisphere ERP P600 grammaticality effect is associated with greater right-hemisphere inhibition: Evidence from right-handers with familial sinistrality. Brain Res 2020; 1738: 146815. doi: 10.1016/j.brainres.2020.146815.

West, J.B. Do climbs to extreme altitude cause brain damage? The Lancet 1986: 2: 387–388. doi: 10.1016/s0140–6736(86)90066–8

Westerhausen, R. Interhemispheric integration after callosotomy: A meta-analysis of Poffenberger and redundant-target paradigms. Neuropsychol Rev. 2023; 33: 872–890. doi: 10.1007/s11065–022–09569–w.

Wheaton, L.A., Bohlhalter, S., Nolte, G., Shibasaki, H., Hattori, N., Fridman, E. et al. Cortico-cortical networks in patients with ideomotor apraxia as revealed by EEG coherence analysis. Neurosci Lett 2008 Mar 12; 433(2): 87–92. doi: 10.1016/j.neulet.2007.12.065.

Whitworth, J.A. Should eponyms be abandoned? No. BMJ. 2007 Sep 1; 335(7617): 425. doi: 10.1136/bmj.39308.380567.AD.

Wigan, A.L. A New View of Insanity. The Duality of the Mind. Longman, Brown, Green and Longmans: London, 1844

Wilkinson, A., Hloch, A., Mueller-Paul, J. and Huber, L. The effect of brumation on memory retention. Sci Rep 2017; 7: 40079. doi: 10.1038/srep40079.

Willems, R.M., Hagoort, P. and Casasanto, D. Body-specific representations of action verbs: neural evidence from right- and left-handers. Psychol Sci 2010 Jan; 21(1): 67–74. doi: 10.1177/0956797609354072.

Willems, R.M., van der Haegen, L., Fisher, S.E. and Francks, C. On the other hand: including left-handers in cognitive neuroscience and neurogenetics. Nat Rev Neurosci 2014; 15: 193–201. doi: 10.1038/nrn3679.

Williamson, V.J. and Stewart, L. Congenital amusia. Handb Clin Neurol 2013; 111: 237–9. doi: 10.1016/B978–0–444–52891–9.00024–5.

Winer, G.A., Cottrell, J.E., Gregg, V., Fournier, J.S. and Bica, L.A. Fundamentally misunderstanding visual perception. Adults' belief in visual emissions. American Psychologist 2002; 57: 417–424.

Wolfle, D. Extrasensory Perception. Science 1956 Jan 6; 123(3184): 7. doi: 10.1126/science.123.3184.7.

Wollenberg, R. Über systematische Störungen der egozentrischen Orientierung bei Geistesgesunden. Archiv für Psychiatrie und Nervenkrankheiten 1925; 74: 337–349.

Wolters, F., Peerdeman, K.J., Evers, A.W.M. Placebo and nocebo effects across symptoms: From pain to fatigue, dyspnea, nausea, and itch. Front Psychiatry 2019; 10: 470. doi: 10.3389/fpsyt.2019.00470. eCollection 2019.

Wong, B., Yin, B. and O'Brien, B. Neurolinguistics: Structure, function, and connectivity in the bilingual brain. Biomed Res Int 2016; 2016: 7069274. doi: 10.1155/2016/7069274

Woollacott, I.O., Fletcher, P.D., Massey, L.A., Pasupathy, A., Rossor, M.N., Caine, D. et al. Compulsive versifying after treatment of transient epileptic amnesia. Neurocase 2015; 21: 548–53. doi: 10.1080/13554794.2014.953178.

World Health Organization. (2019). International Statistical Classification of Diseases and Related Health Problems (11th ed.). https: //icd.who.int/.

Woywodt, A. and Matteson, E. Should eponyms be abandoned? Yes. BMJ 2007; 335: 424. doi: 10.1136/bmj.39308.342639.AD.

Wurm, M.F. and Caramazza, A. Two 'what' pathways for action and object recognition. Trends Cogn Sci 2022; 26: 103–116. doi: 10.1016/j.tics.2021.10.003.

Wurtz, R.H. Recounting the impact of Hubel and Wiesel. J Physiol. 2009; 587: 2817–23. doi: 10.1113/jphysiol.2009.170209.

Yamada, Y., Xue, J., Li, P., Ruiz-Fernández, S., Özdoğru, A.A., Sarı, Ş., Torres, S.C. et al. Where the 'bad' and the 'good' go: A multi-lab direct replication report of Casasanto (2009, Experiment 1). Mem Cognit. 2024 Sep 23. doi: 10.3758/s13421–024–01637–1.

Yang, Y., Raine, A., Lencz, T., Bihrle, S., Lacasse, L. and Colletti, P. Prefrontal white matter in pathological liars. Br J Psychiatry 2005 Oct; 187: 320–5. doi: 10.1192/bjp.187.4.320.

Yoon, J.S., Ericsson, A.K. and Donatelli, D. Effects of 30 years of disuse on exceptional memory performance. Cogn Sci 2018; 42 Suppl 3: 884–903. doi: 10.1111/cogs.12562.

Zäch, P. and Brugger, P. Subjective time in near and far representational space. Cogn Behav Neurol 2008; 21: 8–13. doi: 10.1097/WNN.0b013e31815f237c.

Zamboni, G., Budriesi, C. and Nichelli, P. "Seeing oneself": a case of autoscopy. Neurocase 2005; 11: 212–5. doi: 10.1080/13554790590944799.

Zeki, S. A century of achromatopsia. Brain 1990; 113: 1721–1777. doi: 10.1093/brain/113.6.1721.

Zeki, S. Vision of the Brain. Blackwell Scientific Pubications: Oxford, UK, 1993.

Zeman, A., Monzel, M., Pearson, J., Scholz, C.O. and Simner, J. Definition: Aphantasia. Cortex 2025; 182: 212–213. doi: 10.1016/j.cortex.2024.07.019.

Zhang, L., Meng, J., Li, H., Tang, M., Zhou, Z., Zhou, X. et al. Hippocampal adaptation to high altitude: a neuroanatomic profile of hippocampal subfields in Tibetans and acclimatized Han Chinese residents. Front Neuroanat 2022; 16: 999033. doi: 10.3389/fnana.2022.999033.

Zhang, Y., Lu, H. and Barmann, C.I. Pathogenic bacteria induce aversive olfactory learning in Caenorhabditis elegans. Nature 2005; 438; 179–184. doi: 10.1038/nature04216.

Zhao, J., Shen, G., Nan, G. and Zhao, S. Aphasia due to right hemisphere infarction in a patient with situs inversus after carotid artery stenting: a case report. BMC Neurol 2024; 24: 289. doi: 10.1186/s12883–024–03773–2.

Zihl, J., von Cramon, D. and Mai, N. Selective disturbance of movement vision after bilateral brain damage. Brain 1983 Jun; 106 (Pt 2): 313–40. doi: 10.1093/brain/106.2.313.

Zimmermann, K. and Brugger, P. Signed soliloquy: visible private speech. J Deaf Stud Deaf Educ 2013; 18: 261–70. doi: 10.1093/deafed/ens072.

Zivotofsky, A.Z. Choosing sides: lateralization in line trisection and quadrisection as a function of reading direction and handedness. Brain Res Cogn Brain Res 2004; 20: 206–11. doi: 10.1016/j.cogbrainres.2004.03.001.

Zorzi, M., Priftis, K. and Umilta, C. Brain damage: neglect disrupts the mental number line. Nature 2002; 417: 138–9. doi: 10.1038/417138a.

Züst, M.A., Ruch, S., Wiest, R. and Henke, K. Implicit Vocabulary Learning during Sleep Is Bound to Slow-Wave Peaks. Curr Biol 2019 Feb 18; 29(4): 541–553.e7. doi: 10.1016/j.cub.2018.12.038.

Zwaan, R.A. and Yaxley, R.H. Hemispheric differences in semantic-relatedness judgments. Cognition 2003; 87: B79–86. doi: 10.1016/s0010–0277(02)00235–4.

Index

N

O

P

Q

R

S

T

U

V

W

Z

For Product Safety Concerns and Information please contact our
EU representative GPSR@taylorandfrancis.com Taylor & Francis
Verlag GmbH, Kaufingerstraße 24, 80331 München, Germany